AF356209

Physiology and
Pathophysiology of Placenta

Physiology and Pathophysiology of Placenta

Editor

Giovanni Tossetta

Basel • Beijing • Wuhan • Barcelona • Belgrade • Novi Sad • Cluj • Manchester

Editor
Giovanni Tossetta
Department of Experimental
and Clinical Medicine
Università Politecnica delle Marche
Ancona
Italy

Editorial Office
MDPI
St. Alban-Anlage 66
4052 Basel, Switzerland

This is a reprint of articles from the Special Issue published online in the open access journal *International Journal of Molecular Sciences* (ISSN 1422-0067) (available at: www.mdpi.com/journal/ijms/special_issues/Placenta).

For citation purposes, cite each article independently as indicated on the article page online and as indicated below:

Lastname, A.A.; Lastname, B.B. Article Title. *Journal Name* **Year**, *Volume Number*, Page Range.

ISBN 978-3-7258-0792-5 (Hbk)
ISBN 978-3-7258-0791-8 (PDF)
doi.org/10.3390/books978-3-7258-0791-8

Contents

International Journal of
Molecular Sciences

Editorial

Physiology and Pathophysiology of the Placenta

Giovanni Tossetta 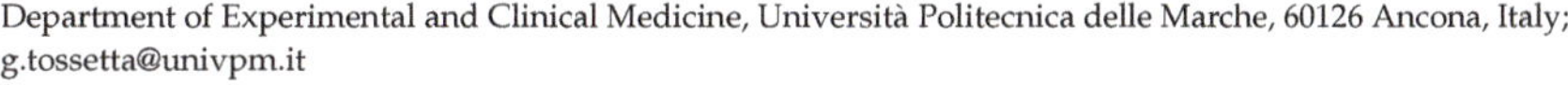

Department of Experimental and Clinical Medicine, Università Politecnica delle Marche, 60126 Ancona, Italy;
g.tossetta@univpm.it

check for
updates

Citation: Tossetta, G. Physiology and Pathophysiology of the Placenta. *Int. J. Mol. Sci.* **2023**, 24, 9066. https://doi.org/10.3390/ijms24109066

Received: 6 May 2023
Accepted: 17 May 2023
Published: 22 May 2023

We are pleased to present this Special Issue of the *International Journal of Molecular Sciences*, entitled "Physiology and Pathophysiology of Placenta".

Placentation is an important and tightly regulated process that ensures the development of placenta, allowing for the normal progression of the fetus. The placenta is an essential organ which plays different and fundamental functions during pregnancy, and its development is regulated by several growth factors, hormones, their receptors and many other types of molecules involved in the regulation of placental cell proliferation, differentiation, migration and invasion [1,2].

The correct function of these processes is tightly regulated by the activation or inhibition of several signalling pathways that regulate the expression of specific genes necessary for a successful pregnancy. The importance of normal placental development becomes evident in the case of impaired placental development, which can lead to significant pregnancy complications such as preeclampsia (PE) [3], fetal growth restriction (FGR) [4], gestational trophoblastic diseases (GTD) [5], preterm delivery [6,7] and gestational diabetes mellitus (GDM) [8]. Pregnancy can also be impaired by exposure to exogenous agents such as bacteria [7], viruses [9], chemicals and natural compounds [9,10] that can alter the normal placental functions, compromising pregnancy outcome. Many of the disorders/pathologies previously mentioned are associated with an increase in maternal and fetal mortality and morbidity, and can lead to life-long health complications for both mother and child.

Important signalling pathways such as Wnt/-catenin, TGF/SMAD, PI3K/AKT/mTOR and JAK/STAT pathways have been reported to be impaired in several pregnancy complications such as PE, GDM and FGR [10–13], which share an inflammatory and oxidative stress condition [6,8]. In addition to the previously mentioned pathologies, viral and bacterial infections during pregnancy can also lead to an increase in inflammatory cytokines that alter the normal function of the placenta and amniotic membranes, causing preterm delivery or significant neonatal complications [7,9,14]. All the pathologies already mentioned cause systemic inflammation (acute or chronic) that turns into endothelial dysfunction. Endothelial dysfunction impairs the normal functionality of the endothelium, and thus can alter the normal function of reproductive organs [15–18].

For these reasons, new and specific biomarkers are necessary in clinical practice to allow an early diagnosis of many of the above-mentioned pregnancy complications, in order to carry out early treatment of the pathology, improving the outcome of the pregnancy or resolving the pathology [8,19,20].

Several natural and synthetic compounds have shown important beneficial effects in treating several diseases. These compounds have also demonstrated important effects in pregnancy complications, suggesting a possible use of these compounds, alone or in combination with classical drugs, to treat these diseases, improving pregnancy outcomes [21,22].

Understanding the mechanisms involved in the regulation of human placenta development in normal and pathological conditions can help to open new perspectives in the treatment of these pregnancy complications.

Thus, the aim of this Special Issue is to provide an overview of the physiology and pathophysiology of the placenta, in order to better understand its development in normal and pathological conditions.

Conflicts of Interest: The authors declare no conflict of interest.

References

1. Tossetta, G.; Avellini, C.; Licini, C.; Giannubilo, S.R.; Castellucci, M.; Marzioni, D. High temperature requirement A1 and fibronectin: Two possible players in placental tissue remodelling. *Eur. J. Histochem.* **2016**, *60*, 2724. [CrossRef] [PubMed]
2. Marzioni, D.; Crescimanno, C.; Zaccheo, D.; Coppari, R.; Underhill, C.B.; Castellucci, M. Hyaluronate and CD44 expression patterns in the human placenta throughout pregnancy. *Eur. J. Histochem.* **2001**, *45*, 131–140. [CrossRef] [PubMed]
3. Tossetta, G.; Fantone, S.; Giannubilo, S.R.; Marinelli Busilacchi, E.; Ciavattini, A.; Castellucci, M.; Di Simone, N.; Mattioli-Belmonte, M.; Marzioni, D. Pre-eclampsia onset and SPARC: A possible involvement in placenta development. *J. Cell Physiol.* **2019**, *234*, 6091–6098. [CrossRef]
4. Cardaropoli, S.; Paulesu, L.; Romagnoli, R.; Ietta, F.; Marzioni, D.; Castellucci, M.; Rolfo, A.; Vasario, E.; Piccoli, E.; Todros, T. Macrophage migration inhibitory factor in fetoplacental tissues from preeclamptic pregnancies with or without fetal growth restriction. *Clin. Dev. Immunol.* **2012**, *2012*, 639342. [CrossRef] [PubMed]
5. Marzioni, D.; Quaranta, A.; Lorenzi, T.; Morroni, M.; Crescimanno, C.; De Nictolis, M.; Toti, P.; Muzzonigro, G.; Baldi, A.; De Luca, A.; et al. Expression pattern alterations of the serine protease HtrA1 in normal human placental tissues and in gestational trophoblastic diseases. *Histol. Histopathol.* **2009**, *24*, 1213–1222. [CrossRef]
6. Cecati, M.; Sartini, D.; Campagna, R.; Biagini, A.; Ciavattini, A.; Emanuelli, M.; Giannubilo, S.R. Molecular analysis of endometrial inflammation in preterm birth. *Cell Mol. Biol. (Noisy-Le-Grand)* **2017**, *63*, 51–57. [CrossRef]
7. Licini, C.; Tossetta, G.; Avellini, C.; Ciarmela, P.; Lorenzi, T.; Toti, P.; Gesuita, R.; Voltolini, C.; Petraglia, F.; Castellucci, M.; et al. Analysis of cell-cell junctions in human amnion and chorionic plate affected by chorioamnionitis. *Histol. Histopathol.* **2016**, *31*, 759–767. [CrossRef]
8. Tossetta, G.; Fantone, S.; Gesuita, R.; Di Renzo, G.C.; Meyyazhagan, A.; Tersigni, C.; Scambia, G.; Di Simone, N.; Marzioni, D. HtrA1 in Gestational Diabetes Mellitus: A Possible Biomarker? *Diagnostics* **2022**, *12*, 2705. [CrossRef]
9. Tossetta, G.; Fantone, S.; Delli Muti, N.; Balercia, G.; Ciavattini, A.; Giannubilo, S.R.; Marzioni, D. Preeclampsia and severe acute respiratory syndrome coronavirus 2 infection: A systematic review. *J. Hypertens.* **2022**, *40*, 1629–1638. [CrossRef]
10. Alijotas-Reig, J.; Esteve-Valverde, E.; Ferrer-Oliveras, R.; Llurba, E.; Gris, J.M. Tumor Necrosis Factor-Alpha and Pregnancy: Focus on Biologics. An Updated and Comprehensive Review. *Clin. Rev. Allergy Immunol.* **2017**, *53*, 40–53. [CrossRef]
11. Villalobos-Labra, R.; Silva, L.; Subiabre, M.; Araos, J.; Salsoso, R.; Fuenzalida, B.; Saez, T.; Toledo, F.; Gonzalez, M.; Quezada, C.; et al. Akt/mTOR Role in Human Foetoplacental Vascular Insulin Resistance in Diseases of Pregnancy. *J. Diabetes Res.* **2017**, *2017*, 5947859. [CrossRef] [PubMed]
12. Zhang, Z.; Wang, X.; Zhang, L.; Shi, Y.; Wang, J.; Yan, H. Wnt/beta-catenin signaling pathway in trophoblasts and abnormal activation in preeclampsia (Review). *Mol. Med. Rep.* **2017**, *16*, 1007–1013. [CrossRef]
13. Li, Y.; Yan, J.; Chang, H.M.; Chen, Z.J.; Leung, P.C.K. Roles of TGF-beta Superfamily Proteins in Extravillous Trophoblast Invasion. *Trends Endocrinol. Metab.* **2021**, *32*, 170–189. [CrossRef] [PubMed]
14. Ikumi, N.M.; Matjila, M. Preterm Birth in Women With HIV: The Role of the Placenta. *Front. Glob Womens Health* **2022**, *3*, 820759. [CrossRef] [PubMed]
15. Mateuszuk, L.; Campagna, R.; Kutryb-Zajac, B.; Kus, K.; Slominska, E.M.; Smolenski, R.T.; Chlopicki, S. Reversal of endothelial dysfunction by nicotinamide mononucleotide via extracellular conversion to nicotinamide riboside. *Biochem. Pharmacol.* **2020**, *178*, 114019. [CrossRef]
16. Szczesny-Malysiak, E.; Stojak, M.; Campagna, R.; Grosicki, M.; Jamrozik, M.; Kaczara, P.; Chlopicki, S. Bardoxolone Methyl Displays Detrimental Effects on Endothelial Bioenergetics, Suppresses Endothelial ET-1 Release, and Increases Endothelial Permeability in Human Microvascular Endothelium. *Oxid. Med. Cell Longev.* **2020**, *2020*, 4678252. [CrossRef]
17. Campagna, R.; Mateuszuk, L.; Wojnar-Lason, K.; Kaczara, P.; Tworzydlo, A.; Kij, A.; Bujok, R.; Mlynarski, J.; Wang, Y.; Sartini, D.; et al. Nicotinamide N-methyltransferase in endothelium protects against oxidant stress-induced endothelial injury. *Biochim. Biophys. Acta Mol. Cell Res.* **2021**, *1868*, 119082. [CrossRef]
18. Zapotoczny, B.; Braet, F.; Kus, E.; Ginda-Makela, K.; Klejevskaja, B.; Campagna, R.; Chlopicki, S.; Szymonski, M. Actin-spectrin scaffold supports open fenestrae in liver sinusoidal endothelial cells. *Traffic* **2019**, *20*, 932–942. [CrossRef] [PubMed]
19. Gesuita, R.; Licini, C.; Picchiassi, E.; Tarquini, F.; Coata, G.; Fantone, S.; Tossetta, G.; Ciavattini, A.; Castellucci, M.; Di Renzo, G.C.; et al. Association between first trimester plasma htra1 level and subsequent preeclampsia: A possible early marker? *Pregnancy Hypertens.* **2019**, *18*, 58–62. [CrossRef]
20. Licini, C.; Avellini, C.; Picchiassi, E.; Mensa, E.; Fantone, S.; Ramini, D.; Tersigni, C.; Tossetta, G.; Castellucci, C.; Tarquini, F.; et al. Pre-eclampsia predictive ability of maternal miR-125b: A clinical and experimental study. *Transl. Res.* **2021**, *228*, 13–27. [CrossRef]

21. Tossetta, G.; Fantone, S.; Giannubilo, S.R.; Marzioni, D. The Multifaced Actions of Curcumin in Pregnancy Outcome. *Antioxidants* **2021**, *10*, 126. [CrossRef] [PubMed]
22. Novakovic, R.; Rajkovic, J.; Gostimirovic, M.; Gojkovic-Bukarica, L.; Radunovic, N. Resveratrol and Reproductive Health. *Life* **2022**, *12*, 294. [CrossRef] [PubMed]

International Journal of
Molecular Sciences

Article

FOXM1 Participates in Trophoblast Migration and Early Trophoblast Invasion: Potential Role in Blastocyst Implantation

Reyna Peñailillo [1,2,†], Victoria Velásquez [1,†], Stephanie Acuña-Gallardo [1,2,3], Felipe García [1,2], Mario Sánchez [4], Gino Nardocci [2,3,5], Sebastián E. Illanes [1,2,3] and Lara J. Monteiro [1,2,3,*]

[1] Program in Biology of Reproduction, Center for Biomedical Research and Innovation (CiiB), Universidad de los Andes, Santiago 7620001, Chile; rpenailillo@uandes.cl (R.P.); vpvelasquez@uc.cl (V.V.); sacuna@uandes.cl (S.A.-G.); fagarcia@uandes.cl (F.G.); sillanes@uandes.cl (S.E.I.)

[2] IMPACT, Center of Interventional Medicine for Precision and Advanced Cellular Therapy, Santiago 7620001, Chile; gnardocci@uandes.cl

[3] School of Medicine, Faculty of Medicine, Universidad de los Andes, Santiago 7620001, Chile

[4] Program in Neuroscience, Centre for Biomedical Research and Innovation (CiiB), Universidad de los Andes, Santiago 7620001, Chile; mesanchez@uandes.cl

[5] Molecular Biology and Bioinformatics Lab, Program in Molecular Biology and Bioinformatics, Centre for Biomedical Research and Innovation (CiiB), Universidad de los Andes, Santiago 7620001, Chile

[*] Correspondence: lmonteiro@uandes.cl

[†] These authors contributed equally to this work.

Citation: Peñailillo, R.; Velásquez, V.; Acuña-Gallardo, S.; García, F.; Sánchez, M.; Nardocci, G.; Illanes, S.E.; Monteiro, L.J. FOXM1 Participates in Trophoblast Migration and Early Trophoblast Invasion: Potential Role in Blastocyst Implantation. *Int. J. Mol. Sci.* **2024**, *25*, 1678. https://doi.org/10.3390/ijms25031678

Academic Editor: Giovanni Tossetta

Received: 30 December 2023
Revised: 21 January 2024
Accepted: 21 January 2024
Published: 30 January 2024

Abstract: Successful implantation requires coordinated migration and invasion of trophoblast cells into a receptive endometrium. Reduced forkhead box M1 (FOXM1) expression limits trophoblast migration and angiogenesis in choriocarcinoma cell lines, and in a rat model, placental FOXM1 protein expression was significantly upregulated in the early stages of pregnancy compared to term pregnancy. However, the precise role of FOXM1 in implantation events remains unknown. By analyzing mice blastocysts at embryonic day (E3.5), we have demonstrated that FOXM1 is expressed as early as the blastocyst stage, and it is expressed in the trophectoderm of the blastocyst. Since controlled oxygen tension is determinant for achieving normal implantation and placentation and a chronic hypoxic environment leads to shallow trophoblast invasion, we evaluated if FOXM1 expression changes in response to different oxygen tensions in the HTR-8/SVneo first trimester human trophoblast cell line and observed that FOXM1 expression was significantly higher when trophoblast cells were cultured at 3% O_2, which coincides with oxygen concentrations in the uteroplacental interface at the time of implantation. Conversely, FOXM1 expression diminished in response to 1% O_2 that resembles a hypoxic environment in utero. Migration and angiogenesis were assessed following FOXM1 knockdown and overexpression at 3% O_2 and 1% O_2, respectively, in HTR-8/SVneo cells. FOXM1 overexpression increased transmigration ability and tubule formation. Using a 3D trophoblast invasion model with trophospheres from HTR-8/SVneo cells cultured on a layer of MATRIGEL and of mesenchymal stem cells isolated from menstrual fluid, we observed that trophospheres obtained from 3D trophoblast invasion displayed higher FOXM1 expression compared with pre-invasion trophospheres. Moreover, we have also observed that FOXM1-overexpressing trophospheres increased trophoblast invasion compared with controls. HTR-8/SVneo-FOXM1-depleted cells led to a downregulation of *PLK4*, *VEGF*, and *MMP2* mRNA expression. Our current findings suggest that FOXM1 participates in embryo implantation by contributing to trophoblast migration and early trophoblast invasion, by inducing transcription activation of genes involved in these processes. Maternal-fetal communication is crucial for trophoblast invasion, and maternal stromal cells may induce higher levels of FOXM1 in trophoblast cells.

Keywords: FOXM1; blastocyst implantation; trophoblast invasion and migration

1. Introduction

Embryo implantation is a tightly regulated process by which a competent blastocyst reaches the maternal endometrium at its most receptive phase (window of implantation). This process begins when the outer layer of the blastocyst, the trophectoderm cells, come into contact and attach to the endometrial epithelium (apposition-attachment) [1–3]. Shortly thereafter, the trophectoderm undergoes differentiation, giving rise to two different regions: (i) syncytiotrophoblast, a multinucleated outer cell layer that penetrates the endometrial epithelial membrane, allowing the embryo's implantation into the endometrium; and (ii) mononuclear cytotrophoblast cells that proliferate to form cell columns that further penetrate the endometrial stroma [1,3,4]. Emerging from the tips of these anchoring villi structures are extravillous cytotrophoblast cells (EVT), known for their highly migratory, proliferative, and invasive characteristics [5–7].

Trophoblast invasion of the uterus involves attachment of EVT to the extracellular matrix and degradation of the matrix as well as the uterine vasculature (endovascular invasion) to ultimately establish the uteroplacental circulation [3,5]. When this physiological equilibrium is disrupted, implantation failure and subsequent spontaneous abortion may occur [3]. Moreover, shallow trophoblast invasion and impaired uterine vasculature remodeling during the early stages of gestation can also perpetuate during pregnancy and lead to pregnancy complications such as preeclampsia, intrauterine growth restriction (IUGR), and premature birth [7–9]. Despite significant progress in reproductive research, efforts are still needed towards a better understanding of the physiological processes initiated during implantation, such as trophoblast invasion and migration, as well as the proteins that might orchestrate them. These two processes are closely related since both require changes in the cell microenvironment and concomitant activation of extracellular proteases [10]. The proteolytic activity of extracellular matrix-degrading metalloproteinases (MMPs) has been largely implicated in the efficiency of trophoblast invasion, specifically in the disruption of the extracellular matrix at the fetal-maternal interface during embryo implantation [11].

We have previously demonstrated, in a model of extravillous trophoblast of choriocarcinoma cell lines, that the forkhead box M1 (FOXM1) transcription factor is an important mediator of angiogenesis and migration, which are key processes involved in early placentation [12]. Additionally, we have also showed in a rat model that placental FOXM1 protein expression decreased as gestational age progresses, further indicating a role for FOXM1 in the early stages of pregnancy [12]. Although our previous results are compelling, the precise role of FOXM1 in implantation and whether FOXM1 is expressed in the pre-implantation blastocyst stage remain unknown. FOXM1 plays a crucial role in a plethora of biological processes, by directly regulating the transcription of downstream target genes involved in cell proliferation, cell cycle progression, angiogenesis, migration, invasion, cell differentiation, and DNA damage repair [13,14]. FOXM1 activity is also required in embryonic and fetal development [15,16]. Indeed, during mice embryogenesis, FOXM1 expression is found in the mesenchymal and epithelial cells of the liver, lung, intestine, renal cortex, and urinary tract; yet, its adult expression pattern is restricted to actively proliferating epithelial cells of the intestine, the spermatocytes and spermatids of the testis, the thymus, and colon [17]. Moreover, Foxm1 knockout mice are characterized by an embryonic lethal phenotype due to severe abnormalities in the development of the heart and liver [18]. In this study, we tested the hypothesis that FOXM1 is expressed in the trophectoderm of the blastocyst and participates in embryo implantation by contributing to trophoblast migration and early trophoblast invasion, by inducing transcription activation of genes involved in these processes.

2. Results

2.1. Foxm1 Is Expressed in the Inner Cell Mass and in the Trophectoderm of Mice Blastocysts

We have previously demonstrated that, in rat placentae, FoxM1 expression decreased as pregnancy progressed from E14.5 to E20.5, suggesting that FOXM1 is important in early placentation events [12]. To further elucidate the role of FOXM1 in the earlier stages of pregnancy, we first sought to indagate whether it is expressed as early as the stage of the pre-implantation blastocyst. Indeed, as seen in Figure 1A, mice embryos in stage E3.5 (early blastocysts) express *Foxm1* mRNA. Mice ovarian tissue was used as the positive control, and a band of *Foxm1* can be appreciated. *Gapdh* was included as the blastocyst housekeeping gene control. Regarding the localization of Foxm1 within the blastocyst, we observed that Foxm1 is expressed not only in the inner cell mass but also in the outer layer of the blastocyst cells (Figure 1B), namely the trophectoderm, an epithelial monolayer of cells, that differentiates into trophoblast cells and directly attaches to and invades the receptive endometrium to establish the placenta. These results strongly suggest that FOXM1 is involved in implantation and early placentation.

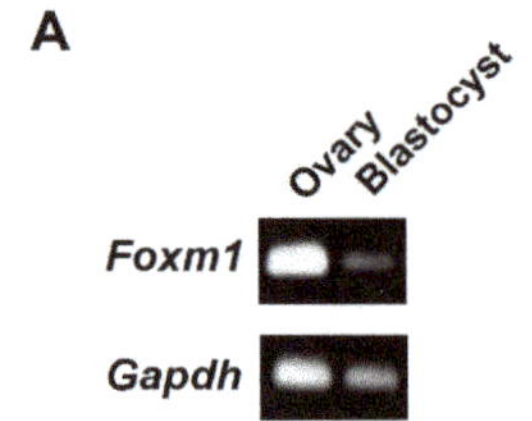

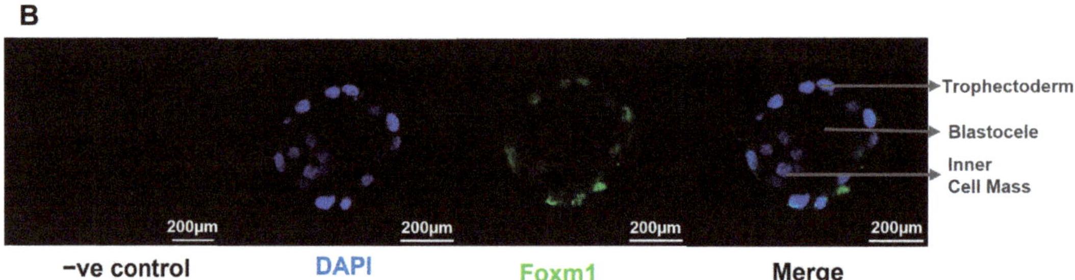

Figure 1. Characterization of Foxm1 in blastocyst. Blastocysts and ovaries were collected from C57BL/6 female mice at embryonic day 3.5 (E3.5). n = 4 mice. (**A**) Representative image of 14 blastocysts and 2 ovaries that were used to isolate RNA and amplified by qRT-PCR to evaluate the expression of *Foxm1*. PCR product was run on 1% agarose gel and visualized using a UV transilluminator. Bands correspond to the 101 bp *Foxm1* mRNA template and to the 83 bp *Gapdh* mRNA template that was incorporated as housekeeping gene control. (**B**) Blastocysts were pipetted into drops onto a glass slide and stained for FOXM1 with Alexa Fluor 488 anti-mouse sera (green) and counterstained with DAPI (blue) to visualize the nuclei. Single plane images were acquired with Leica TCS SP8 with a 20× air (N.A. = 0.5) objective. −ve control, negative control: no primary antibody was added. N.A., numerical aperture.

2.2. FOXM1 Expression Peaks at Time of Implantation in a Human First Trimester Trophoblast Cell Line

To explore whether FOXM1 plays a role in implantation, we analyzed its expression in a cell model of human first trimester trophoblasts, the HTR-8/SVneo cell line. It has been postulated that, in human pregnancy, the oxygen concentration within the uterine surface typically ranges from 2.5 to 5% O_2 (<20 mmHg) until the eighth week [19]. After the placenta establishes connections with maternal vasculature, that is around 12 weeks, the oxygen concentrations rise until around 8.5% (~60 mmHg). This remains steady until birth [8,19–21]. Having that controlled oxygen tension is determinant to achieve normal implantation and placentation, and since a chronic hypoxic environment leads to shallow trophoblast invasion, it is relevant to evaluate if FOXM1 expression changes in response to different oxygen tensions in first trimester human trophoblast cells. Thus, following exposure of HTR-8/SVneo cells to 21%, 8%, 3%, and 1% O_2, we observed that both FOXM1 protein (Figure 2A) and mRNA expression (Figure 2B) were significantly ($p = 0.0002$) higher when trophoblast cells were cultured at 3% O_2, which coincides with oxygen concentrations in the uteroplacental interface at the time of implantation. Conversely, we observed FOXM1 expression diminishing in response to 1% O_2 that resembles a hypoxic environment in utero. VEGF was included as a putative control target of FOXM1 [12,22]. However, we did not observe the same expression trend as with FOXM1.

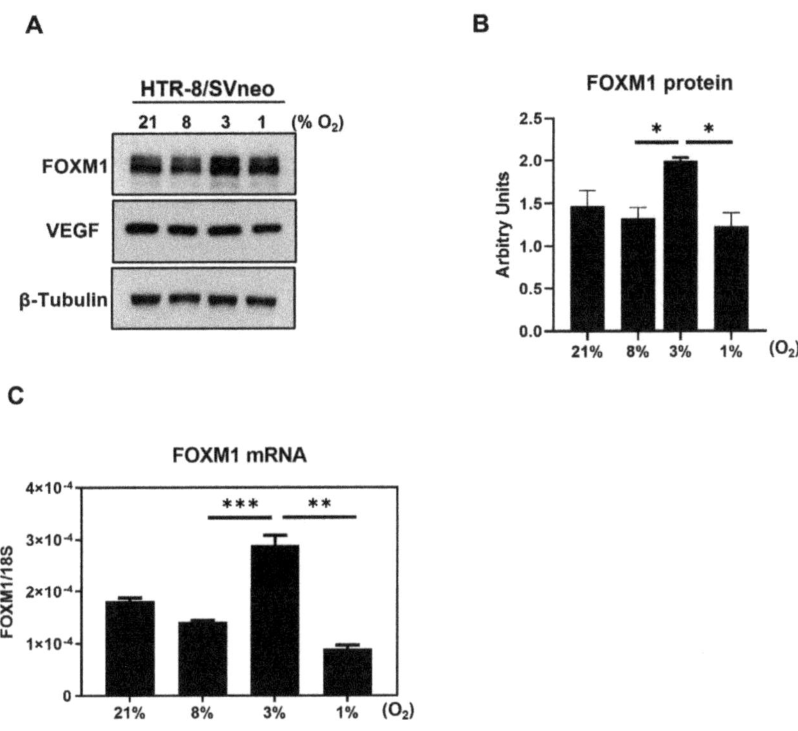

Figure 2. Expression of FOXM1 in the presence of different oxygen percentages. HTR-8/SVneo cells were exposed to 21%, 8%, 3%, and 1% O_2 for 24 h. (**A**) Cells were then collected and analyzed by western blot to determine the protein expression levels of FOXM1, VEGF, and β-Tubulin (loading control) (representative image of 3 independent experiments is shown). (**B**) FOXM1 band density was quantified and normalized to β-Tubulin. (**C**) qRT-PCR was performed to determine *FOXM1* mRNA transcript levels normalized to *18S* mRNA expression. Results are the mean ± SEM of three independent experiments in duplicate. Statistical analysis was performed using Student's *t*-tests. * $p \leq 0.05$, ** $p \leq 0.01$, *** $p \leq 0.001$, significant. SEM, standard error of the mean.

2.3. Overexpression of FOXM1 in HTR-8/SVneo Cell Line Induces Transmigration and Tubule Formation

To investigate if FOXM1 is important for peri-implantation events, we performed gain and loss of function assays in the first trimester human trophoblast cells and assessed whether changes in FOXM1 expression influence migration, angiogenesis, and the invasion ability of trophoblast cells. Since FOXM1 expression increased at 3% O_2 and diminished at 1% O_2, we knocked down FOXM1 at 3% O_2 and overexpressed FOXM1 at 1% O_2. As seen in Figure 3A,B, FOXM1 protein was efficiently silenced and overexpressed, respectively. Following FOXM1 silencing, a significantly lower ($p = 0.0001$) number of migrating cells was observed compared with the non-specific control cells (Figure 3C). However, HTR-8/SVneo cells depleted for FOXM1 only demonstrated a modest but non-significant difference in the tubule network formation compared to those cells transfected with the NS-siRNA control (Figure 3E). Concomitantly, FOXM1 overexpression significantly increased the transmigration ability (Figure 3D, $p = 0.0001$) and tubule formation (Figure 3F; nodes, $p = 0.0015$; junctions, $p = 0.0008$; meshes, $p = 0.0014$) of HTR-8/SVneo cells compared to empty-vector transfection. Nonetheless, we did not observe significant changes in the ability of HTR-8/SVneo cells to invade following overexpression or silencing of FOXM1, as determined by a transwell invasion assay (Supplementary Figure S1). Altogether, these results suggest that FOXM1 is involved in early placentation events, such as transmigration and angiogenesis.

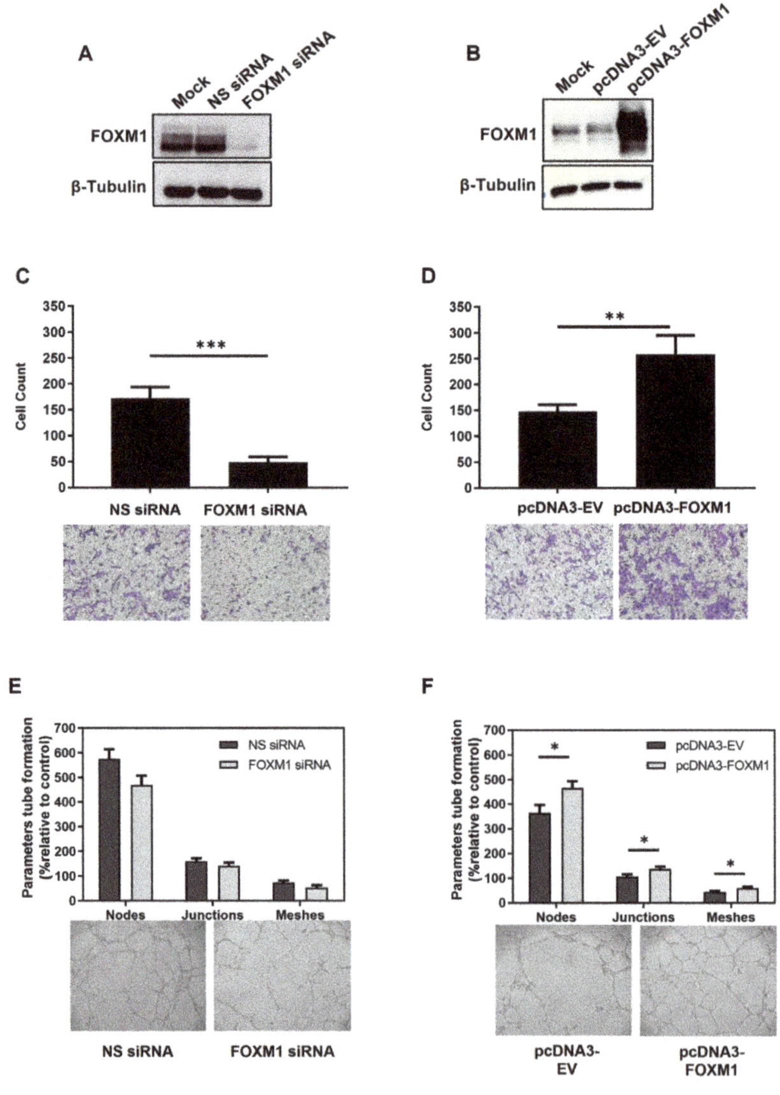

Figure 3. Effects of FOXM1 knockdown and overexpression in the migration and angiogenesis of HTR-8/SVneo cells. (**A**) HTR-8/SVneo cells maintained at 3% O_2 were transfected with mock (transfection reagent only), non-specific (NS) siRNA or FOXM1 siRNA for 24 h. (**B**) HTR-8/SVneo cells maintained at 1% O_2 were transfected with mock (transfection reagent only), pcDNA3-empty

vector (EV) (control) or pcDNA3-FOXM1 for 24 h. Cells were trypsinized and analyzed for western blotting to confirm: A. FOXM1 silencing and B. FOXM1 overexpression. Fifty thousand remaining cells from A and B were re-seeded in a millicell insert (pore 8 µm) with media with reduced serum (0.5% FBS) and placed on a well of a 48-well plate containing media supplemented with 10% FBS. Transmigration capacity was evaluated after 12 h of incubation, following (**C**) FOXM1 knockdown and following (**D**) FOXM1 overexpression. The represented data are averages of three independent experiments ± SEM. The percentage of migrated cells was calculated: number of cells after/before scraping × 100 (average of the 5 fields). Representative images are shown in the lower panel. Remaining cells from (**A**) and (**B**) were re-seeded onto 96-well plates pre-coated with MATRIGEL to evaluate tubule formation after 6 h of incubation, following (**E**) FOXM1 knockdown and following (**F**) FOXM1 overexpression. Number of nodes, junctions, and meshes were analyzed using ImageJ-Angiogenesis Analyzer-HUVEC Phase Contrast software v1.0.c. Nodes are the intersection of 3 segments, junctions are the combinations of several nodes, and meshes are networks in which nodes are linked together. Data are presented as mean values of three independent experiments ± SEM. Representative images show tube formation networks after 6 h (lower panel). Statistical analysis was conducted using Student's *t*-tests. * $p \leq 0.05$; ** $p \leq 0.01$; *** $p \leq 0.001$, significant. All images were captured with the Primo Vert microscope (Zeiss, Jena, Germany) with a 4× (N.A. = 0.10, W.D. = 12 mm) objective using the AxioCam ERc5s camera (Zeiss, Jena, Germany). SEM, standard error of the mean. N.A., numerical aperture.

2.4. FOXM1 Is Involved in Early Invasion of the Trophoblast

During implantation, trophoblast invasion of the uterine lining depends on the communication between the trophectoderm and the maternal decidua and involves the migration and invasion of the trophoblast cells from out of the blastocyst and through various layers of the endometrium, accompanied by the degradation of the extracellular matrix and stroma [5,10]. In order to understand whether FOXM1 is involved in trophoblast invasion during implantation, we used an in vitro 3D trophoblast invasion model [23,24] that mimics the processes of migration and early invasion of the trophoblast and assessed the expression of FOXM1 in trophoblast cells. This assay consists of generating, in vitro, blastocyst like-structures (trophospheres) and co-culturing them onto a layer of MATRIGEL and mesenchymal stem cells isolated from menstrual fluid (MenSCs), mimicking the extracellular matrix and the endometrial stromal cells, respectively (Figure 4A). To resemble the physiological hormones and O_2 conditions within the endometrium during the mid-secretory phase of the menstrual cycle, MenSCs were subjected to E2 for 24 h followed by E2 and P4 for an additional 24 h at 5% O_2 (endometrial mimic), prior to the co-culture with trophospheres. Following 72 h of the co-culture, we observed the formation of projections that penetrated the MATRIGEL in a radial orientation (Figure 4B, top image), which are trophoblast cells that have migrated from the compact blastocyst-like spheroid and invaded the MATRIGEL. In contrast, we hardly observed trophoblast invasion when the trophospheres were cultured only with MATRIGEL, demonstrating that MATRIGEL without MenSCs does not provide the necessary signals to promote trophoblast migration and invasion. Interestingly, we found that FOXM1 transcript levels of the trophospheres that were co-cultured with MenSCs and MATRIGEL were significantly higher compared to trophospheres before 3D invasion assay (Figure 4C, $p = 0.02$), indicating that FOXM1 may be required during early trophoblast invasion of the maternal decidua. Additionally, trophospheres that were generated with pcDNA3-FOXM1 cells exhibited a significant increase ($p = 0.03$) in the area invaded by the trophosphere, compared to those cells transfected with pcDNA3-empty vector control (Figure 5). These results further suggest that FOXM1 plays a role in early invasion of the trophoblast and that this process depends on communication mechanisms between maternal decidual cells and the trophoblast cells of the blastocyst.

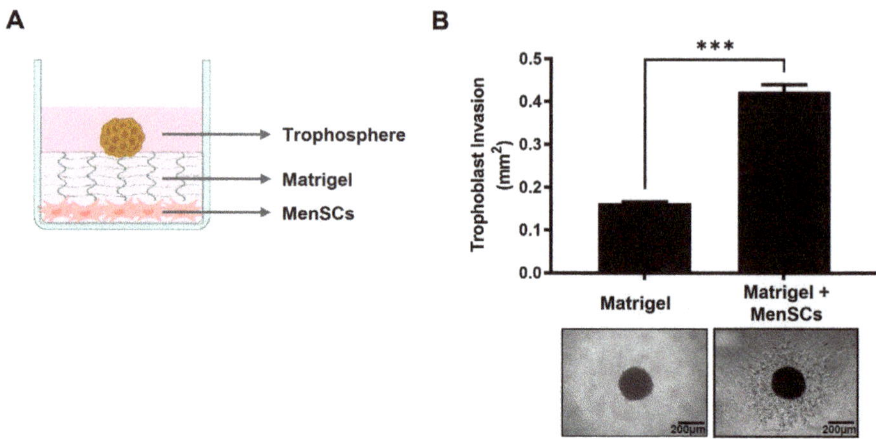

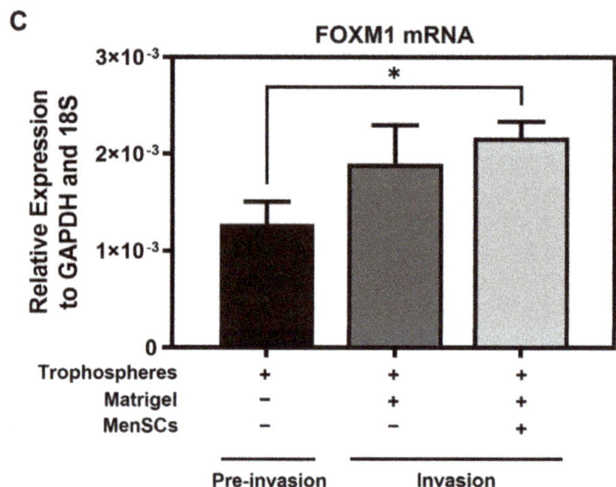

Figure 4. FOXM1 expression is upregulated when trophoblast cells are invading into the matrix. (**A**) Schematic representation of the components of the 3D trophoblast invasion model: MenSCs on the bottom of the plate as a monolayer of endometrial stromal cells, MATRIGEL mimicking the extracellular matrix, and on top is the HTR-8/SVneo trophosphere that is used to mimic the trophectoderm of the blastocyst (image created with BioRender.com). (**B**) HTR-8/SVneo trophospheres were transferred to 8 individual wells of a 96-well plate that contained MATRIGEL alone or a monolayer of MenSCs (n = 4) (previously subjected to endometrial mimic) seeded below the MATRIGEL matrix for 72 h. Representative images of trophoblast migration from the trophospheres and invading into the MATRIGEL were captured at day 3 with an Olympus CKX41 microscope and an Axiocam 208 color camera (10× objective lens, N.A. = 0.25, W.D. = 10.5 mm). (**C**) Bars represent: in black: trophospheres immediately after formation, in dark grey: trophospheres in MATRIGEL, and in light grey: trophospheres in MATRIGEL with MenSCs after 3D invasion assay. Trophospheres were collected and recovered from MATRIGEL, and FOXM1 mRNA was analyzed using qRT-PCR and normalized to *GAPDH* and *18S* housekeeping genes. Results are expressed as mean ± SEM of eight trophospheres and n = 4 MenSCs. Statistical analysis was conducted using Student's t-tests. * $p \leq 0.05$; *** $p \leq 0.001$, significant. MenSCs, mesenchymal stem cells isolated from menstrual fluid; SEM, standard error of the mean. N.A., numerical aperture.

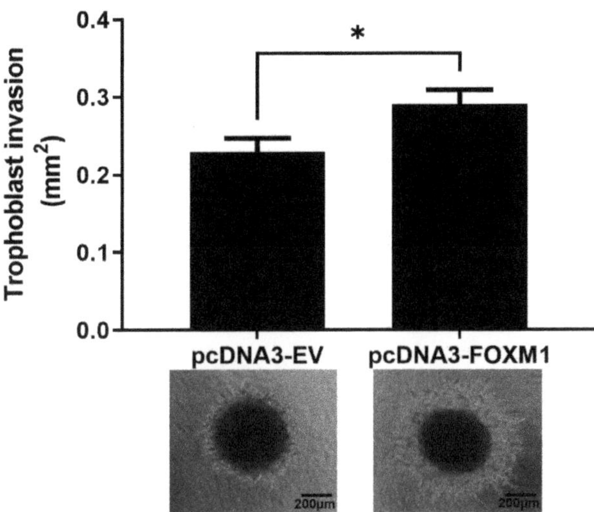

Figure 5. Trophoblast invasion increases in FOXM1-overexpressed trophospheres. HTR-8/SVneo cells were transfected either with pcDNA3-empty vector (EV) (control) or pcDNA3-FOXM1 for 24 h prior to trophosphere generation. Trophospheres were then transferred onto MATRIGEL + MenSCs and trophoblast invasion was evaluated after 48 h. Phase contrast images were captured by the Olympus CKX41 and Axiocam 208 camera (10× objective lens, N.A. = 0.25, W.D. = 10.5). Trophoblast invasion was quantified as area (mm^2) of trophosphere + invasive trophoblasts (projections that emerge from the trophosphere) by using the ImageJ software v1.54h. Columns are the mean of 2 independent experiments with 6 trophospheres per condition; bars ± SEM. Statistical analysis was conducted using Student's *t*-tests. * $p \leq 0.05$, significant. SEM, standard error of the mean. N.A., numerical aperture.

2.5. FOXM1 Modulates PKL4, VEGF, and MMP2 Transcript Expression in HTR8/SVneo Cells

Fox proteins have been reported to bind to the core consensus sequence AAA(C/T)A [25] of their targets´ promoters in order to promote transcription. This consensus sequence is frequently referred to as forkhead responsive element (FHRE). Since FOXM1 may be modulating migration and invasion of trophoblast cells, we seek to identify potential FOXM1 targets whereby it could be mediating these processes. Thus, we analyzed chromatin immunoprecipitation followed by sequencing (ChIP-seq) to investigate the genome-wide chromatin binding mechanisms used by FOXM1. Data for FOXM1 ChIP-seq experiments were obtained from the NCBI GEO database using the accession number GSE32465 [26]. We downloaded three FOXM1 ChIP-seq experiments conducted on ECC1 (human endometrium adenocarcinoma cell line), MCF-7 (human breast adenocarcinoma cell line), and SK-N-SH (human neuroblastoma cell line) cell lines and visualized them using the Interactive Genomic Viewer (IGV). As shown in Figure 6A, FOXM1 binds largely to the proximal promoter regions of *PLK4* and *VEGF* of the different cancer cell lines analyzed (ECC1, MCF-7, and SK-N-SH). To validate these findings in HTR-8/Svneo cells, we evaluated the transcription expression of *VEGF* and *PLK4* following knockdown and overexpression of FOXM1 in HTR-8/Svneo cells (Figure 7A,B). In concordance with the in silico analysis, the depletion of FOXM1 led to a downregulation of *VEGF* ($p = 0.03$) and *PLK4* ($p = 0.01$) mRNA expression, and a significant increase in the transcription levels of these genes was detected following ectopic expression of FOXM1 (*VEGF*, $p = 0.005$; *PLK4*, $p = 0.009$). Since MMP2 has been identified to be secreted by the cytotrophoblast cells that are responsible in digesting the major constituents of the endometrial matrix [27] and has been demonstrated to be a direct target of FOXM1 in retinoblastoma cells [28], we also evaluated the mRNA expression of *MMP2* following gain ($p = 0.03$) and loss ($p = 0.03$) of FOXM1 expression. Interestingly, we observed that MMP2 also followed the same pattern of FOXM1 (Figure 7A,B). Altogether, these results provide evidence that FOXM1 binds to the promotors and directly regulates the transcription of *VEGF*, *MMP2*, and *PLK4* genes, which have important roles in angiogenesis, invasion, and trophoblast differentiation, respectively.

A

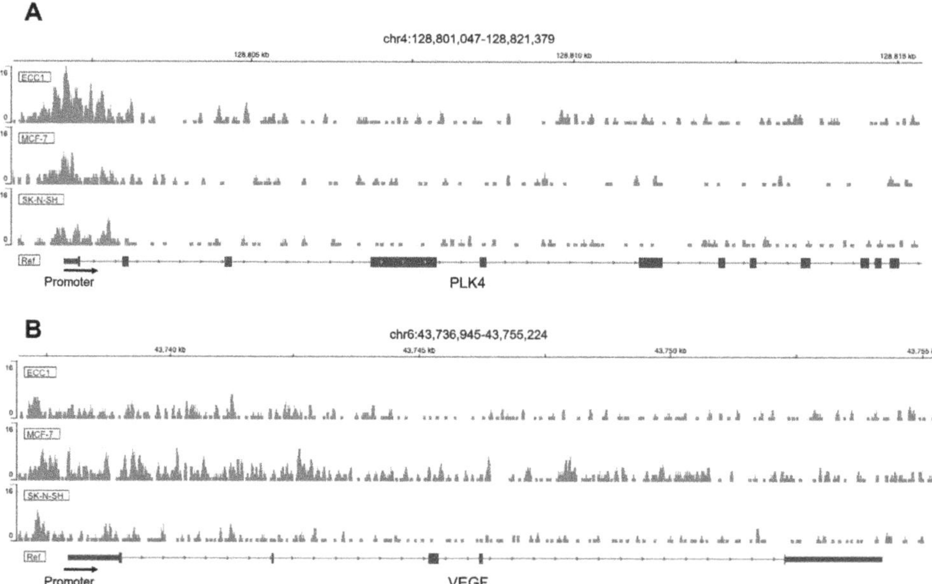

B

Figure 6. ChIP-Seq enrichment of FOXM1 in 3 different cell lines. Genome binding of FOXM1 transcription factor to (**A**) PLK4 and (**B**) VEGF genes in different cell lines (ECC1, MCF-7, and SK-N-SH). The *x*-axis represents genomic coordinates and the *y*-axis for each track is auto-scaled to the highest peak in each chromosome region shown.

A

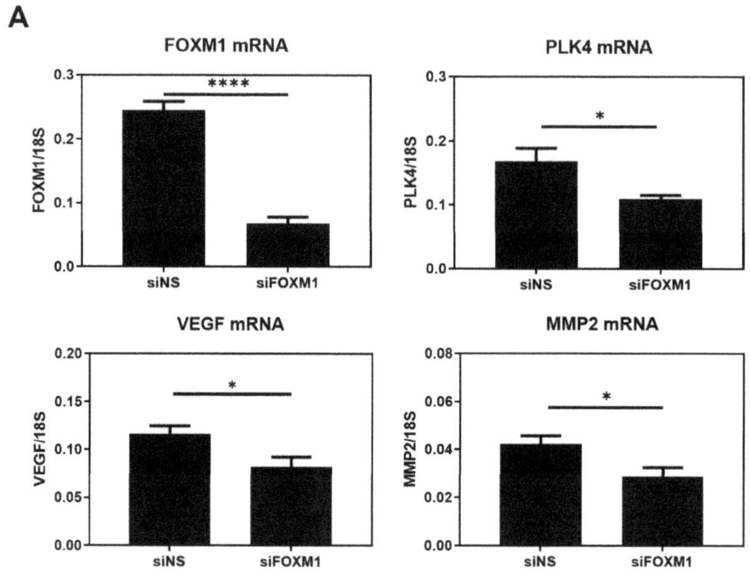

Figure 7. *Cont.*

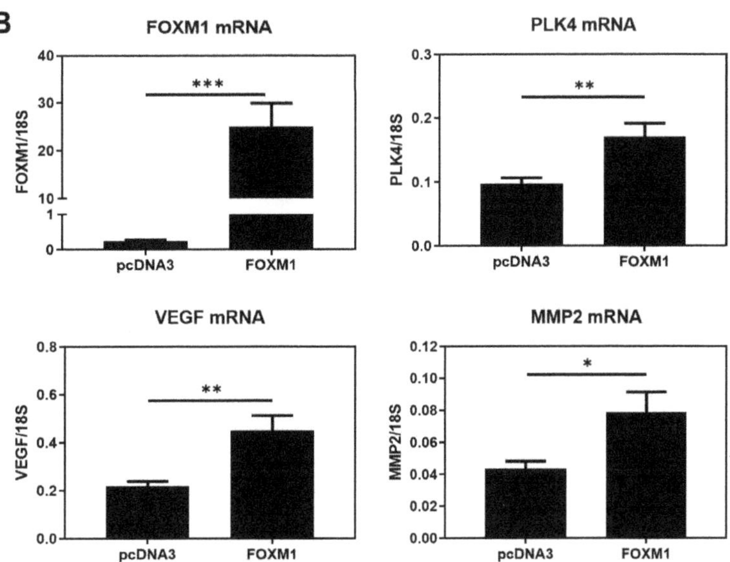

Figure 7. FOXM1 modulates *VEGF*, *PLK4*, and *MMP2* transcript levels in HTR-8/Svneo cells. (**A**) HTR-8/Svneo cells cultured at 3% O_2 were either transfected with NS siRNA or with FOXM1 siRNA specific pool. (**B**) HTR-8/Svneo cells cultured at 1% O_2 were transfected with pcDNA3-empty vector (EV) (control) or pcDNA3-FOXM1. Twenty-four hours post-transfection, qRT-PCR analysis was conducted to examine *FOXM1*, *MMP2*, *VEGF*, and *PLK* mRNA expression. All qRT-PCR results were normalized to *18S* mRNA expression. Columns are the mean of 3 independent experiments in duplicate; bars $\pm$ SEM. Statistical significance was performed using Student's *t*-tests. * $p \leq 0.05$; ** $p \leq 0.01$; *** $p \leq 0.001$; **** $p \leq 0.0001$; significant. SEM, standard error of the mean.

3. Discussion

Implantation and early development of the placenta are highly regulated processes that are pivotal for a successful pregnancy. Indeed, there is large body of evidence showing that the impairment of these processes contributes to the etiopathology of pregnancy complications such as preeclampsia, intrauterine growth restriction, and preterm labor. An improved understanding of the molecular mechanisms involved in normal implantation and early trophoblast development will improve the ability of clinicians to understand and treat these pregnancy disorders. Using a model of choriocarcinoma cell lines that exhibits an extravillous trophoblast phenotype, we showed that FOXM1 is an important mediator of key processes involved in early placentation including angiogenesis and migration [12]. Here, we further demonstrate that, in a model of first trimester trophoblast cells, FOXM1 expression significantly increases in response to 3% O_2, which coincides with the oxygen concentration within the uterine surface at implantation, and 1% oxygen, resembling a hypoxic environment, diminished FOXM1 expression [19]. Moreover, we have observed, for the first time, that FOXM1 is expressed as early as the blastocyst stage in mice embryos and that it is not only is expressed in the inner cell mass but also in the trophectoderm, highlighting the potential role of this protein in the early phases of trophoblast differentiation and blastocyst implantation. This process begins when the trophectoderm cells of the blastocyst attach to the endometrial epithelium, start to migrate out of the blastocyst, and invade the endometrium [1–3]. The detection of FOXM1 in the inner cell mass was not surprising since FOXM1 has been shown to be required for embryonic and fetal development [15,16]. Indeed, during embryogenesis, FOXM1 is expressed to ensure correct development of both epithelial and mesenchymal tissues [17].

Additionally, FOXM1 has been demonstrated to be essential for embryonic development of mice pulmonary vasculature [18].

In the cancer scenario, it has been demonstrated that FOXM1 induces tumor proliferation, migration, invasion, and angiogenesis [14], which are processes shared by trophoblast cells during the peri-implantation phase. Accordingly, we have also demonstrated that FOXM1 is necessary for the migration and angiogenesis of the human HTR8/Svneo first trimester trophoblast cell line while not influencing its proliferative ability (Supplementary Figure S2). Moreover, neither silencing nor overexpression of FOXM1 led to statistically significant effects on the ability of HTR8/Svneo cells to invade MATRIGEL, as assessed by transwell invasion assays (Supplementary Figure S1).

Successful implantation requires effective maternal-embryonic communication [2–4,27]. Using an in vitro 3D trophoblast invasion model, we have confirmed that trophoblast cells were only able to migrate from the trophosphere and invade through the MATRIGEL in the presence of MenSCs of endometrial origin, which was pre-treated with the endometrial mimic, suggesting that MenSCs secrete important factors that trigger the motility and invasion of the trophoblast cells, as described previously [10,23,24]. More importantly, FOXM1 mRNA expression was significantly higher in those trophoblast cells that invaded greater into the MATRIGEL, i.e., the trophospheres that were co-cultured with MATRIGEL and MenSCs compared to pre-invasion trophospheres. Together, these results suggest that trophoblast invasion strongly relies on maternal-fetal communication and that maternal decidual cells may induce trophoblast cells to express higher levels of FOXM1 that subsequently activate genes involved in trophoblast migration and invasion. Mitochondria or extracellular vesicle transfer could be studied as potential communication mechanisms. To further validate that FOXM1 is indeed required for trophoblast invasion, we demonstrated that FOXM1-overexpressed trophospheres exhibited a significant increase in the area invaded by the trophosphere, compared to those cells transfected with the pcDNA3-empty vector control. It is worth mentioning that spheroid formation from HTR-8/Svneo morphologically resembles a blastocyst and expresses higher stemness markers, NANOG and SOX2, and lower CDH1 epithelial marker, in comparison with its adherent counterpart [23].

As a transcription factor, it is plausible that FOXM1 exerts its effects by directly binding the promoters and inducing the transcription of genes involved in trophoblast differentiation, migration, invasion, and angiogenesis. Thus, we propose that *PLK4*, *VEGF*, and *MMP2* are FOXM1 downstream targets involved in these processes. Invasion and migration are closely related mechanisms that involve the ability of the cells to move in response to a stimulus and to advance through the extracellular matrix within a tissue by activating extracellular proteases [29]. The proteolytic activity of extracellular matrix-degrading metalloproteinases (MMPs) has been largely implicated in the efficiency of trophoblast invasion, specifically in disrupting the extracellular matrix at the fetal-maternal interface during embryo implantation [11]. Specifically, MMP2 has been identified to be secreted by the cytotrophoblast cells that are responsible in digesting the major constituents of the endometrial matrix [27]. Indeed, MMP2 can be directly regulated by FOXM1 in human retinoblastoma Y-79 cells [28]. Consistently, a large-scale microarray study of the transcriptome of the rat placenta throughout mid-late gestation also reveals the concomitant expression of FOXM1 and angiogenic genes, including VEGFA, MMP14, Caveolin-1, and Angiopoietin-4 [30]. Moreover, VEGF has also been demonstrated to be a direct target of FOXM1 in breast cancer cells [22]. Polo-like kinase four (PLK4) is a member of the serine/threonine kinase family that plays an important role in cell cycle regulation, by participating in centriole duplication and maintaining mitotic accuracy in normal cells. Its deregulation, henceforth, has been associated with a prominent role in cancer and metastasis [31,32]. PLK4 has also been demonstrated to be involved in the differentiation of trophoblast stem cells into trophoblast giant cells during placental development [31,33]. This evidence, together with the ChIP-Seq data, indicates that *VEFG*, *PLK4*, and *MMP2* are potential downstream targets of FOXM1, through which it promotes trophoblast differentiation, migration, invasion, and angiogenesis during early placentation. To support this conjecture, we have also shown that the mRNA expression of these genes is significantly downregulated following silencing of FOXM1 and significantly upregulated following FOXM1 overexpression in HTR-8/Svneo cells. Further studies using ChIP are required to validate that *VEFG*, *PLK4*, and *MMP2* are downstream targets of FOXM1 in HTR-8/Svneo cells.

Our results demonstrate, for the first time, that FOXM1 is present in the trophectoderm of mice blastocysts. By using a 3D in vitro model of implantation, we have confirmed that maternal-fetal communication is crucial for trophoblast invasion and that maternal stromal cells may induce trophoblast cells to express higher levels of FOXM1, which subsequently activates the transcription of its downstream targets genes that are important for successful implantation: *PLK4*, *VEGF*, and *MMP2*.

4. Materials and Methods

4.1. Cell Culture

The human HTR8/Svneo first trimester trophoblast cell line was purchased from the American Type Culture Collection (CRL-3271; Lot #70016636, ATCC, Manassas, VA, USA) and cultured in RPMI-1640 medium (GE Healthcare, Piscataway, NJ, USA), supplemented with 10% heat-inactivated fetal bovine serum (FBS), 1 mmol/L sodium pyruvate, and 1% penicillin/streptomycin (P/S) (all Gibco, ThermoFisher, Waltham, MA, USA). Cells were maintained at 37 °C in a humidified incubator with 5% CO_2. For hypoxia experiments, cells were seeded in the appropriate cell culture plastic wells and maintained at 21%, 8%, 3%, or 1% O_2 for 24 h to 72 h prior to experiments. Hypoxic chamber C-474 equipped with Pro-Ox 110 oxygen controlling regulator (BioSpherix, Parish, NY, USA) was used.

4.2. Embryo Collection

C57BL/6 mice were purchased from Universidad de Chile. Animals were maintained in the animal facility of Universidad de los Andes at 25 °C and 12 h light:dark cycling and fed standard chow and water ad libitum. Protocols were conducted in agreement with the National Research Council (NRC) publication Guide for the Care and Use of Laboratory Animals (eighth edition, The National Academies Press, Washington, DC, USA). The animal study was reviewed and approved by the Ethical Scientific Committees of Universidad de los Andes, Santiago, Chile. For blastocyst collection, female mice were superovulated by intraperitoneal injections with 5 IU of pregnant mares' serum gonadotropin (PMSG) (Novormon, Buenos Aires, Argentina) and 5 IU of human chorionic gonadotropin (HCG) (Sigma-Aldrich, St Louis, MO, USA) 48 h later. Immediately after HCG injection, female mice were mated 1:2 to male mice of the same strain. Mating was evaluated by inspection of the vaginal plug on the following day. The presence of sperm plug was considered as embryonic day 0.5 (E0.5) of pregnancy. Pregnant females were euthanized at E3.5 and subsequently, blastocysts were obtained. The procedure involved initial anesthesia with ketamine:xylazine, followed by euthanasia by cervical dislocation. Embryos were collected by flushing the uterus with M2 medium (Sigma-Aldrich, St Louis, MO, USA). Then, freshly harvested embryos were washed once in M2, followed by three washes with M2 medium containing polyvinyl pyrrolidone (PVP; 6 mg/mL, H6PVP Sigma-Aldrich, USA) and three washes with 1X phosphate-buffered saline (PBS) (Cytiva, Logan, UT, USA). Embryos and oocytes with signs of disintegration were classified as degenerated embryos or oocytes. At least 14 embryos at the blastocyst stage were used for qRT-PCR. The blastocysts were transferred to a tube with a minimal volume (1-2 µL) of PBS, snap-frozen in liquid nitrogen, and stored at −80 °C. For blastocyst differential labelling of cell lineages by immunofluorescence, embryos were washed and used immediately.

4.3. Immunofluorescent Staining

Immunofluorescence staining of the blastocyst was carried out as described previously [34]. After embryo collection, blastocysts (E3.5) were washed three times in M2 medium, and the *zona pellucida* was removed using acidified Tyrodes solution (Sigma-Aldrich, USA), at pH 2.3 at 37 °C for 30–60 s, followed by extensive washing in M2. Blastocysts were fixed with 4%

PFA (ThermoFisher, Rockford, IL, USA) in PBS for 15 min at room temperature. The fixative was removed, and the embryos were washed three times with 1X PBS and then permeabilized with 0.2% Triton X-100 (Sigma-Aldrich, USA) in 1X PBS for 10 min at room temperature. Embryos were washed again three times with PBS containing 0.1% (*v/v*) Tween 20 (Winkler, Santiago, Chile) (PBST), followed by blocking with 1% BSA (Rockland, Philadelphia, PA, USA) and 5% goat serum G9023 (Sigma-Aldrich, St. Louis, MO, USA) in PBST for 1 h at room temperature. FOXM1 G-5 (Santa Cruz Biotechnology, Santa Cruz, CA, USA) primary antibody in PBST, 1% BSA, 1% goat serum, and glycine at 0.05 M (Winkler, Chile) were added overnight at 4 °C. The next day, the embryos were washed three times with PBST with 1% BSA, followed by secondary antibody Alexa Fluor 488 anti-mouse (ThermoFisher, Eugene, OR, USA) incubation for 1 h at room temperature in a dark environment. Final PBST with BSA washes were performed. A drop of Fluoromount-G (Invitrogen, Carlsbad CA, USA) mounting medium with 4′,6-diamidino-2-phenylindole (DAPI) (Vector Laboratories, Burlingame, CA, USA) was added to each drop of blastocysts before addition of a coverslip and sealing with clear nail varnish. After drying, slides were stored at 4 °C. Images were collected with a Leica TCS SP8 confocal microscope (Leica Microsystems, Mannhein, Germany) with 20× objective (N.A. 0.5) using LAS-X software v3.0 (Leica Microsystems, Mannhein, Germany).

4.4. Western Blot Analysis

Cells were harvested by trypsinization and whole cell lysates prepared as described previously [35]. Primary antibodies, FOXM1 D12D5 (Cell Signaling, Danvers, MA, USA), β-Tubulin H-235 (Santa Cruz Biotechnology, USA), and VEFG-A ab46154 (Abcam, Cambridge, MA, USA), were detected using horseradish peroxidase-linked anti-rabbit or anti-mouse conjugates as appropriate (KPL, Gaithersburg, MD, USA), and proteins were visualized using enhanced chemiluminescence (ECL) Pierce ECL Western Blotting Substrate detection system (ThermoFisher, USA) with X-ray films. The obtained images were analyzed using ImageJ software v1.54h (National Institutes of Health, Bethesda, MD, USA). All samples were normalized for protein loading using β-tubulin.

4.5. RNA Isolation and Quantitative Real-Time PCR

Total RNA was extracted from HTR8/SVneo cells with the TRIzol Reagent (Ambion, Carlsbad CA, USA) according to the manufacturer's protocols. RNA from the pooled blastocyst ($n = 14$) was isolated using PureLink RNA Micro Kit (Invitrogen, Carlsbad CA, USA) according to the manufacturer's protocols. RNA concentration and quality were evaluated using the NanoDrop One spectrophotometer (ThermoFisher, Madison WI, USA). Before cDNA synthesis, total RNA was treated with DNase I (Invitrogen, Carlsbad CA, USA). Complementary DNA generated by SuperScript II Reverse Transcription System (Invitrogen, Carlsbad CA, USA), according to the manufacturer's instructions, was analyzed using quantitative real-time PCR (qRT-PCR) in the Stratagene Mx3000P system (Agilent Technologies, Santa Clara, CA, USA), using Brilliant III SYBR Green qPCR Master Mix (Agilent Technologies, USA). The qRT-PCR was set to 95 °C for 10 min for enzyme activation, followed by 35 cycles of denaturation and primer annealing/extension consisting of 95 °C for 15 s, 60 °C for 15 s, and 72 °C for 15 s, respectively. After the PCR runs, a dissociation curve was generated to confirm the absence of nonspecific amplification. Transcription levels were quantified using the $2^{-\Delta\Delta CT}$ method. 18S ribosomal RNA or GAPDH housekeeping genes were used to normalize input complementary DNA. The following gene-specific primers were used for mouse: Foxm1-sense: 5′-GGACATCTACACTTGGATTGAGG-3′ and Foxm1-antisense: 5′-TGTCATGGAGAGAAAGGTTGTG-3′; Gapdh-sense: 5′-AGTGGCAAAGTGGAGATT-3′ and Gapdh-antisense 5′-GTGGAGTCATACTGGAACA-3′; and for human HTR8/SVneo cells: FOXM1-sense: 5′-TGCAGCTAGGGATGTGAATCTTC-3′ and FOXM1-antisense: 5′-GGAGCCCAGTCCATCAGAACT-3′; VEGF-sense: 5′-TATGCGGATCAAACCTCACC-3′ and VEGF-antisense: 5′-CTTGTCTTGCTCTATCTTTCTTTGG-3′; MMP2-sense: 5′-TGTG-ACGCCACGTGACAAG-3′ and MMP2-antisense: 5′-CCAGTATTCATTCCCTGCAAAGA-

3′; PLK-4-sense: 5′-GATAGACCACCCTCACCTACT-3′ and PLK-4-antisense: 5′-CTGTACA-AACCTGGAAGCATATTG-3′; 18S-sense: 5′-GCCGCTAGAGGTGAAATTCTTGGA-3′ and 18S-antisense: 5′-ATCGCCAGTCGGCATCGTTTAT-3′; GAPDH-sense: GTCAGGGTCTCT-CTCTTCCT and GAPDH-antisense: GCTCTCCTCTGACTTGAACA.

4.6. Transient Gene Silencing and Overexpression

For gene silencing, HTR8/SVneo cells were transiently transfected with ON-TARGETplus SMARTpool siRNA (Dharmacon ThermoFisher, Lafayette, CO, USA), using Oligofectamine (Invitrogen, Carlsbad, CA, USA) and Opti-MEM (Gibco, ThermoFisher, Waltham, MA, USA) according to the manufacturer's instructions. siRNA FOXM1 (L-009762-00-0005) and the non-specific (NS) control siRNA (D-001810-10-05) SMARTpools were used. RNA interference experiments were carried out at 3% O_2. For gene overexpression, HTR8/SVneo cells were transfected with either the pcDNA3-empty vector or pcDNA3-FOXM1 [36] using FuGENE 6 Transfection Reagent (Promega, Madison, WI, USA) in a 3:1 ratio (μL of FuGENE: μg of DNA) according to the manufacturer's recommendations. The overexpression experiments were performed at 1% O_2.

4.7. Transmigration Assay

HTR8/SVneo cells transiently transfected with either FOXM1 siRNA or control NS siRNA, or pcDNA3-FOXM1 or empty vector EV-pcDNA3, under different oxygen tensions were used to examine cell migration in vitro. Following 24 h of transfection, 50,000 HTR-8/SVneo cells were seeded in a millicell insert (pore 8 μm, 12 mm, Millipore, Billerica, MA, USA) with 400 μL RPMI with reduced serum (0.5% FBS). Inserts were placed on 500 μL of RPMI supplemented with 10% FBS in 48-well plates. Transmigration capacity was evaluated following 12 h incubation either at 3 or 1% O_2. Briefly, the insert was washed with 1X PBS, fixed with cold methanol for 2 min, and stained with 0.5% crystal violet (Winkler, Santiago, Chile). Cells inside the inserts were scraped with cotton swabs moistened with 1X PBS to ensure that only migrated cells were analyzed. Five fields were captured for each insert at 4× objective magnification (N.A. = 0.10, W.D. = 10.5 mm) before and after scraping under an inverted microscope (Primo Vert, Zeiss, Jena, Germany), using the AxioCam ERc5s camera (Zeiss, Jena, Germany). Images were analyzed with AxioVision Rel analysis software v4.8 (Zeiss, Jena, Germany). The percentage of migrated cells was calculated as follows: number of cells after/before scraping × 100 (average of the 5 fields). The experiments were performed in duplicate.

4.8. MATRIGEL-Based Tube Forming Assay

HTR8/SVneo cells transiently transfected with either FOXM1 siRNA or control NS siRNA, or pcDNA3-FOXM1 or empty vector EV-pcDNA3, under different oxygen tensions were resuspended in the RPMI medium with 2% FBS or in the endothelial growth media-2 (EGM-2, Bullet Kit, Lonza, Verviers, Belgium) (positive control) and seeded in triplicate in 96-well plates pre-coated with 70 μL growth factor reduced phenol-red free MATRIGEL matrix (Corning Life Sciences, Union City, CA, USA). Cells were incubated at 37 °C and under 3 or 1% O_2 (accordingly) for 6 h and tube-like structures were examined with a phase-contrast microscope Primo Vert (Zeiss, Jena, Germany). One image per well was captured using an AxioCam ERc5s camera (Zeiss, Jena, Germany). Number of nodes, junctions, and meshes were analyzed using ImageJ-Angiogenesis Analyzer-HUVEC Phase Contrast software v1.0.c.

4.9. Trophosphere Formation

Trophoblast spheres (Trophospheres) are blastocyst like-structures formed in vitro using adherent HTR8/SVneo cells, as described previously [23]. In brief, 20,000 HTR8/SVneo cells were suspended in 200 μL of supplemented RPMI media and placed into each well of an ultra-low attachment 96-well plate (Costar, Kennebunk, ME, USA). Following centrifugation at 300× g for 5 min, cells were incubated for 72 h at 37 °C in a humidified atmosphere hypoxia chamber with 5% O_2 and 5% CO_2. The trophospheres were washed with 1X PBS and used for 3D invasion assays. The differential cellular characteristics of trophoblast cells as a 3D model or monolayer were previously described in detail [23].

4.10. 3D Trophoblast Invasion Assay

To mimic the structure of the endometrium, mesenchymal stem cells isolated from menstrual fluid of healthy donors, who had not used hormonal contraceptives for at least three months, were isolated and cultured as described in [23,24]. All experiments were performed using MenSCs at early passages (P) P3 to P7 ($n = 4$). Menstrual fluid was self-collected by consenting donors following their written informed consent to participate in this study according to a protocol reviewed and approved by the Ethical Scientific Committee of the Universidad de los Andes, Santiago, Chile. To achieve the 3D invasion assay [23], growth factor reduced phenol-red free MATRIGEL (Corning Life Sciences, Union City, CA, USA) was mixed with Dulbecco's Modified Eagle Medium (DMEM) without phenol red (Mediatech Inc., Manassas, VA, USA) containing 10% charcoal-stripped FBS and 1% P/S in a 1:1 ratio and added onto 3000 MenSCs that had been previously seeded in a 96-well plate and subjected to 213 pg/mL 17β-estradiol (E2) for 24 h, followed by 146 pg/mL E2 and 11 ng/mL progesterone (P4) for an additional 24 h at 5% O_2 (endometrial mimic), in order to resemble the physiological hormonal and O_2 conditions within the endometrium, throughout the menstrual cycle [23]. The plate was incubated for 30 min at 37 °C to allow the MATRIGEL to solidify. A single trophosphere was placed on each well containing MenSCs and MATRIGEL, and 150 μL of endometrial mimic media were added to embed the sphere. Plates were incubated at 37 °C in a humidified atmosphere hypoxia chamber with 5% O_2 and 5% CO_2. Trophosphere invasion was evaluated after 72 h. Phase contrast images were captured by the contrast microscope Olympus CKX41 and Axiocam 208 camara (10× objective lens, N.A. = 0.25, W.D. = 10.5 mm) and AxioVision Rel software v4.8 (Zeiss, Jena, Germany). The invasion level (area) was quantified by using ImageJ software v1.51j8, and the trophospheres were collected for qRT-PCR assay. For mRNA analysis of the trophospheres following 3D invasion assay, the trophospheres were recovered from MATRIGEL and incubated with 300 μL of Cell Recovery Solution (Corning Life Sciences, Bedford, MA, USA) for 15 min on ice, followed by centrifugation at 400× *g* for 5 min at 4 °C. The supernatant was removed and the trophospheres were washed with PBS, followed with centrifugation at 350× *g* for 2 min at 4 °C. PBS was removed and the trophospheres were lysed with TRIzol for RNA extraction.

4.11. Statistical Analyses

Statistical comparisons between two means were performed using Student's *t*-tests. Results were deemed statistically significant when $p \leq 0.05$. Data normality was tested using the D'Agostino-Pearson test. For analysis of 2 groups, unpaired *t*-test was used. Graphing and statistical analyses were performed using GraphPad Prism 9.0.

Supplementary Materials: The following supporting information can be downloaded at: https://www.mdpi.com/article/10.3390/ijms25031678/s1.

Author Contributions: Conceptualization, L.J.M. and S.E.I.; methodology, R.P., S.A.-G., V.V., F.G. and M.S.; writing—original draft preparation, R.P., V.V., M.S. and L.J.M.; writing—review and editing, R.P., V.V., S.A.-G., G.N., S.E.I. and L.J.M.; visualization, R.P., V.V., G.N. and L.J.M.; supervision, L.J.M.; funding acquisition, L.J.M. and S.E.I. All authors have read and agreed to the published version of the manuscript.

Funding: This research was funded by ANID-FONDECYT de Iniciación #11181249 (to LJM) and ANID-FONDECYT Regular #1230932 (to LJM); ANID-FONDECYT Regular #1201851 (to SEI); ANID-FONDECYT de Iniciación #11190998 (to GN), ANID-FONDECYT Postdoctorado #3230201 (to RP); and ANID-BASAL funding for Scientific and Technological Center of Excellence, IMPACT, #FB210024 (to LJM, GN and SEI).

Institutional Review Board Statement: The animal study protocol was approved by the Ethical Scientific Committee of Universidad de los Andes, Santiago, Chile #CEC201874 (date of approval: 9 November 2018).

Informed Consent Statement: Menstrual fluid was self-collected by consenting donors following their written informed consent to participate in this study according to a protocol approved by the Ethical Scientific Committee of the Universidad de los Andes, Santiago, Chile #CEC202025 (date of approval: 22 April 2020).

Data Availability Statement: Data are contained within the article and Supplementary Materials.

Conflicts of Interest: The authors declare no conflicts of interest.

References

1. Kim, S.-M.; Kim, J.-S. A Review of Mechanisms of Implantation. *Dev. Reprod.* **2017**, *21*, 351–359. [CrossRef]
2. Norwitz, E.R. Defective implantation and placentation: Laying the blueprint for pregnancy complications. *Reprod. BioMed. Online* **2006**, *13*, 591–599. [CrossRef]
3. Norwitz, E.R.; Schust, D.J.; Fisher, S.J. Implantation and the survival of early pregnancy. *N. Engl. J. Med.* **2001**, *345*, 1400–1408. [CrossRef] [PubMed]
4. Cha, J.; Sun, X.; Dey, S.K. Mechanisms of implantation: Strategies for successful pregnancy. *Nat. Med.* **2012**, *18*, 1754–1767. [CrossRef] [PubMed]
5. Lyall, F. Mechanisms regulating cytotrophoblast invasion in normal pregnancy and pre-eclampsia. *Aust. N. Z. J. Obstet. Gynaecol.* **2006**, *46*, 266–273. [CrossRef] [PubMed]
6. Perez-Sepulveda, A.; España-Perrot, P.P.; Norwitz, E.R.; Illanes, S.E. Metabolic pathways involved in 2-methoxyestradiol synthesis and their role in preeclampsia. *Reprod. Sci.* **2013**, *20*, 1020–1029. [CrossRef] [PubMed]
7. Chaiworapongsa, T.; Chaemsaithong, P.; Yeo, L.; Romero, R. Pre-eclampsia part 1: Current understanding of its pathophysiology. *Nat. Rev. Nephrol.* **2014**, *10*, 466–480. [CrossRef] [PubMed]
8. Burton, G.; Caniggia, I. Hypoxia: Implications for implantation to delivery—A workshop report. *Placenta* **2001**, *22* (Suppl. A), S63–S65. [CrossRef]
9. Hannan, N.J.; Paiva, P.; Dimitriadis, E.; Salamonsen, L.A. Models for study of human embryo implantation: Choice of cell lines? *Biol. Reprod.* **2010**, *82*, 235–245. [CrossRef] [PubMed]
10. You, Y.; Stelzl, P.; Joseph, D.N.; Aldo, P.B.; Maxwell, A.J.; Dekel, N.; Liao, A.; Whirledge, S.; Mor, G. TNF-alpha Regulated Endometrial Stroma Secretome Promotes Trophoblast Invasion. *Front. Immunol.* **2021**, *12*, 737401. [CrossRef]
11. Ferretti, C.; Bruni, L.; Dangles-Marie, V.; Pecking, A.; Bellet, D. Molecular circuits shared by placental and cancer cells, and their implications in the proliferative, invasive and migratory capacities of trophoblasts. *Hum. Reprod. Update* **2007**, *13*, 121–141. [CrossRef]
12. Monteiro, L.J.; Cubillos, S.; Sanchez, M.; Acuña-Gallardo, S.; Venegas, P.; Herrera, V.; Lam, E.W.-F.; Varas-Godoy, M.; Illanes, S.E. Reduced FOXM1 Expression Limits Trophoblast Migration and Angiogenesis and Is Associated with Preeclampsia. *Reprod. Sci.* **2019**, *26*, 580–590. [CrossRef]
13. Kalathil, D.; John, S.; Nair, A.S. FOXM1 and Cancer: Faulty Cellular Signaling Derails Homeostasis. *Front. Oncol.* **2020**, *10*, 626836. [CrossRef]
14. Koo, C.-Y.; Muir, K.W.; Lam, E.W.-F. FOXM1: From cancer initiation to progression and treatment. *Biochim. Biophys. Acta* **2012**, *1819*, 28–37. [CrossRef]
15. Wierstra, I. The transcription factor FOXM1 (Forkhead box M1): Proliferation-specific expression, transcription factor function, target genes, mouse models, and normal biological roles. *Adv. Cancer Res.* **2013**, *118*, 97–398.
16. Bella, L.; Zona, S.; de Moraes, G.N.; Lam, E.W.-F. FOXM1: A key oncofoetal transcription factor in health and disease. *Semin. Cancer Biol.* **2014**, *29*, 32–39. [CrossRef] [PubMed]
17. Ye, H.; Kelly, T.F.; Samadani, U.; Lim, L.; Rubio, S.; Overdier, D.G.; Roebuck, K.A.; Costa, R.H. Hepatocyte nuclear factor 3/fork head homolog 11 is expressed in proliferating epithelial and mesenchymal cells of embryonic and adult tissues. *Mol. Cell. Biol.* **1997**, *17*, 1626–1641. [CrossRef] [PubMed]
18. Kim, I.-M.; Ramakrishna, S.; Gusarova, G.A.; Yoder, H.M.; Costa, R.H.; Kalinichenko, V.V. The forkhead box m1 transcription factor is essential for embryonic development of pulmonary vasculature. *J. Biol. Chem.* **2005**, *280*, 22278–22286. [CrossRef] [PubMed]
19. Rodesch, F.; Simon, P.; Donner, C.; Jauniaux, E. Oxygen measurements in endometrial and trophoblastic tissues during early pregnancy. *Obstet. Gynecol.* **1992**, *80*, 283–285. [PubMed]
20. Jauniaux, E.; Watson, A.L.; Hempstock, J.; Bao, Y.-P.; Skepper, J.N.; Burton, G.J. Onset of maternal arterial blood flow and placental oxidative stress. A possible factor in human early pregnancy failure. *Am. J. Pathol.* **2000**, *157*, 2111–2122. [CrossRef] [PubMed]
21. Palmer, K.; Saglam, B.; Whitehead, C.; Stock, O.; Lappas, M.; Tong, S. Severe early-onset preeclampsia is not associated with a change in placental catechol O-methyltransferase (COMT) expression. *Am. J. Pathol.* **2011**, *178*, 2484–2488. [CrossRef] [PubMed]

22. Karadedou, C.T.; Gomes, A.R.; Chen, J.; Petkovic, M.; Ho, K.-K.; Zwolinska, A.K.; Feltes, A.; Wong, S.Y.; Chan, K.Y.K.; Cheung, Y.-N.; et al. FOXO3a represses VEGF expression through FOXM1-dependent and -independent mechanisms in breast cancer. *Oncogene* **2012**, *31*, 1845–1858. [CrossRef] [PubMed]

23. Peñaililло, R.; Acuña-Gallardo, S.; García, F.; Monteiro, L.J.; Nardocci, G.; Choolani, M.A.; Kemp, M.W.; Romero, R.; Illanes, S.E. Mesenchymal Stem Cells-Induced Trophoblast Invasion Is Reduced in Patients with a Previous History of Preeclampsia. *Int. J. Mol. Sci.* **2022**, *23*, 9071. [CrossRef]

24. You, Y.; Stelzl, P.; Zhaing, Y.; Porter, J.; Liu, H.; Liao, A.; Aldo, P.B.; Mor, G. Novel 3D in vitro models to evaluate trophoblast migration and invasion. *Am. J. Reprod. Immunol.* **2019**, *81*, e13076. [CrossRef] [PubMed]

25. Myatt, S.S.; Lam, E.W. The emerging roles of forkhead box (Fox) proteins in cancer. *Nat. Rev. Cancer* **2007**, *7*, 847–859. [CrossRef] [PubMed]

26. Gertz, J.; Savic, D.; Varley, K.E.; Partridge, E.C.; Safi, A.; Jain, P.; Cooper, G.M.; Reddy, T.E.; Crawford, G.E.; Myers, R.M. Distinct properties of cell-type-specific and shared transcription factor binding sites. *Mol. Cell* **2013**, *52*, 25–36. [CrossRef] [PubMed]

27. Emiliani, S.; Delbaere, A.; Devreker, F.; Englert, Y. Embryo-maternal interactive factors regulating the implantation process: Implications in assisted reproductive. *Reprod. Biomed. Online* **2005**, *10*, 527–540. [CrossRef] [PubMed]

28. Zhu, X.; Yu, M.; Wang, K.; Zou, W.; Zhu, L. FoxM1 affects adhesive, migratory, and invasive abilities of human retinoblastoma Y-79 cells by targeting matrix metalloproteinase 2. *Acta Biochim. Biophys. Sin.* **2020**, *52*, 294–301. [CrossRef]

29. Pijuan, J.; Barceló, C.; Moreno, D.F.; Maiques, O.; Sisó, P.; Marti, R.M.; Macià, A.; Panosa, A. In Vitro Cell Migration, Invasion, and Adhesion Assays: From Cell Imaging to Data Analysis. *Front. Cell Dev. Biol.* **2019**, *7*, 107. [CrossRef]

30. Vaswani, K.; Hum, M.W.-C.; Chan, H.-W.; Ryan, J.; Wood-Bradley, R.J.; Nitert, M.D.; Mitchell, M.D.; Armitage, J.A.; Rice, G.E. The effect of gestational age on angiogenic gene expression in the rat placenta. *PLoS ONE* **2013**, *8*, e83762. [CrossRef]

31. Maniswami, R.R.; Prashanth, S.; Karanth, A.V.; Koushik, S.; Govindaraj, H.; Mullangi, R.; Rajagopal, S.; Jegatheesan, S.K. PLK4: A link between centriole biogenesis and cancer. *Expert Opin. Ther. Targets* **2018**, *22*, 59–73. [CrossRef]

32. Garvey, D.R.; Chhabra, G.; Ndiaye, M.A.; Ahmad, N. Role of Polo-Like Kinase 4 (PLK4) in Epithelial Cancers and Recent Progress in its Small Molecule Targeting for Cancer Management. *Mol. Cancer Ther.* **2021**, *20*, 632–640. [CrossRef]

33. Tanenbaum, M.E.; Medema, R.H. Cell fate in the Hand of Plk4. *Nat. Cell Biol.* **2007**, *9*, 1127–1129. [CrossRef]

34. Penailillo, R.S.; Eckert, J.J.; Burton, M.A.; Burdge, G.C.; Fleming, T.P.; Lillycrop, K.A. High maternal folic acid intake around conception alters mouse blastocyst lineage allocation and expression of key developmental regulatory genes. *Mol. Reprod. Dev.* **2021**, *88*, 261–273. [CrossRef] [PubMed]

35. Perez-Sepulveda, A.; Monteiro, L.J.; Dobierzewska, A.; España-Perrot, P.P.; Venegas-Araneda, P.; Guzmán-Rojas, A.M.; González, M.I.; Palominos-Rivera, M.; Irarrazabal, C.E.; Figueroa-Diesel, H.; et al. Placental Aromatase Is Deficient in Placental Ischemia and Preeclampsia. *PLoS ONE* **2015**, *10*, e0139682. [CrossRef] [PubMed]

36. Monteiro, L.J.; Khongkow, P.; Kongsema, M.; Morris, J.R.; Man, C.; Weekes, D.; Koo, C.Y.; Gomes, A.R.; Pinto, P.H.; Varghese, V.; et al. The Forkhead Box M1 protein regulates BRIP1 expression and DNA damage repair in epirubicin treatment. *Oncogene* **2013**, *32*, 4634–4645. [CrossRef] [PubMed]

International Journal of
Molecular Sciences

Review

Circulating miRNAs and Preeclampsia: From Implantation to Epigenetics [†]

Stefano Raffaele Giannubilo [1], Monia Cecati [1,*], Daniela Marzioni [2] and Andrea Ciavattini [1]

[1] Department of Clinical Sciences, Università Politecnica delle Marche, 60020 Ancona, Italy;
     s.giannubilo@univpm.it (S.R.G.); a.ciavattini@staff.univpm.it (A.C.)
[2] Department of Experimental and Clinical Medicine, Università Politecnica delle Marche, 60126 Ancona, Italy;
     d.marzioni@staff.univpm.it
*   Correspondence: moniacecati@gmail.com; Tel.: +39-071-596-2214
†   This paper is dedicated to the memory of Prof. Andrea L. Tranquilli.

Abstract: In this review, we comprehensively present the literature on circulating microRNAs (miRNAs) associated with preeclampsia, a pregnancy-specific disease considered the primary reason for maternal and fetal mortality and morbidity. miRNAs are single-stranded non-coding RNAs, 20–24 nt long, which control mRNA expression. Changes in miRNA expression can induce a variation in the relative mRNA level and influence cellular homeostasis, and the strong presence of miRNAs in all body fluids has made them useful biomarkers of several diseases. Preeclampsia is a multifactorial disease, but the etiopathogenesis remains unclear. The functions of trophoblasts, including differentiation, proliferation, migration, invasion and apoptosis, are essential for a successful pregnancy. During the early stages of placental development, trophoblasts are strictly regulated by several molecular pathways; however, an imbalance in these molecular pathways can lead to severe placental lesions and pregnancy complications. We then discuss the role of miRNAs in trophoblast invasion and in the pathogenesis, diagnosis and prediction of preeclampsia. We also discuss the potential role of miRNAs from an epigenetic perspective with possible future therapeutic implications.

Keywords: miRNA; preeclampsia; trophoblast; placenta; epigenetic

Citation: Giannubilo, S.R.; Cecati, M.; Marzioni, D.; Ciavattini, A. Circulating miRNAs and Preeclampsia: From Implantation to Epigenetics. *Int. J. Mol. Sci.* **2024**, *25*, 1418. https://doi.org/10.3390/ijms25031418

Academic Editor: Robert Garfield

Received: 28 December 2023
Revised: 16 January 2024
Accepted: 23 January 2024
Published: 24 January 2024

1. Introduction

The placenta is a transitory organ, necessary for development in utero in humans and mammals [1–3]. The important role of this organ is highlighted when normal placental development is compromised, leading to important pregnancy complications, such as preeclampsia (PE) [4,5], fetal growth restriction (FGR) [6–8], gestational diabetes mellitus (GDM) [9], preterm birth (PTB) [9,10] and gestational trophoblastic disease (GTD) [11,12].

Preeclampsia is a hypertensive disorder associated with pregnancy that affects 5–7% of pregnancies and is responsible for over 70,000 maternal deaths each year [13]. Parenteral or oral drug therapies are available and administered to preeclamptic women to stabilize the mother and fetus and reduce the risk of failure of several maternal organs, such as the liver, kidney and brain [14]. Nevertheless, to date, delivery at an optimal time is the only reliable treatment for preeclampsia [15].

Reduced arterial compliance and peripheral vasoconstriction characterize the preeclamptic condition from the early stages [16]; however, the exact pathophysiology of preeclampsia is still unknown. Probably, the initial defect in placentation and vascularization in the placenta bed due to poor trophoblast invasion of the spiral arteries [17–19] leads to an inappropriate activation of the immune system [20], as it occurs in recurrent miscarriage [21]. It is known that oxidative stress and inflammation are involved in endothelial dysfunction [22–27], a key process underlying several diseases including preeclampsia [28–32]. Hypoxia due to inadequate spiral artery remodeling compromises the placental endothelial function [33], which results in an oxidative stress condition [34,35], made worse by an increased production of nitric oxide [36,37]. Continued oxidative stress leads to chronic inflammation, which induces a premature senescence of preeclamptic placental tissue, as suggested by the short length of telomeric DNA [38].

The inability of preeclamptic placental tissue to regain a condition of equilibrium leads to an inevitable dysregulation of several metabolic pathways [39,40]. The understanding of this pathology is still elusive, although different studies tried to identify candidate genes involved in preeclampsia onset, measuring their expression in placental tissue [41,42].

Since preeclampsia can cause maternal and neonatal morbidity and mortality [43], several studies have been focused on finding efficient markers able to predict preeclampsia onset to improve the treatment of these pregnancies [5,44–49].

The presence of circulating RNAs released from many body tissues, the placenta included, offers a potential tool to indirectly observe any pathologies in real time, starting from the onset and during development. In this way, it possible to associate the physiological changes in the circulating RNA variation level in preeclamptic subjects with respect to healthy ones [50]. Circulating RNAs contain fragments that are transcribed but not translated called non-coding RNAs (ncRNAs). Bioinformatic analysis estimates that a quote of the human genome equal to 75–80% is transcribed into RNA molecules. Nevertheless, just 2% of transcribed RNA molecules are converted into proteins [51]. Usually, ncRNAs are divided into four different classes: PIWI-interacting RNAs (piRNAs), circular RNAs (circRNAs), long ncRNAs (lncRNAs) with a length over 200 nucleotides and small ncRNAs (known also as miRNAs) with a length under 200 nucleotides.

The name piRNAs was derived from their ability to bind to PIWI family proteins. They are between 24 and 30 nucleotides in length, and they are known to participate in the regulation of cell proliferation, apoptosis and the cell cycle [52]. LncRNAs and circRNAs are more than 200 nucleotides in length but have different shapes. As suggested by their name, lncRNAs are linear unlike circRNAs, which are ringlike. Both lncRNAs and circRNAs are transcribed from regions located at exons, introns, intergenic regions or $5'/3'$-untranslational regions and interact with DNA, RNA and proteins by folding themselves into complicated second structures [53]. LncRNAs and circRNAs can modulate gene expression through several strategies. They can prevent mRNA degradation, avoiding the miRNA, by bind to the targeted mRNA. They can regulate gene expression by modulating transcription factors to tie to promoters [54]. In addition, they can exert a regulative action on several signaling pathways, acting as a scaffold to regulate the interactions between proteins [55]. miRNAs are small ncRNAs, which control mRNA expression. Changes in miRNA expression can induce a variation in the relative mRNA level and influence cellular homeostasis [56,57]. Many miRNAs remain in the cell where they were assembled, whereas others can be released outside (known as circulating miRNAs) under normal and pathological conditions [58]. The strong presence of circulating miRNAs in all body fluids makes them useful biomarkers of several diseases, cancer included [44,59,60]. Circulating miRNAs are present in gastric juices, saliva and urine [61–63], but plasma and serum represent the source most frequently used to collect them in healthy and non-healthy subjects [64].

This review summarized the miRNAs recovered from the peripheral maternal blood circulation that resulted in a dysregulated preeclamptic pregnancy from its onset through to its development. The authors also focused on the epigenetic regulations of miRNA expression as a possible tool to reduce the risk of cardiovascular diseases in women who experienced preeclamptic pregnancy.

2. Materials and Methods

During preparation of this scoping review (September to December 2023), we searched PubMed using terms containing "miRNA", "pregnancy", "preeclampsia", "hypertension", and "Placenta". Only studies on circulating miRNAs were included. We also identified relevant studies via 'snowballing'/citation chasing that were relevant for background information. All studies were assessed based on their quality of reporting, bias in participant selection, result presentation or author conflicts of interest.

3. miRNA Biogenesis, Mechanisms of Export and Annotation Criteria

miRNAs are single-stranded non-coding RNAs, 20–24 nt long, which originate from the primary miRNA transcript (pri-miRNA) [65]. In the nucleus, an RNase III endonuclease (named Drosha) combined to DGCR8 (DiGeorge critical region 8) protein (named Pasha) begins the maturation of the pri-miRNA, liberating a ~60–70-nt stem loop intermediate (pre-miRNA). The complex Exportin-5–Ran-GTP moves the pre-miRNA precursor hairpin from the nucleus to the cytoplasm. The RNase Dicer associated with the transactivation response element RNA-binding protein (TRBP) digests the pre-miRNA to find its mature length. Finally, the passenger strand of the mature miRNA is digested, while the functional strand is loaded with Argonaute (Ago2) proteins into the RNA-induced silencing complex (RISC). RISC silences the mRNA target through mRNA deadenylation, translational repression or cleavage driven by the functional strand of the mature miRNA, which recognizes 3′-untranslated region (3′-UTR) [66,67]. Although the 3′UTR sequence is the most prevalent site for the recognition of mRNA target, miRNAs can also interact with 5′-untranslated region (5′-UTR) [68] (Figure 1).

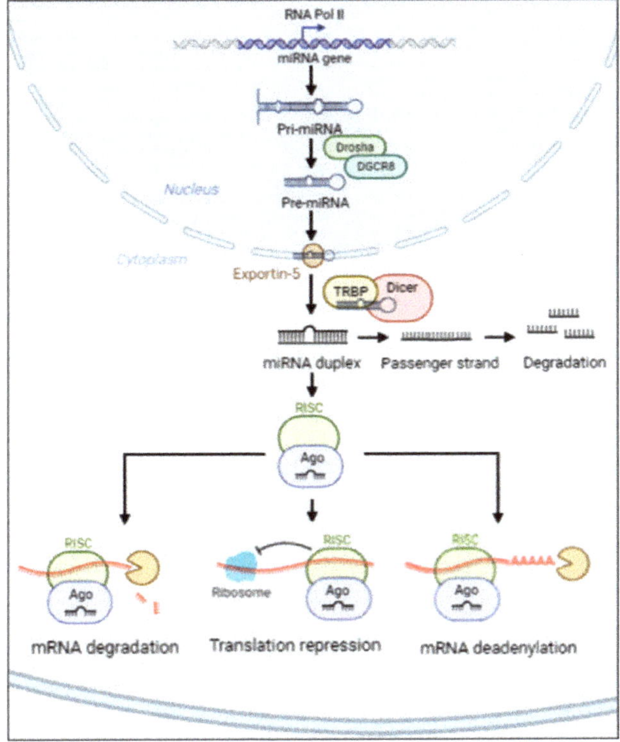

Figure 1. Schematic overview of miRNA biosynthesis. The miRNA pathway produces pri-miRNA transcripts from miRNA genes encoded in exonic, intronic or intergenic regions. In the nucleus, Drosha and DGCR8 digest the pri-miRNA into pre-miRNA. The pre-miRNA is driven into the cytoplasm by the complex Exportin-5–Ran-GTP. Once the mature length is obtained, the functional strand is loaded with Ago2 protein into RISC complex which silences the mRNA's target through mRNA deadenylation, translational repression or degradation.

In recent years, circulating miRNAs have gained the attention of activity research because they function in the same manner as classical signaling, made of growth factors, hormones and cytokines [69]. Circulating miRNAs are surprisingly stable and able to carry out their suppressive function against mRNA target into the recipient cells. Mitchell et al. [70] and Chen et al. [71] demonstrated that miRNAs isolated from human serum and plasma were resistant to activity of RNase. Contrarily, exogenous miRNAs synthetized to be added to plasma and serum resulted in rapid degradation, suggesting that the integrity of endogenous miRNA in human biological fluids depends on something different than their structure or short length [70].

Experimental data demonstrated that miRNAs can be released outside the cell under four different forms: (a) miRNAs combined to Argonaute2 protein (Ago2) [72,73], (b) miRNAs tied to RNA-binding protein nucleophosmin (NPM1) [74], (c) miRNAs tied to high-density lipoprotein (HDL) [75], and (d) miRNAs closed within extracellular vesicles (EVs) such as exosomes and microvesicles. Notably, the presence of one form of miRNA does not exclude the presence of another extracellular form of miRNAs, as suggested by results in all cell lines or human biological samples tested [76].

The circulating miRNAs detected in plasma, and not included in EV, are mostly bound to proteins. The Ago2 protein not only takes part in miRNA maturation via RISC complex but also exercises a shielding effect on extracellular miRNAs against RNases. Particularly, Arroyo et al. suggested that the entire Ago2–miRNA complex may be able to modulate gene expression in recipient cells [73].

Like Ago2, NPM1 also has a protective role for circulating miRNAs. Wang et al. hypothesized that NPM1 is involved not only in the packaging and release of miRNAs outside the cell but that it remains bound to them in the peripheral circulation [74]. Nevertheless, the biological function of circulating miRNAs associated with both Ago2 and NPM1 proteins in pregnancy and pregnancy-related diseases is still unknown.

HDL is one of the five different types of lipoproteins, including chylomicrons, very-low-density lipoprotein (VLDL), intermediate-density lipoprotein (IDL) and low-density lipoprotein (LDL). HDLs have a micellular constitution and organization, like most polar lipids, and a mass composed of 50–60% proteins, mostly represented by apolipoproteins A-I (APOA1) [74]. It is widely known that HDLs also vehicle miRNAs, not only lipids and proteins [75]. The load of miRNAs in HDFs is different in composition and concentration between normal and pathological subjects and equally attributable to a specific disease [77–79]. It has been demonstrated that the miRNAs-HDL complex is internalized only if the recipient cell expresses the scavenger receptor BI (SR-BI) [75]. Unfortunately, currently, no evidence is available in the literature about the biological importance of this mechanism in pregnancy and especially in the onset of pathological pregnancies.

EVs are small lipid particles released from most human cell types, both malignant and healthy [80]. Interestingly, EVs have also been suggested as a promising class of therapeutic agents [81]. EVs mediate intercellular communication by moving heterogeneous molecules (i.e., DNA, RNAs, miRNAs, proteins and lipids). The recognition of miRNAs sorted for cargo into EVs relies on heterogeneous nuclear ribonucleoprotein A2B1 (hnRNPA2B1) or synaptotagmin-binding cytoplasmic RNA interaction protein (SYNCRIP). A paper published by Villarroya-Beltri et al. demonstrated that the selective choice of miRNAs to be released outside by EVs depends on the recognition of special sequences named EXOmotifs (e.g., GGAG or AGG) in miRNAs by hnRNPA2B1 or SYNCRIP [82]. The membranous protein caveolin-1 has been identified as necessary to drive the miRNAs-hnRNPA2B1/SYNCRIP complex into EVs [83]. Once formed, EVs and their cargo can be liberated, either directly from the plasma membrane or grouped into multivesicular bodies and then released [81].

A passive release of miRNAs can occur in a state of chronic inflammation, apoptosis or necrosis, leading to cell lysis. In this case, the miRNAs are released into biological fluids associated with Ago proteins [72]. A quantitative analysis of the passive release of miRNAs has not yet been carried out.

miRNAs are identified by a code made of the "miR" prefix and a unique identifying number. miRNAs belonging to certain species are indicated with an additional three-letter prefix; for example, hsa-miR-124 is a miRNA belonging to the human species (the prefix hsa means Homo Sapiens). The identifying numbers are given sequentially, meaning that miR-124 was discovered and named before miR-456. miRNAs with nearly identical sequences, except for one or two nucleotides, can be given with an additional lowercase letter [84]. Let-7 is a fundamental miRNA tumor suppressor. In humans, the let-7 family includes 12 different members (let-7a-1, 7a-2, 7a-3, 7b, 7c, 7d, 7e, 7f-1, 7f-2, 7g, 7i and miR-98) [85]. The chromosome 19 miRNA cluster (C19MC) includes approximately 8% of all known human miRNA genes. C19MC is only expressed in the placenta and in undifferentiated cells [86].

4. miRNAs, Implantation and Preeclampsia

The possibility that miRNAs may play a crucial role in the pathogenesis of preeclampsia stems from the consideration that these molecules actively participate in the implantation processes of pregnancy and, on the other hand, that preeclampsia finds its pathophysiological basis in the very early stages of pregnancy. Nucleic acids have been detected in uterine fluid and, more specifically, in EVs containing miRNAs, suggesting a role of miRNAs in embryo–endometrial communication [87]. Using biopsy material and modern transcriptomics, it has also been shown that miRNAs are dysregulated in the endometrium of women with recurrent implantation failure [88]. Trophoblast migration and invasion, and cellular adaptations in the physiological changes underlying gestation, involve EVs as key modulators. Endometrial luminal epithelial cells [89] and proper communication between these cells determine the success or failure of pregnancy [90]. Exosome nanovesicles can transfer information (e.g., hsa-miR-30b, hsa-miR-200c, hsamiR-17 and hsa-miR-106a, miRNAs involved in endometrial receptivity and implantation) from the endometrium to the blastocyst, thereby promoting implantation [91,92]. In addition, extracellular vesicles secreted by the endometrium are internalized by the embryo to enhance adhesion and invasion [93], mediating embryo–endometrial communication [94]. During the early invasion phase (8 to 10 days after conception), the cytotrophoblast differentiates into an extravillous interstitial trophoblast and invades the decidua. At this point, feto–maternal communication occurs between the extravillous trophoblast and the decidualized endometrial stroma [95]. One of the most probable hypotheses to describe the etiology of preeclampsia is based on a failure of extravillous trophoblasts to invade the uterine spiral arteries in the placental bed. Insufficient placental vascular remodeling induces placental hypoperfusion, which is critical for the pathogenesis of preeclampsia. Whitley et al. showed that first-trimester extravillous trophoblasts from pregnancies with high uterine artery resistance were inherently more sensitive to apoptotic stimuli, which may be associated with reduced remodeling of the maternal spiral arteries [96]. Placental insufficiency has also been associated with abnormal levels of extracellular fetal DNA, mRNA transcripts and circulating C19MC microRNAs (miR-516b-5p, miR-517-5p, miR-520a-5p, miR-525-5p and miR-526a) [97]. Exosomes secreted by cytotrophoblasts, which express placenta-specific miRNAs, including syncytin-2, have been implicated in embryo implantation through the promotion of Treg differentiation and suppression of the nuclear factor B signaling pathway and, thus, the immune and inflammatory response [98]. The specific loading of miRNAs in maternal plasma exosomes obtained in the first trimester of pregnancy in women who developed preeclampsia [99] suggests a potential role of miRNAs in the pathogenesis of preeclampsia as early as the first trimester. For example, miR-1269 mediates the downregulation of the tumor suppressor gene forkhead box O1 (FOXO1) [100]. Because FOXO1 is a protein involved in the stromal decidualization of the endometrium and the implantation process [101], alterations in its expression can have deleterious consequences in pregnancy, such as preeclampsia. The expression level of miR-33b-3p in patients with preeclampsia is significantly different from that of healthy pregnant women. It has been reported that transforming growth factor beta-1 (TGFB1), a target of miR-33b-3p, may play a role in the pathogenesis of preeclampsia by preventing the differentiation of trophoblasts toward an invasive phenotype [102]. In other experimental studies, it was seen that the migratory and invasive abilities of trophoblastic cells were significantly inhibited by miR-486-5p. Rho GTPase-activating protein 5 (ARHGAP5) is the most abundant GTPase-activating protein (GAP) of the small Rho GTPase family. This protein plays a role in the regulation of actin cytoskeleton-based mechanisms, thereby influencing cell migration and invasion [103]. The possible pathogenesis of preeclampsia could depend on increased miR-486-5p, which modulates the expression of ARHGAP5 in trophoblast cells [104]. The X-linked inhibitor of apoptosis protein (XIAP) is an inhibitory protein of apoptosis (IAP) [105]; the expression of XIAP in trophoblasts is decreased in preeclampsia, which may be associated with increased apoptosis in the placenta [106]. Serum miR-515-5p was significantly increased in patients with preeclampsia compared with normal pregnant women. Functionally, the overexpression of miR-515-5p suppressed the proliferation and invasion of trophoblastic cells, as observed in HTR-8/SVneo trophoblastic cells. Using luciferase reporter assays, XIAP was identified as a direct target gene of miR-515-5p, and the overexpression of miR-515-5p was found to decrease the level of XIAP in HTR-8/SVneo cells [107]. Of particular interest is the relationship between miRNAs and oxygenation. Early in gestation (<12 weeks), the placenta develops in a low-oxygen environment (<20 mmHg), which is necessary for successful placentation [108]. The insufficient syncytization of cytotrophoblastic villus cells results in suboptimal placental perfusion and, thus, chronic hypoxia, which is a hallmark of preeclampsia. Placental renin–angiotensin system (RAS) activity depends, in part, on the prevailing oxygen tension, and the latter controls the levels of placental miRNAs that regulate placental RAS expression. The suppression of miRNAs targeting placental RAS early in gestation is partly responsible for the increase in RAS expression during this period to promote placental development [109]. In other studies in this area, it has been reported that miR455-3P acts as a rheostat that restrains the hypoxia response that might otherwise prevent the differentiation of the cytotrophoblast into syncytiotrophoblast: reduced miR455 expression is causally linked to the development of preeclampsia [110].

5. miRNAs and Clinical Preeclampsia

This section reports results from different studies about the possibility of using miRNAs as a tool for the prediction of preeclampsia onset and severity. Significantly dysregulated miRNAs are reported in Table 1.

Table 1. miRNAs significantly dysregulated in clinical diagnosis of preeclampsia.

miRNAs	Diagnosed PE	Early Onset PE	Late Onset PE	Mild PE	Severe PE	Prediction of PE	Ref.
miR-15a-5p						↑	[111]
	↑						[112]
miR-15b					↓		[113]
miR-16	↓						[114]
			↓				[115]
miR-16-5p	↓						[116]
miR-17-5p	↓						[117]
miR-18a						↓	[118]
					↓		[113]
			↓				[115]
miR-19b1						↓	[118]
					↓		[113]

Table 1. *Cont.*

miRNAs	Diagnosed PE	Early Onset PE	Late Onset PE	Mild PE	Severe PE	Prediction of PE	Ref.
miR-21			↑				[119]
					↑		[113]
miR-21-5p			↑				[120]
miR-22-5p						↓	[111]
miR-24-3p	↑						
miR-23b-5p						↓	[121]
miR-24	↑						[122]
miR-26a	↑						[122]
miR-29a				↑			[123]
	↑						[124]
					↑		[113]
	↑						[125]
miR-29a-3p	↑						[117]
miR-31		↑					[119]
miR-31-5p	↑						[126]
miR-92a-1						↓	[118]
miR-92-a-1-3p						↑	[118]
miR-93-5p						↓	[111]
miR-99-5p						↓	[121]
miR-103	↑						[122]
miR-106a						↑	[111]
miR-106b	↑						[127]
miR-125b						↑	[111]
						↑	[44]
	↑						[125]
miR-125a-5p	↑						[125]
miR-126		↓					[128]
miR-126-3p						↓	[111]
miR-130a-3p						↑	[111]
miR-130b	↑						[122]
miR-132						↑	[129]
miR-132-3p	↑						[117]
miR-133b						↑	[129]
miR-136	↑						[130]
	↑						[125]

Table 1. *Cont.*

miRNAs	Diagnosed PE	Early Onset PE	Late Onset PE	Mild PE	Severe PE	Prediction of PE	Ref.
miR-141				↓			[131]
				↑			[123]
miR-144				↓	↓		[123]
					↓		[113]
miR-146a						↑	[129]
miR-149			↓				[115]
miR-151a-3p						↑	[132]
miR-152	↓						[133]
miR-155					↑		[113]
	↑						[134]
	↑						[135]
miR-155-5p	↑						[126]
			↑				[120]
miR-181a	↑						[122]
	↑						[122]
	↑						[136]
miR-185	↓						[125]
miR-186	↑						[136]
miR-191-5p						↑	[111]
	↓						[117]
miR-195-5p	↑						[137]
miR-196b	↓						[138]
miR197-3p	↑						[117]
miR-200c	↑						[114]
miR-204-3p						↓	[111]
miR-206			↑				[139]
	↑						[140]
miR-210	↑						[133]
	↑						[117]
						↑	[118]
				↑	↑		[141]
					↑		[113]
	↑						[142]
	↓						[116]
						↑	[129]
miR-214-3p	↑						[126]

Table 1. *Cont.*

miRNAs	Diagnosed PE	Early Onset PE	Late Onset PE	Mild PE	Severe PE	Prediction of PE	Ref.
miR-215					↑		[113]
miR-218-5p	↑						[117]
miR-221				↓			[131]
miR-218-5p	↑						[117]
miR-223	↓						[125]
miR-302b-3p	↓						[117]
miR-320c	↓						[125]
miR-326	↑						[127]
miR-328	↑						[117]
miR-342-3p	↑						[122]
miR-363			↓				[115]
miR-365a-3p						↓	[111]
miR-374a-5p						↑	[111]
miR-375	↑						[117]
miR-424			↓				[115]
miR-494	↑						[130]
miR-495	↑						[130]
miR-510	↓						[143]
miR-512-3p						↑	[144]
miR-515-5p		↑					[145]
miR-515-5p	↑						[107]
miR-516-5p	↑						[146]
miR-516a-5p		↑					[146]
miR-516b		↑					[145]
miR-517	↑						[146]
miR-517						↑	[147]
miR-517-5p						↑	[148]
miR-517-5p						↓	[149]
mirR-517b	↑						[125]
miR-517c	↑						[125]
miR-518b					↑		[113]
miR-518b			↑				[115]
miR-518e	↑						[125]
miR-518f3p						↑	[144]
miR-519a	↑						[125]
miR-519d	↑						[125]

Table 1. *Cont.*

miRNAs	Diagnosed PE	Early Onset PE	Late Onset PE	Mild PE	Severe PE	Prediction of PE	Ref.
miR-520-5p		↑					[145]
miR-520a	↑						[146]
miR-520a-5p						↓	[149]
miR-520c-3p						↑	[144]
miR-520d-3p						↑	[144]
miR-520g	↑						[125]
miR-520h		↑					[145]
miR-520h	↑						[125]
miR-521	↑						[125]
miR-525	↑						[146]
miR-525-5p		↑					[145]
miR-525-5p						↓	[149]
miR-526		↑					[145]
miR-526						↑	[147]
miR-559-5p						↓	[111]
miR-526a	↑						[146]
miR-542-3p	↑						[125]
miR-573						↑	[132]
miR-574-5p	↑						[122]
miR-574-5p						↑	[111]
miR-628-3p						↑	[132]
miR-650					↑		[113]
miR-885-5p	↑						[150]
miR-942						↓	[151]
miR-1229p						↑	[129]
miR-1233						↑	[152]
miR-1244						↑	[129]
miR-1260	↓						[125]
miR-1272	↓						[125]
miR-1283			↓				[115]
miR-1290-3p	↓						[126]
miR-1323		↑					[145]
miR-4264-5p						↓	[111]
Let-7a-5p						↑	[111]

Table 1. *Cont.*

miRNAs	Diagnosed PE	Early Onset PE	Late Onset PE	Mild PE	Severe PE	Prediction of PE	Ref.
Let-7a	⬆						[125]
Let-7d	⬇						[125]
Let-7f	⬇						[125]
Let-7f-1	⬆						[125]

Abbreviations: PE: preeclampsia. ⬆ Up-regulated; ⬇ Down-regulated.

5.1. miRNAs and Diagnosed Preeclampsia

One of the first demonstrations of the possibility of using circulating miRNAs as a marker for preeclamptic pregnancy was by Gunel et al. They measured the expression level of miR-210 and miR-152 from the maternal plasma of both healthy and preeclamptic pregnant women. The results demonstrated an upregulation for miR-210 and a downregulation for miR-152 in women with preeclamptic pregnancy [133]. In the study of Campos et al., a lower expression of circulating miR-196b in maternal plasma was correlated to a preeclamptic condition in pregnant women [138]. Both miR-195-5p [137] and miR-885-5p [150] overexpression was correlated to a preeclamptic condition in women. Investigating the plasma of women with diagnosed pregnancy complicated by preeclampsia, Sheng et al. observed an upregulation for miR-206 [140]. The study realized by Akgor et al. analyzed the circulating level of miRNAs in women with diagnosed pregnancy complicated by preeclampsia. The authors observed a panel of miRNAs, many of which are already known as potential biomarkers for the non-invasive diagnosis of preeclampsia. However, they observed, for the first time, an upregulation of two novel miRNAs (miR-191-5p and miR-197) associated with the preeclamptic condition [117]. Ayoub et al. found an upregulation for both miR-186 and miR-181a in the serum of women with pregnancy complicated by preeclampsia [136]. The dysregulated expression of circulating miRNAs depends on an epigenetic mechanism, as observed by Sekar et al. They found an increased level of miR-510 in the plasma of women with preeclamptic pregnancy correlated to a decreased methylation status of its promoter [143]. Also, C19MC has a correlation with the preeclamptic condition. Using next-generation sequencing technology, Yang et al. observed the presence of 22 different circulating dysregulated miRNAs (15 miRNAs were up-expressed and 7 miRNAs were down-expressed) in the serum of women with diagnosed preeclampsia. Among upregulated miRNAs, Yang et al. also observed three miRNAs belonging to C19MC that means miR-517, miR-518 and miR-519 [125]. The C19MC miRNAs also have been associated with the severity of the preeclamptic condition: the upregulation of miR-516-5p, miR-517, miR-520a, miR-525 and miR-526a was correlated to clinical signs of preeclampsia, requirements for the delivery and Doppler ultrasound parameters [146]. Activity research focused not only on the identification of deregulated miRNAs involved in preeclamptic pregnancy but also on the metabolic pathways that they take part in. Wu et al. not only observed dysregulated miRNAs (miR-24, miR-26a, miR-103, miR-130b, miR-181a, miR-342-3p and miR-574-5p) in the plasma of women affected by preeclampsia but revealed that they were also associated with specific signaling pathways such as the regulation of transcription and the cell cycle [122]. Khaliq's study demonstrated that the expression level of miR-29a and miR-181a was increased in the serum of women with pregnancy complicated by preeclampsia and correlated to AKT/PI3K in the insulin signaling pathway [124]. A significant upregulation of both miR-106b and miR-326 was associated with an imbalance in T-helper type 17 (Th17)/regulatory T (Treg) cells in women affected by preeclampsia [127]. Circulating miRNAs in pregnancy can modulate several pathways, VEGF included. A study realized by Witvrouwen et al. investigated the expression level of circulating VEGF-related miRNAs in the blood samples recovered from women with diagnosed preeclampsia. The authors demonstrated that the downregulation of miR-16 in patients with preeclampsia can impair the endothelial function. Moreover, the overexpression of circulating miR-200c was increased in preeclamptic subjects and correlated with higher arterial stiffness [114]. A recent study by Nunode et al. observed a significant upregulation in the maternal serum from preeclamptic patients for miR-515-5p. In addition, bioinformatics prediction suggested a potential role of miR-515-5p in placental development: miR-515-5p could suppress trophoblast cell invasion, inhibiting the X-linked inhibitor of apoptosis protein (XIAP) which promotes cell migration by enhancing epithelial–mesenchymal transition (EMT) [107]. Kim et al. carried out an interesting study, which led to the identification of several dysregulated miRNAs associated with some clinical hallmarks of preeclampsia. In the serum of women with diagnosed preeclampsia, the expression level of miR-31-5p, miR-155-5p and miR-214-3p was significantly increased, whereas miR-1290-3p was significantly downregulated. The authors suggested that a dysregulation in the expression level of the above-mentioned miRNAs was associated with the clinical features of preeclampsia (such as hypoxia, inflammation and decreased estrogen levels) [126]. The results obtained by Kim et al. were confirmed in a study carried out by Wang et al. The authors suggested that an increased expression of miR-155 is associated with clinical manifestations of preeclampsia [135]. Also, miRNAs delivered by circulating exosomes resulted as promising biomarkers in the detection of the preeclamptic condition. Motawi et al. observed an upregulation for miRNA-136, miRNA-494 and miRNA-495 isolated from exosomes in the blood samples of women with a diagnosed preeclamptic condition [130]. Wang et al. studied the role of pregnancy-associated exosomes and their miRNA cargo miR-15a-5p in preeclamptic pathology. They isolated exosomes from the peripheral whole blood of women with pregnancy affected by preeclampsia and verified an upregulation for miR-15a-5p with respect to women with uncomplicated pregnancy [112]. Ntsethe et al. characterized the presence of exosomes in the serum of women with pregnancy complicated by preeclampsia. Exosome levels were higher in pathological pregnancies than uncomplicated ones. Both miR-155 and miR-222 moved by exosomes in the peripheral circulation were dysregulated. In fact, miR-155 was upregulated and miR-222 downregulated in the serum of women with complicated pregnancy with respect to controls [134]. Also, Aharon et al. investigated miRNAs present in exosomes. Particularly, they analyzed the blood samples of women with diagnosed preeclamptic pregnancy, isolating before exosomes and then characterizing the miRNA content. The results demonstrated a significant downregulation for miR-16-5p and miR-210 in subjects with pregnancy affected by preeclampsia with respect to healthy pregnant controls [116]. Also, the Ago-bound miR-210 was increased in the peripheral circulation of women with pregnancy complicated by preeclampsia [142].

Conversely, Luque et al. investigated miRNA circulating in maternal plasma. Evidence demonstrated that the investigated miRNAs (miR-192, miR-125b, miR-143, miR-126, miR-221, miR-942 and miR-127) were not a useful tool to predict preeclampsia, considering that their serum levels demonstrated no significant differences between the preeclampsia and control groups [153]. In the same way, Let7a-5p is not associated with preeclampsia because its expression in maternal plasma is not significantly different between uncomplicated and preeclamptic pregnancy [154]. Gunel et al. observed that the circulating level of miR-195 is not significantly different between preeclamptic and normotensive plasma samples, although miR-195 was downregulated in preeclamptic placenta samples [155].

5.2. miRNAs and Onset of Preeclampsia

miRNAs have been investigated to distinguish between the early (before 34 weeks of pregnancy) and late (after 34 weeks of pregnancy) onset of preeclampsia.

Miura et al. studied the expression level of all ten miRNAs belonging to C19MC, isolated from maternal blood samples at 27–34 weeks of gestation. The authors demonstrated that, except for miR-518b and miR-519d, the remaining miRNAs were upregulated in women with an early onset of preeclampsia compared with women with a late onset of preeclampsia [145]. Dong et al. confirmed the usefulness of miRNAs as a biomarker in the detection of preeclampsia onset. Particularly, they measured the expression level of both miR-21 and miR-31 in the maternal plasma of preeclamptic women. The results demonstrated that an increase in circulating miR-31 was associated with early onset preeclampsia; meanwhile, an increase in circulating miR-21 was related to late-onset preeclampsia [119]. Decreased levels of circulating miR-126 were associated with early onset preeclampsia compared to gestation-matched controls [128]. Kolkova et al. verified the expression of miRNAs isolated from the venous blood of women with pregnancy complicated by preeclampsia. Subjects affected by preeclampsia were divided into subgroups considering the preeclampsia onset (early or late). A significant difference in circulating miRNA was observed for miR-21-5p and miR-155-5p in preeclamptic pregnancies, compared to women with uncomplicated pregnancy. The overexpression of these two miRNAs was observed in subjects with a late onset of preeclampsia compared to healthy pregnancies [120]. Whigham et al. performed a case–control study using blood samples from pregnant women. They measured the miRNA expression levels at 28 and 36 weeks of gestation from subjects who developed preeclampsia after 36 weeks' gestation, comparing the results with levels in gestation-matched blood samples from a cohort of randomly selected controls. Notably, the expression level of miR-18a, miR-363, miR-1283, miR-149, miR-16 and miR-424 was significantly reduced in subjects who developed preeclampsia at 36 weeks' gestation [115]. Akehurst et al. measured the miR-206 in the maternal plasma of women at 28 weeks of gestation and observed that miR-206 was upregulated in the maternal plasma of women who later developed preeclampsia [139]. Also, the miR-518b belonging to C19MC is a good biomarker for predicting pregnancy complicated by preeclampsia. Jelena et al. demonstrated that women at 24–38 weeks of gestation with an upregulation of miR-518b in plasma later developed preeclamptic pregnancy [115].

Research activity has led to the identification of miRNAs being added to the list of potential predictors for the severity of preeclampsia. Pan et al. analyzed the expression level of miRNAs in the plasma of women with normal pregnancy and pregnancy complicated by mild preeclampsia. Plasma was collected before and after parturition. They demonstrated that the parturition influenced the expression of miRNAs in the plasma of the same women and that the expression level of miR-141 and miR-221 was different between normal and preeclamptic plasma, both before and after parturition [131]. Li et al. demonstrated that the differential expression of circulating miRNAs investigated was related to the severity of the preeclamptic condition. In fact, both miR-141 and miR-29a are significantly overexpressed in the plasma of women affected by mild preeclampsia. On the contrary, miR-144 was significantly downregulated in the plasma of women affected by mild preeclampsia and severe preeclampsia with respect to uncomplicated pregnancy [123]. Birò et al. highlighted an upregulation for miR-210 in women with pregnancy complicated by severe and mild preeclampsia upon the measured level of expression in maternal plasma. However, the miR-210 expression level was not significantly different between the mild and severe preeclampsia groups [141]. Jairajpuri et al. suggested that miR-215, miR-155, miR-650, miR-210, miR-21, miR-518b and miR-29a were upregulated, and miR-18a, miR-19b1, miR-144 and miR-15b were downregulated in pregnancy complicated by severe preeclampsia versus mild preeclampsia [113].

5.3. miRNAs and Prediction of Preeclampsia

The ability to predict preeclampsia is a major challenge in contemporary obstetrics, and resources are now focused on the first trimester of pregnancy, where prophylactic strategies can help reduce the incidence of this disorder [156] (Figure 2).

Combined tests, such as the measurement of mean arterial pressure (MAP), the ratio of soluble Fms-like tyrosine kinase-1 to placental growth factor (sFlt-1/PlGF) and the uterine artery pulsatility index (UTPI), are already widely validated [157]. The strategy of combining biochemical and biophysical data stems from the consideration that it is unlikely that preeclampsia can be detected early by a single predictive parameter with sufficient accuracy to be clinically useful. The association between miRNAs and biophysical parameters was evaluated by demonstrating a negative correlation between miR-942 levels and maternal blood pressure and between miR-143 levels and the uterine artery Doppler pulsatility index [153].

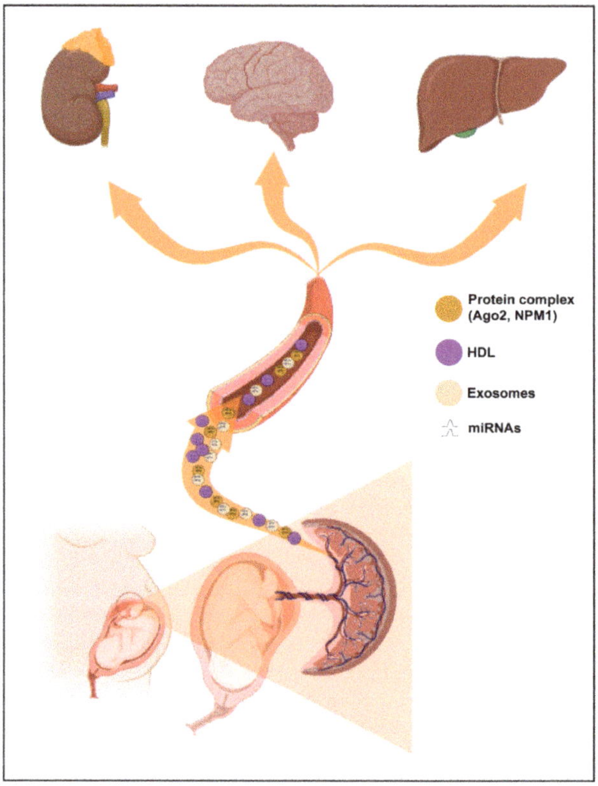

Figure 2. Trafficking routes of circulating miRNAs during pregnancy. Circulating miRNAs may be delivered to maternal circulation and affect various cell events in maternal targeting organs (e.g., kidney, brain and liver).

A study by Luque et al. [153] indicated that the assessment of maternal serum microR-NAs at the end of the first trimester of pregnancy does not appear to have any predictive value for early preeclampsia. Other studies, however, have strongly supported the need for more detailed exploration of microRNAs in the maternal circulation since they represent potential biomarkers for pregnancy-related complications [146,152,158,159]. There is a need to consider that increased placental dysfunction may stimulate the gradual release of placental mediators, including miRNAs, into the maternal circulation, leading to further diffuse maternal vascular damage and increased differential profiles of circulating miRNAs only at later stages of gestation. Studies by Winger et al. have shown that, in the first trimester of pregnancy, miRNA profiling in maternal peripheral blood mononuclear cells can successfully predict adverse outcomes, such as preeclampsia and miscarriage [129]. Circulating levels of miR-942 were lower at mid-pregnancy (12–20 weeks' gestation) in women with preeclampsia than in the control group [151]. miR-942 might play a role in preeclampsia through ENG, an endothelial growth factor with anti-angiogenic properties [160].

Ura et al. suggested that the overexpression of miR-1233 measured at the early stages of gestation (12–14 weeks) might be a potential marker to distinguish women who later developed severe preeclampsia in the third trimester of pregnancy [152]. A subsequent study investigated the expression level of global microRNAs in first-trimester (12–14 weeks) plasma obtained from women who subsequently developed preeclampsia (at or after 34 weeks of gestation) compared to uncomplicated pregnancies. The results demonstrated that both miR-23b-5p and miR-99b-5p were downregulated in subjects with pregnancy complicated by later preeclampsia compared to the controls [121]. The panel of miRNAs that are useful in the diagnosis of preeclampsia is enriched by a study that was carried out in 2020. Li et al. investigated biological samples from women at the 12–13th week of gestation and subsequently tested the results in vitro. Analysis of the plasma samples recovered from women who later developed preeclampsia revealed a downregulation for 16 miRNAs with respect to women with uncomplicated pregnancy in the same period of gestation. Notably, the miR-125b was associated in vitro with a strong inhibition of trophoblast invasion and with the compromised activity of endothelial cells [111]. Xu et al. measured the expression level of several miRNAs in the maternal plasma of women at two different gestational weeks (15–18 weeks and 35–38 weeks). The expression level of miR-18a, miR-19b1 and miR-92a1 was reduced, while miR-210 was upregulated in the plasma of patients affected by preeclampsia compared to those in normal controls at both gestational stages [118]. In a retrospective nested cohort case–control study, 34 subjects (16 women who later developed preeclampsia and 18 women with uncomplicated pregnancy) were enrolled. Patients were invited to donate serum samples at 12, 16 and 20 weeks of gestation and at the time of preeclampsia diagnosis. The authors demonstrated that miR-628-3p was the earliest miRNA to be deregulated both at 12 and at 20 weeks of gestation and so it can be considered as a strong biomarker in preeclampsia diagnosis. Also, miR-151a-3p, miR-573 and miR-628-3p were upregulated at 16 and 20 weeks of gestation and, in the authors' judgement, should be added to the panel of potential biomarkers for predicting pregnancy complicated by preeclampsia [132]. miR-125b was found to be significantly upregulated in the plasma of women who later developed PE, when the logical value of plasma miR-125b expression levels at the beginning of pregnancy combined with maternal age and BMI in predicting preeclampsia was 0.85 [44].

Also, miRNAs belonging to C19MC have been investigated to verify their potential role as predictors for the preeclampsia onset. A nested case–control study of a longitudinal cohort enrolled women at 10 to 13 gestational weeks. The analysis of maternal plasma demonstrated a higher level of expression for miR-517-5p in women who later developed preeclamptic pregnancy [148]. Different results emerged using microRNAs from exosomes in the plasma of women at 10–13 weeks of gestation. In fact, measurements of the expression level for miRNAs belonging to C19MC demonstrated that the downregulation of miR-517-5p, miR-520a-5p and miR-525-5p was associated with the occurrence of preeclampsia [149]. Kondracka et al. examined blood samples from women in the first trimester of gestation for the expression level of miRNAs belonging to C19MC. They reported an upregulation for miR-517 but also for miR-526 in pregnant women who later developed preeclampsia but made no significant suggestions about the possible use of miR-517-5p as a predictor of preeclampsia [147]. Analyzing the maternal serum at 12, 16 and 20 weeks of gestation, Martinez-Fierro et al. observed that increased circulating levels of 512-3p, 518f3p, 520c-3p and 520d-3p were associated with a later occurrence of preeclampsia [144]. A meta-analysis of a total of 20 studies from 8 articles including 273 patients with preeclampsia and 343 healthy pregnancies showed that circulating miRNAs could be a useful screening tool to diagnose and predict preeclampsia, with a sensitivity of 0.88 (95% CI: 0.80–0.93), a specificity of 0.87 (95% CI: 0.78–0.92) and a diagnostic odds ratio of 50.24 (95% CI: 21.28–118.62) [161]. Another meta-analysis of 14 articles, which included cyclic RNAs and miRNAs, reported a pooled AUC value of 0.86 (pooled sensitivity = 71%; pooled specificity = 84%) and a diagnostic odds ratio of 13 (95% CI: 11–19) [162].

6. miRNAs, Preeclampsia and Epigenetics

Preeclampsia, in addition to being one of the most frequent causes of maternal and fetal morbidity and mortality in pregnancy, has long-term negative implications for both mother and offspring [163]. Epidemiological studies indicate that women who experience preeclampsia during pregnancy have an increased vascular and metabolic risk later, as do the children of preeclamptic mothers [164]. These epidemiological considerations underlie epigenetic studies of preeclamptic disease. Although preeclampsia is a very complex disease, a great deal of evidence confirms that endothelial dysfunction is a central feature of pathogenesis and a factor that epigenetically may lead to an increased cardiovascular risk in later life. Epigenetics, or how the environment influences gene expression without altering the DNA sequence, is one of the mechanisms by which gestational hypoxia enables adaptive responses to change in the placental environment in preeclampsia. Epigenetic modifications are one of the potential mechanisms, including aberrant miRNA expression, through which the exposure to an altered environment in utero results in the development of chronic disease. The actions of miRNAs, DNA methylation and histone modification are the three most studied epigenetic processes [165]. In vitro studies have shown that miRNA expression is modulated by hypoxia, cell signaling pathways and epigenetic modifications through promoter methylation [166]. The downregulation of miRNAs also results from the hypermethylation of promoter regions [167,168]. Hypomethylation of the miR-141-3p promoter has been reported to increase the expression of miR-141-3p, which, in turn, induces inflammasome formation, a decreased expression of MMP2/9 and the inhibition of trophoblast proliferation and invasion [169]. From this point of view, the degree of miRNA methylation might have an epigenetic effect. From the perspective of endothelial dysfunction with possible epigenetic spillover, maternal and cord-derived endothelial progenitor cells (EPCs) from preeclamptic pregnancies show an aberrant miRNAs profile compared with healthy pregnancies [170]. EPCs are essential for maintaining a healthy endothelium throughout an individual's lifetime. Decreased cell numbers and colony-forming units of maternal EPCs are described as a sign of impaired endothelial repair capacity in preeclampsia [171]. The importance of studying miRNA changes in the epigenetic domain stems from the possible therapeutic developments in preeclampsia and women's future cardiovascular risk. The relevance of miRNAs in vascular neovascularization has been demonstrated by several knockdown approaches of enzymes involved in miRNA biogenesis [172]. Because miRNAs are known to be critical in the fine regulation and maintenance of the physiological balance of the vascular endothelium, they are targets of miRNA-based therapies through reprogramming endothelial cells.

7. Conclusions

Several non-coding RNAs and, thus, miRNAs are differentially expressed in the pathophysiology of women's health in general [173] and in pregnancy and the placenta in particular. In recent years, the number of identified miRNAs has significantly increased; however, the exact mechanisms remain to be elucidated, mainly because of the cell-specific functions exhibited by many miRNAs. A deeper understanding of miRNAs and their relationships with gene modifications will help in determining the mechanism by which these molecules contribute to placental development. The identification of miRNAs that may act as potential non-invasive biomarkers for the prediction of pregnancy outcomes in the first trimester, especially among high-risk women, may have implications for research, identifying signaling pathways for further investigation and clinical implications, facilitating early diagnosis and timely interventions. This challenge is not easy to address since preeclampsia is probably not a single disease but may present in several different forms, and there are many difficulties in finding, measuring and reproducing miRNA results. From this review, we see the possibility that, in the near future, molecular biology may, through the application of gene therapy on miRNAs, intervene in the pathological basis of preeclampsia from the very early stages of placental implantation and development.

Author Contributions: M.C. and S.R.G. conceptualized and drafted the manuscript; A.C. and D.M. revised and supervised the manuscript. M.C. and S.R.G. developed the table and figures and reviewed and edited the manuscript. All authors have read and agreed to the published version of the manuscript.

Funding: This research received no external funding.

Institutional Review Board Statement: Review and ethics approval were waived for this study because of local Ethics Committee rules that do not require evaluation for literature review manuscripts.

Informed Consent Statement: Not applicable.

Data Availability Statement: No new data were created or analyzed in this study. Data sharing is not applicable to this article.

Conflicts of Interest: The authors declare no conflicts of interest.

References

1. Tossetta, G.; Avellini, C.; Licini, C.; Giannubilo, S.R.; Castellucci, M.; Marzioni, D. High temperature requirement A1 and fibronectin: Two possible players in placental tissue remodelling. *Eur. J. Histochem.* **2016**, *60*, 2724. [CrossRef] [PubMed]
2. Huppertz, B. The anatomy of the normal placenta. *J. Clin. Pathol.* **2008**, *61*, 1296–1302. [CrossRef] [PubMed]
3. Tossetta, G.; Fantone, S.; Busilacchi, E.M.; Di Simone, N.; Giannubilo, S.R.; Scambia, G.; Giordano, A.; Marzioni, D. Modulation of matrix metalloproteases by ciliary neurotrophic factor in human placental development. *Cell Tissue Res.* **2022**, *390*, 113–129. [CrossRef]
4. Tossetta, G.; Fantone, S.; Giannubilo, S.R.; Marinelli Busilacchi, E.; Ciavattini, A.; Castellucci, M.; Di Simone, N.; Mattioli-Belmonte, M.; Marzioni, D. Pre-eclampsia onset and SPARC: A possible involvement in placenta development. *J. Cell Physiol.* **2019**, *234*, 6091–6098. [CrossRef] [PubMed]
5. Gesuita, R.; Licini, C.; Picchiassi, E.; Tarquini, F.; Coata, G.; Fantone, S.; Tossetta, G.; Ciavattini, A.; Castellucci, M.; Di Renzo, G.C.; et al. Association between first trimester plasma htra1 level and subsequent preeclampsia: A possible early marker? *Pregnancy Hypertens.* **2019**, *18*, 58–62. [CrossRef]
6. Cardaropoli, S.; Paulesu, L.; Romagnoli, R.; Ietta, F.; Marzioni, D.; Castellucci, M.; Rolfo, A.; Vasario, E.; Piccoli, E.; Todros, T. Macrophage migration inhibitory factor in fetoplacental tissues from preeclamptic pregnancies with or without fetal growth restriction. *Clin. Dev. Immunol.* **2012**, *2012*, 639342. [CrossRef]
7. Marzioni, D.; Todros, T.; Cardaropoli, S.; Rolfo, A.; Lorenzi, T.; Ciarmela, P.; Romagnoli, R.; Paulesu, L.; Castellucci, M. Activating protein-1 family of transcription factors in the human placenta complicated by preeclampsia with and without fetal growth restriction. *Placenta* **2010**, *31*, 919–927. [CrossRef]
8. Todros, T.; Marzioni, D.; Lorenzi, T.; Piccoli, E.; Capparuccia, L.; Perugini, V.; Cardaropoli, S.; Romagnoli, R.; Gesuita, R.; Rolfo, A.; et al. Evidence for a role of TGF-beta1 in the expression and regulation of alpha-SMA in fetal growth restricted placentae. *Placenta* **2007**, *28*, 1123–1132. [CrossRef]
9. Tossetta, G.; Fantone, S.; Gesuita, R.; Di Renzo, G.C.; Meyyazhagan, A.; Tersigni, C.; Scambia, G.; Di Simone, N.; Marzioni, D. HtrA1 in Gestational Diabetes Mellitus: A Possible Biomarker? *Diagnostics* **2022**, *12*, 2705. [CrossRef]
10. Giannubilo, S.R.; Licini, C.; Picchiassi, E.; Tarquini, F.; Coata, G.; Fantone, S.; Tossetta, G.; Ciavattini, A.; Castellucci, M.; Giardina, I.; et al. First trimester HtrA1 maternal plasma level and spontaneous preterm birth. *J. Matern. Fetal Neonatal Med.* **2022**, *35*, 780–784. [CrossRef]
11. Capparuccia, L.; Marzioni, D.; Giordano, A.; Fazioli, F.; De Nictolis, M.; Busso, N.; Todros, T.; Castellucci, M. PPARgamma expression in normal human placenta, hydatidiform mole and choriocarcinoma. *Mol. Hum. Reprod.* **2002**, *8*, 574–579. [CrossRef] [PubMed]
12. Crescimanno, C.; Marzioni, D.; Paradinas, F.J.; Schrurs, B.; Muhlhauser, J.; Todros, T.; Newlands, E.; David, G.; Castellucci, M. Expression pattern alterations of syndecans and glypican-1 in normal and pathological trophoblast. *J. Pathol.* **1999**, *189*, 600–608. [CrossRef]
13. Dimitriadis, E.; Rolnik, D.L.; Zhou, W.; Estrada-Gutierrez, G.; Koga, K.; Francisco, R.P.V.; Whitehead, C.; Hyett, J.; da Silva Costa, F.; Nicolaides, K.; et al. Pre-eclampsia. *Nat. Rev. Dis. Primers* **2023**, *9*, 8. [CrossRef] [PubMed]
14. Odigboegwu, O.; Pan, L.J.; Chatterjee, P. Use of Antihypertensive Drugs During Preeclampsia. *Front. Cardiovasc. Med.* **2018**, *5*, 50. [CrossRef] [PubMed]
15. Mol, B.W.J.; Roberts, C.T.; Thangaratinam, S.; Magee, L.A.; de Groot, C.J.M.; Hofmeyr, G.J. Pre-eclampsia. *Lancet* **2016**, *387*, 999–1011. [CrossRef]
16. Thadhani, R.; Ecker, J.L.; Kettyle, E.; Sandler, L.; Frigoletto, F.D. Pulse pressure and risk of preeclampsia: A prospective study. *Obstet. Gynecol.* **2001**, *97*, 515–520. [CrossRef]
17. Fantone, S.; Mazzucchelli, R.; Giannubilo, S.R.; Ciavattini, A.; Marzioni, D.; Tossetta, G. AT-rich interactive domain 1A protein expression in normal and pathological pregnancies complicated by preeclampsia. *Histochem. Cell Biol.* **2020**, *154*, 339–346. [CrossRef]
18. Tossetta, G.; Fantone, S.; Giannubilo, S.R.; Ciavattini, A.; Senzacqua, M.; Frontini, A.; Marzioni, D. HTRA1 in Placental Cell Models: A Possible Role in Preeclampsia. *Curr. Issues Mol. Biol.* **2023**, *45*, 3815–3828. [CrossRef]

19. Fantone, S.; Giannubilo, S.R.; Marzioni, D.; Tossetta, G. HTRA family proteins in pregnancy outcome. *Tissue Cell* **2021**, *72*, 101549. [CrossRef]

20. Ali, S.M.; Khalil, R.A. Genetic, immune and vasoactive factors in the vascular dysfunction associated with hypertension in pregnancy. *Expert. Opin. Ther. Targets* **2015**, *19*, 1495–1515. [CrossRef]

21. Cecati, M.; Emanuelli, M.; Giannubilo, S.R.; Quarona, V.; Senetta, R.; Malavasi, F.; Tranquilli, A.L.; Saccucci, F. Contribution of adenosine-producing ectoenzymes to the mechanisms underlying the mitigation of maternal-fetal conflicts. *J. Biol. Regul. Homeost. Agents* **2013**, *27*, 519–529. [PubMed]

22. Ferreira, L.B.; Williams, K.A.; Best, G.; Haydinger, C.D.; Smith, J.R. Inflammatory cytokines as mediators of retinal endothelial barrier dysfunction in non-infectious uveitis. *Clin. Transl. Immunol.* **2023**, *12*, e1479. [CrossRef] [PubMed]

23. Mateuszuk, L.; Campagna, R.; Kutryb-Zajac, B.; Kus, K.; Slominska, E.M.; Smolenski, R.T.; Chlopicki, S. Reversal of endothelial dysfunction by nicotinamide mononucleotide via extracellular conversion to nicotinamide riboside. *Biochem. Pharmacol.* **2020**, *178*, 114019. [CrossRef] [PubMed]

24. Tossetta, G.; Piani, F.; Borghi, C.; Marzioni, D. Role of CD93 in Health and Disease. *Cells* **2023**, *12*, 1778. [CrossRef] [PubMed]

25. Campagna, R.; Vignini, A. NAD(+) Homeostasis and NAD(+)-Consuming Enzymes: Implications for Vascular Health. *Antioxidants* **2023**, *12*, 376. [CrossRef]

26. Campagna, R.; Mateuszuk, L.; Wojnar-Lason, K.; Kaczara, P.; Tworzydlo, A.; Kij, A.; Bujok, R.; Mlynarski, J.; Wang, Y.; Sartini, D.; et al. Nicotinamide N-methyltransferase in endothelium protects against oxidant stress-induced endothelial injury. *Biochim. Biophys. Acta Mol. Cell Res.* **2021**, *1868*, 119082. [CrossRef]

27. Szczesny-Malysiak, E.; Stojak, M.; Campagna, R.; Grosicki, M.; Jamrozik, M.; Kaczara, P.; Chlopicki, S. Bardoxolone Methyl Displays Detrimental Effects on Endothelial Bioenergetics, Suppresses Endothelial ET-1 Release, and Increases Endothelial Permeability in Human Microvascular Endothelium. *Oxid. Med. Cell Longev.* **2020**, *2020*, 4678252. [CrossRef]

28. Piani, F.; Tossetta, G.; Cara-Fuentes, G.; Agnoletti, D.; Marzioni, D.; Borghi, C. Diagnostic and Prognostic Role of CD93 in Cardiovascular Disease: A Systematic Review. *Biomolecules* **2023**, *13*, 910. [CrossRef]

29. Ristovska, E.C.; Genadieva-Dimitrova, M.; Todorovska, B.; Milivojevic, V.; Rankovic, I.; Samardziski, I.; Bojadzioska, M. The Role of Endothelial Dysfunction in the Pathogenesis of Pregnancy-Related Pathological Conditions: A Review. *Pril (Makedon. Akad. Nauk. Umet. Odd. Med. Nauki)* **2023**, *44*, 113–137. [CrossRef]

30. Shen, J.; San, W.; Zheng, Y.; Zhang, S.; Cao, D.; Chen, Y.; Meng, G. Different types of cell death in diabetic endothelial dysfunction. *Biomed. Pharmacother.* **2023**, *168*, 115802. [CrossRef]

31. Tossetta, G.; Fantone, S.; Delli Muti, N.; Balercia, G.; Ciavattini, A.; Giannubilo, S.R.; Marzioni, D. Preeclampsia and severe acute respiratory syndrome coronavirus 2 infection: A systematic review. *J. Hypertens.* **2022**, *40*, 1629–1638. [CrossRef] [PubMed]

32. Fantone, S.; Tossetta, G.; Di Simone, N.; Tersigni, C.; Scambia, G.; Marcheggiani, F.; Giannubilo, S.R.; Marzioni, D. CD93 a potential player in cytotrophoblast and endothelial cell migration. *Cell Tissue Res.* **2022**, *387*, 123–130. [CrossRef] [PubMed]

33. Cevher Akdulum, M.F.; Demirdag, E.; Arik, S.I.; Safarova, S.; Erdem, M.; Bozkurt, N.; Erdem, A. Is the First-Trimester Systemic Immune-Inflammation Index Associated With Preeclampsia? *Cureus* **2023**, *15*, e44063. [CrossRef] [PubMed]

34. Tossetta, G.; Fantone, S.; Piani, F.; Crescimanno, C.; Ciavattini, A.; Giannubilo, S.R.; Marzioni, D. Modulation of NRF2/KEAP1 Signaling in Preeclampsia. *Cells* **2023**, *12*, 1545. [CrossRef] [PubMed]

35. Mazzanti, L.; Cecati, M.; Vignini, A.; D'Eusanio, S.; Emanuelli, M.; Giannubilo, S.R.; Saccucci, F.; Tranquilli, A.L. Placental expression of endothelial and inducible nitric oxide synthase and nitric oxide levels in patients with HELLP syndrome. *Am. J. Obstet. Gynecol.* **2011**, *205*, 236.e1-7. [CrossRef] [PubMed]

36. Fantone, S.; Ermini, L.; Piani, F.; Di Simone, N.; Barbaro, G.; Giannubilo, S.R.; Gesuita, R.; Tossetta, G.; Marzioni, D. Downregulation of argininosuccinate synthase 1 (ASS1) is associated with hypoxia in placental development. *Hum. Cell* **2023**, *36*, 1190–1198. [CrossRef] [PubMed]

37. Cecati, M.; Giannubilo, S.R.; Saccucci, F.; Sartini, D.; Ciavattini, A.; Emanuelli, M.; Tranquilli, A.L. Potential Role of Placental Klotho in the Pathogenesis of Preeclampsia. *Cell Biochem. Biophys.* **2016**, *74*, 49–57. [CrossRef]

38. Maynard, S.E.; Karumanchi, S.A. Angiogenic factors and preeclampsia. *Semin. Nephrol.* **2011**, *31*, 33–46. [CrossRef]

39. Van Niekerk, G.; Christowitz, C.; Engelbrecht, A.M. Insulin-mediated immune dysfunction in the development of preeclampsia. *J. Mol. Med.* **2021**, *99*, 889–897. [CrossRef]

40. Leavey, K.; Benton, S.J.; Grynspan, D.; Kingdom, J.C.; Bainbridge, S.A.; Cox, B.J. Unsupervised Placental Gene Expression Profiling Identifies Clinically Relevant Subclasses of Human Preeclampsia. *Hypertension* **2016**, *68*, 137–147. [CrossRef]

41. Benton, S.J.; Leavey, K.; Grynspan, D.; Cox, B.J.; Bainbridge, S.A. The clinical heterogeneity of preeclampsia is related to both placental gene expression and placental histopathology. *Am. J. Obstet. Gynecol.* **2018**, *219*, 604.e1–604.e25. [CrossRef]

42. Rana, S.; Lemoine, E.; Granger, J.P.; Karumanchi, S.A. Preeclampsia: Pathophysiology, Challenges, and Perspectives. *Circ. Res.* **2019**, *124*, 1094–1112. [CrossRef] [PubMed]

43. Licini, C.; Avellini, C.; Picchiassi, E.; Mensa, E.; Fantone, S.; Ramini, D.; Tersigni, C.; Tossetta, G.; Castellucci, C.; Tarquini, F.; et al. Pre-eclampsia predictive ability of maternal miR-125b: A clinical and experimental study. *Transl. Res.* **2021**, *228*, 13–27. [CrossRef] [PubMed]

44. Liu, Z.; Pei, J.; Zhang, X.; Wang, C.; Tang, Y.; Liu, H.; Yu, Y.; Luo, S.; Gu, W. APOA1 Is a Novel Marker for Preeclampsia. *Int. J. Mol. Sci.* **2023**, *24*, 16363. [CrossRef] [PubMed]
45. Piani, F.; Agnoletti, D.; Baracchi, A.; Scarduelli, S.; Verde, C.; Tossetta, G.; Montaguti, E.; Simonazzi, G.; Degli Esposti, D.; Borghi, C.; et al. Serum uric acid to creatinine ratio and risk of preeclampsia and adverse pregnancy outcomes. *J. Hypertens.* **2023**, *41*, 1333–1338. [CrossRef]
46. Soobryan, N.; Reddy, K.; Ibrahim, U.H.; Moodley, J.; Kumar, A.; Mackraj, I. Identification of gene signature markers in gestational hypertension and early-onset pre-eclampsia. *Placenta* **2023**, *145*, 1–8. [CrossRef]
47. Piani, F.; Tossetta, G.; Fantone, S.; Agostinis, C.; Di Simone, N.; Mandala, M.; Bulla, R.; Marzioni, D.; Borghi, C. First Trimester CD93 as a Novel Marker of Preeclampsia and Its Complications: A Pilot Study. *High. Blood Press. Cardiovasc. Prev.* **2023**, *30*, 591–594. [CrossRef]
48. Qin, S.; Sun, N.; Xu, L.; Xu, Y.; Tang, Q.; Tan, L.; Chen, A.; Zhang, L.; Liu, S. The Value of Circulating microRNAs for Diagnosis and Prediction of Preeclampsia: A Meta-analysis and Systematic Review. *Reprod. Sci.* **2022**, *29*, 3078–3090. [CrossRef]
49. Ping, Z.; Feng, Y.; Lu, Y.; Ai, L.; Jiang, H. Integrated analysis of microRNA and mRNA expression profiles in Preeclampsia. *BMC Med. Genom.* **2023**, *16*, 309. [CrossRef]
50. Kimura, T. Non-coding Natural Antisense RNA: Mechanisms of Action in the Regulation of Target Gene Expression and Its Clinical Implications. *Yakugaku Zasshi* **2020**, *140*, 687–700. [CrossRef]
51. Zeng, Q.; Wan, H.; Zhao, S.; Xu, H.; Tang, T.; Oware, K.A.; Qu, S. Role of PIWI-interacting RNAs on cell survival: Proliferation, apoptosis, and cycle. *IUBMB Life* **2020**, *72*, 1870–1878. [CrossRef] [PubMed]
52. Wang, N.; Yu, Y.; Xu, B.; Zhang, M.; Li, Q.; Miao, L. Pivotal prognostic and diagnostic role of the long non-coding RNA colon cancer-associated transcript 1 expression in human cancer (Review). *Mol. Med. Rep.* **2019**, *19*, 771–782. [CrossRef] [PubMed]
53. Zhao, W.; An, Y.; Liang, Y.; Xie, X.W. Role of HOTAIR long noncoding RNA in metastatic progression of lung cancer. *Eur. Rev. Med. Pharmacol. Sci.* **2014**, *18*, 1930–1936. [PubMed]
54. Shields, E.J.; Petracovici, A.F.; Bonasio, R. lncRedibly versatile: Biochemical and biological functions of long noncoding RNAs. *Biochem. J.* **2019**, *476*, 1083–1104. [CrossRef] [PubMed]
55. Carthew, R.W.; Sontheimer, E.J. Origins and Mechanisms of miRNAs and siRNAs. *Cell* **2009**, *136*, 642–655. [CrossRef] [PubMed]
56. Macfarlane, L.A.; Murphy, P.R. MicroRNA: Biogenesis, Function and Role in Cancer. *Curr. Genom.* **2010**, *11*, 537–561. [CrossRef] [PubMed]
57. Schwarzenbach, H.; Gahan, P.B. MicroRNA Shuttle from Cell-To-Cell by Exosomes and Its Impact in Cancer. *Noncoding RNA* **2019**, *5*, 28. [CrossRef] [PubMed]
58. Avellini, C.; Licini, C.; Lazzarini, R.; Gesuita, R.; Guerra, E.; Tossetta, G.; Castellucci, C.; Giannubilo, S.R.; Procopio, A.; Alberti, S.; et al. The trophoblast cell surface antigen 2 and miR-125b axis in urothelial bladder cancer. *Oncotarget* **2017**, *8*, 58642–58653. [CrossRef]
59. Filipow, S.; Laczmanski, L. Blood Circulating miRNAs as Cancer Biomarkers for Diagnosis and Surgical Treatment Response. *Front. Genet.* **2019**, *10*, 169. [CrossRef]
60. Cui, L.; Zhang, X.; Ye, G.; Zheng, T.; Song, H.; Deng, H.; Xiao, B.; Xia, T.; Yu, X.; Le, Y.; et al. Gastric juice MicroRNAs as potential biomarkers for the screening of gastric cancer. *Cancer* **2013**, *119*, 1618–1626. [CrossRef]
61. Park, N.J.; Zhou, H.; Elashoff, D.; Henson, B.S.; Kastratovic, D.A.; Abemayor, E.; Wong, D.T. Salivary microRNA: Discovery, characterization, and clinical utility for oral cancer detection. *Clin. Cancer Res.* **2009**, *15*, 5473–5477. [CrossRef] [PubMed]
62. Hanke, M.; Hoefig, K.; Merz, H.; Feller, A.C.; Kausch, I.; Jocham, D.; Warnecke, J.M.; Sczakiel, G. A robust methodology to study urine microRNA as tumor marker: microRNA-126 and microRNA-182 are related to urinary bladder cancer. *Urol. Oncol.* **2010**, *28*, 655–661. [CrossRef] [PubMed]
63. Schwarzenbach, H.; Hoon, D.S.; Pantel, K. Cell-free nucleic acids as biomarkers in cancer patients. *Nat. Rev. Cancer* **2011**, *11*, 426–437. [CrossRef] [PubMed]
64. Lee, Y.; Jeon, K.; Lee, J.T.; Kim, S.; Kim, V.N. MicroRNA maturation: Stepwise processing and subcellular localization. *EMBO J.* **2002**, *21*, 4663–4670. [CrossRef] [PubMed]
65. Winter, J.; Jung, S.; Keller, S.; Gregory, R.I.; Diederichs, S. Many roads to maturity: microRNA biogenesis pathways and their regulation. *Nat. Cell Biol.* **2009**, *11*, 228–234. [CrossRef]
66. Mayya, V.K.; Duchaine, T.F. Ciphers and Executioners: How 3′-Untranslated Regions Determine the Fate of Messenger RNAs. *Front. Genet.* **2019**, *10*, 6. [CrossRef]
67. Broughton, J.P.; Lovci, M.T.; Huang, J.L.; Yeo, G.W.; Pasquinelli, A.E. Pairing beyond the Seed Supports MicroRNA Targeting Specificity. *Mol. Cell* **2016**, *64*, 320–333. [CrossRef]
68. Chen, Y.M.; Zheng, Y.L.; Su, X.; Wang, X.Q. Crosstalk Between MicroRNAs and Circular RNAs in Human Diseases: A Bibliographic Study. *Front. Cell Dev. Biol.* **2021**, *9*, 754880. [CrossRef]
69. Mitchell, P.S.; Parkin, R.K.; Kroh, E.M.; Fritz, B.R.; Wyman, S.K.; Pogosova-Agadjanyan, E.L.; Peterson, A.; Noteboom, J.; O'Briant, K.C.; Allen, A.; et al. Circulating microRNAs as stable blood-based markers for cancer detection. *Proc. Natl. Acad. Sci. USA* **2008**, *105*, 10513–10518. [CrossRef]
70. Chen, X.; Ba, Y.; Ma, L.; Cai, X.; Yin, Y.; Wang, K.; Guo, J.; Zhang, Y.; Chen, J.; Guo, X.; et al. Characterization of microRNAs in serum: A novel class of biomarkers for diagnosis of cancer and other diseases. *Cell Res.* **2008**, *18*, 997–1006. [CrossRef]

71. Turchinovich, A.; Weiz, L.; Langheinz, A.; Burwinkel, B. Characterization of extracellular circulating microRNA. *Nucleic Acids Res.* **2011**, *39*, 7223–7233. [CrossRef] [PubMed]
72. Arroyo, J.D.; Chevillet, J.R.; Kroh, E.M.; Ruf, I.K.; Pritchard, C.C.; Gibson, D.F.; Mitchell, P.S.; Bennett, C.F.; Pogosova-Agadjanyan, E.L.; Stirewalt, D.L.; et al. Argonaute2 complexes carry a population of circulating microRNAs independent of vesicles in human plasma. *Proc. Natl. Acad. Sci. USA* **2011**, *108*, 5003–5008. [CrossRef] [PubMed]
73. Wang, K.; Zhang, S.; Weber, J.; Baxter, D.; Galas, D.J. Export of microRNAs and microRNA-protective protein by mammalian cells. *Nucleic Acids Res.* **2010**, *38*, 7248–7259. [CrossRef] [PubMed]
74. Vickers, K.C.; Palmisano, B.T.; Shoucri, B.M.; Shamburek, R.D.; Remaley, A.T. MicroRNAs are transported in plasma and delivered to recipient cells by high-density lipoproteins. *Nat. Cell Biol.* **2011**, *13*, 423–433. [CrossRef] [PubMed]
75. Ortiz-Quintero, B. Cell-free microRNAs in blood and other body fluids, as cancer biomarkers. *Cell Prolif.* **2016**, *49*, 281–303. [CrossRef] [PubMed]
76. Simionescu, N.; Niculescu, L.S.; Sanda, G.M.; Margina, D.; Sima, A.V. Analysis of circulating microRNAs that are specifically increased in hyperlipidemic and/or hyperglycemic sera. *Mol. Biol. Rep.* **2014**, *41*, 5765–5773. [CrossRef]
77. Axmann, M.; Meier, S.M.; Karner, A.; Strobl, W.; Stangl, H.; Plochberger, B. Serum and Lipoprotein Particle miRNA Profile in Uremia Patients. *Genes* **2018**, *9*, 533. [CrossRef]
78. Watts, G.F.; Barrett, P.H. High-density lipoprotein metabolism in familial hypercholesterolaemia: Significance, mechanisms, therapy. *Nutr. Metab. Cardiovasc. Dis.* **2002**, *12*, 36–41.
79. Valadi, H.; Ekstrom, K.; Bossios, A.; Sjostrand, M.; Lee, J.J.; Lotvall, J.O. Exosome-mediated transfer of mRNAs and microRNAs is a novel mechanism of genetic exchange between cells. *Nat. Cell Biol.* **2007**, *9*, 654–659. [CrossRef]
80. Sabanovic, B.; Piva, F.; Cecati, M.; Giulietti, M. Promising Extracellular Vesicle-Based Vaccines against Viruses, Including SARS-CoV-2. *Biology* **2021**, *10*, 94. [CrossRef]
81. Villarroya-Beltri, C.; Gutierrez-Vazquez, C.; Sanchez-Cabo, F.; Perez-Hernandez, D.; Vazquez, J.; Martin-Cofreces, N.; Martinez-Herrera, D.J.; Pascual-Montano, A.; Mittelbrunn, M.; Sanchez-Madrid, F. Sumoylated hnRNPA2B1 controls the sorting of miRNAs into exosomes through binding to specific motifs. *Nat. Commun.* **2013**, *4*, 2980. [CrossRef] [PubMed]
82. Lee, H.; Li, C.; Zhang, Y.; Zhang, D.; Otterbein, L.E.; Jin, Y. Caveolin-1 selectively regulates microRNA sorting into microvesicles after noxious stimuli. *J. Exp. Med.* **2019**, *216*, 2202–2220. [CrossRef] [PubMed]
83. Ambros, V.; Bartel, B.; Bartel, D.P.; Burge, C.B.; Carrington, J.C.; Chen, X.; Dreyfuss, G.; Eddy, S.R.; Griffiths-Jones, S.; Marshall, M.; et al. A uniform system for microRNA annotation. *RNA* **2003**, *9*, 277–279. [CrossRef] [PubMed]
84. Chirshev, E.; Oberg, K.C.; Ioffe, Y.J.; Unternaehrer, J.J. Let-7 as biomarker, prognostic indicator, and therapy for precision medicine in cancer. *Clin. Transl. Med.* **2019**, *8*, 24. [CrossRef] [PubMed]
85. Donker, R.B.; Mouillet, J.F.; Chu, T.; Hubel, C.A.; Stolz, D.B.; Morelli, A.E.; Sadovsky, Y. The expression profile of C19MC microRNAs in primary human trophoblast cells and exosomes. *Mol. Hum. Reprod.* **2012**, *18*, 417–424. [CrossRef]
86. Ng, Y.H.; Rome, S.; Jalabert, A.; Forterre, A.; Singh, H.; Hincks, C.L.; Salamonsen, L.A. Endometrial exosomes/microvesicles in the uterine microenvironment: A new paradigm for embryo-endometrial cross talk at implantation. *PLoS ONE* **2013**, *8*, e58502. [CrossRef] [PubMed]
87. Rekker, K.; Altmae, S.; Suhorutshenko, M.; Peters, M.; Martinez-Blanch, J.F.; Codoner, F.M.; Vilella, F.; Simon, C.; Salumets, A.; Velthut-Meikas, A. A Two-Cohort RNA-seq Study Reveals Changes in Endometrial and Blood miRNome in Fertile and Infertile Women. *Genes* **2018**, *9*, 574. [CrossRef]
88. Aplin, J.D.; Kimber, S.J. Trophoblast-uterine interactions at implantation. *Reprod. Biol. Endocrinol.* **2004**, *2*, 48. [CrossRef]
89. Desrochers, L.M.; Bordeleau, F.; Reinhart-King, C.A.; Cerione, R.A.; Antonyak, M.A. Microvesicles provide a mechanism for intercellular communication by embryonic stem cells during embryo implantation. *Nat. Commun.* **2016**, *7*, 11958. [CrossRef]
90. Cretoiu, D.; Xu, J.; Xiao, J.; Suciu, N.; Cretoiu, S.M. Circulating MicroRNAs as Potential Molecular Biomarkers in Pathophysiological Evolution of Pregnancy. *Dis. Markers* **2016**, *2016*, 3851054. [CrossRef]
91. Liu, L.; Guo, J.; Gao, W.; Gao, M.; Ma, X. Research progress in the role of non-coding RNAs and embryo implantation. *Zhong Nan Da Xue Xue Bao Yi Xue Ban* **2023**, *48*, 1377–1387. [PubMed]
92. Gurung, S.; Greening, D.W.; Catt, S.; Salamonsen, L.; Evans, J. Exosomes and soluble secretome from hormone-treated endometrial epithelial cells direct embryo implantation. *Mol. Hum. Reprod.* **2020**, *26*, 510–520. [CrossRef] [PubMed]
93. Das, M.; Kale, V. Extracellular vesicles: Mediators of embryo-maternal crosstalk during pregnancy and a new weapon to fight against infertility. *Eur. J. Cell Biol.* **2020**, *99*, 151125. [CrossRef] [PubMed]
94. Atay, S.; Gercel-Taylor, C.; Suttles, J.; Mor, G.; Taylor, D.D. Trophoblast-derived exosomes mediate monocyte recruitment and differentiation. *Am. J. Reprod. Immunol.* **2011**, *65*, 65–77. [CrossRef] [PubMed]
95. Whitley, G.S.; Dash, P.R.; Ayling, L.J.; Prefumo, F.; Thilaganathan, B.; Cartwright, J.E. Increased apoptosis in first trimester extravillous trophoblasts from pregnancies at higher risk of developing preeclampsia. *Am. J. Pathol.* **2007**, *170*, 1903–1909. [CrossRef] [PubMed]
96. Reddy, A.; Zhong, X.Y.; Rusterholz, C.; Hahn, S.; Holzgreve, W.; Redman, C.W.; Sargent, I.L. The effect of labour and placental separation on the shedding of syncytiotrophoblast microparticles, cell-free DNA and mRNA in normal pregnancy and pre-eclampsia. *Placenta* **2008**, *29*, 942–949. [CrossRef]

97. Czernek, L.; Duchler, M. Exosomes as Messengers Between Mother and Fetus in Pregnancy. *Int. J. Mol. Sci.* **2020**, *21*, 4264. [CrossRef]

98. Truong, G.; Guanzon, D.; Kinhal, V.; Elfeky, O.; Lai, A.; Longo, S.; Nuzhat, Z.; Palma, C.; Scholz-Romero, K.; Menon, R.; et al. Oxygen tension regulates the miRNA profile and bioactivity of exosomes released from extravillous trophoblast cells—Liquid biopsies for monitoring complications of pregnancy. *PLoS ONE* **2017**, *12*, e0174514. [CrossRef]

99. Yang, X.W.; Shen, G.Z.; Cao, L.Q.; Jiang, X.F.; Peng, H.P.; Shen, G.; Chen, D.; Xue, P. MicroRNA-1269 promotes proliferation in human hepatocellular carcinoma via downregulation of FOXO1. *BMC Cancer* **2014**, *14*, 909. [CrossRef]

100. Fan, W.; Li, S.W.; Li, W.H.; Wang, Y.; Gong, Y.; Ma, Q.H.; Luo, S. FOXO1 expression and regulation in endometrial tissue during the menstrual cycle and in early pregnancy decidua. *Gynecol. Obstet. Invest.* **2012**, *74*, 56–63. [CrossRef]

101. Kamali Simsek, N.; Benian, A.; Sevgin, K.; Ergun, Y.; Goksever Celik, H.; Karahuseyinoglu, S.; Gunel, T. Microrna analysis of human decidua mesenchymal stromal cells from preeclampsia patients. *Placenta* **2021**, *115*, 12–19. [CrossRef] [PubMed]

102. Zandvakili, I.; Lin, Y.; Morris, J.C.; Zheng, Y. Rho GTPases: Anti- or pro-neoplastic targets? *Oncogene* **2017**, *36*, 3213–3222. [CrossRef]

103. Taga, S.; Hayashi, M.; Nunode, M.; Nakamura, N.; Ohmichi, M. miR-486-5p inhibits invasion and migration of HTR8/SVneo trophoblast cells by down-regulating ARHGAP5. *Placenta* **2022**, *123*, 5–11. [CrossRef] [PubMed]

104. Holcik, M.; Korneluk, R.G. XIAP, the guardian angel. *Nat. Rev. Mol. Cell Biol.* **2001**, *2*, 550–556. [CrossRef] [PubMed]

105. Arroyo, J.; Price, M.; Straszewski-Chavez, S.; Torry, R.J.; Mor, G.; Torry, D.S. XIAP protein is induced by placenta growth factor (PLGF) and decreased during preeclampsia in trophoblast cells. *Syst. Biol. Reprod. Med.* **2014**, *60*, 263–273. [CrossRef] [PubMed]

106. Nunode, M.; Hayashi, M.; Nagayasu, Y.; Sawada, M.; Nakamura, M.; Sano, T.; Fujita, D.; Ohmichi, M. miR-515-5p suppresses trophoblast cell invasion and proliferation through XIAP regulation in preeclampsia. *Mol. Cell Endocrinol.* **2023**, *559*, 111779. [CrossRef] [PubMed]

107. Jauniaux, E.; Gulbis, B.; Burton, G.J. Physiological implications of the materno-fetal oxygen gradient in human early pregnancy. *Reprod. Biomed. Online* **2003**, *7*, 250–253. [CrossRef]

108. Wang, Y.; Lumbers, E.R.; Arthurs, A.L.; Corbisier de Meaultsart, C.; Mathe, A.; Avery-Kiejda, K.A.; Roberts, C.T.; Pipkin, F.B.; Marques, F.Z.; Morris, B.J.; et al. Regulation of the human placental (pro)renin receptor-prorenin-angiotensin system by microRNAs. *Mol. Hum. Reprod.* **2018**, *24*, 453–464. [CrossRef]

109. Lalevee, S.; Lapaire, O.; Buhler, M. miR455 is linked to hypoxia signaling and is deregulated in preeclampsia. *Cell Death Dis.* **2014**, *5*, e1408. [CrossRef]

110. Gunel, T.; Zeybek, Y.G.; Akcakaya, P.; Kalelioglu, I.; Benian, A.; Ermis, H.; Aydinli, K. Serum microRNA expression in pregnancies with preeclampsia. *Genet. Mol. Res.* **2011**, *10*, 4034–4040. [CrossRef]

111. Campos, C.B.; Marques, T.M.; Pereira, R.W.; Sandrim, V.C. Reduced circulating miR-196b levels is associated with preeclampsia. *Pregnancy Hypertens.* **2014**, *4*, 11–13. [CrossRef] [PubMed]

112. Sandrim, V.C.; Eleuterio, N.; Pilan, E.; Tanus-Santos, J.E.; Fernandes, K.; Cavalli, R. Plasma levels of increased miR-195-5p correlates with the sFLT-1 levels in preeclampsia. *Hypertens. Pregnancy* **2016**, *35*, 150–158. [CrossRef]

113. Sandrim, V.C.; Luizon, M.R.; Palei, A.C.; Tanus-Santos, J.E.; Cavalli, R.C. Circulating microRNA expression profiles in pre-eclampsia: Evidence of increased miR-885-5p levels. *BJOG* **2016**, *123*, 2120–2128. [CrossRef] [PubMed]

114. Sheng, C.; Zhao, Y.; Zhu, L. Down-regulation of EDN1 gene expression by circulating miR-206 is associated with risk of preeclampsia. *Medicine* **2020**, *99*, e20319. [CrossRef] [PubMed]

115. Akgor, U.; Ayaz, L.; Cayan, F. Expression levels of maternal plasma microRNAs in preeclamptic pregnancies. *J. Obstet. Gynaecol.* **2021**, *41*, 910–914. [CrossRef] [PubMed]

116. Ayoub, S.E.; Shaker, O.G.; Aboshama, R.A.; Etman, M.K.; Khalefa, A.A.; Khamiss Abd Elguaad, M.M.; Zaki, O.M.; Ali, D.Y.; Hemeda, N.F.; Amin, A.; et al. Expression profile of LncRNA ANRIL, miR-186, miR-181a, and MTMR-3 in patients with preeclampsia. *Noncoding RNA Res.* **2023**, *8*, 481–486. [CrossRef] [PubMed]

117. Sekar, D.; Lakshmanan, G.; Mani, P.; Biruntha, M. Methylation-dependent circulating microRNA 510 in preeclampsia patients. *Hypertens. Res.* **2019**, *42*, 1647–1648. [CrossRef]

118. Yang, Q.; Lu, J.; Wang, S.; Li, H.; Ge, Q.; Lu, Z. Application of next-generation sequencing technology to profile the circulating microRNAs in the serum of preeclampsia versus normal pregnant women. *Clin. Chim. Acta* **2011**, *412*, 2167–2173. [CrossRef]

119. Hromadnikova, I.; Kotlabova, K.; Ondrackova, M.; Kestlerova, A.; Novotna, V.; Hympanova, L.; Doucha, J.; Krofta, L. Circulating C19MC microRNAs in preeclampsia, gestational hypertension, and fetal growth restriction. *Mediat. Inflamm.* **2013**, *2013*, 186041. [CrossRef]

120. Wu, L.; Zhou, H.; Lin, H.; Qi, J.; Zhu, C.; Gao, Z.; Wang, H. Circulating microRNAs are elevated in plasma from severe preeclamptic pregnancies. *Reproduction* **2012**, *143*, 389–397. [CrossRef]

121. Khaliq, O.P.; Murugesan, S.; Moodley, J.; Mackraj, I. Differential expression of miRNAs are associated with the insulin signaling pathway in preeclampsia and gestational hypertension. *Clin. Exp. Hypertens.* **2018**, *40*, 744–751. [CrossRef] [PubMed]

122. Eghbal-Fard, S.; Yousefi, M.; Heydarlou, H.; Ahmadi, M.; Taghavi, S.; Movasaghpour, A.; Jadidi-Niaragh, F.; Yousefi, B.; Dolati, S.; Hojjat-Farsangi, M.; et al. The imbalance of Th17/Treg axis involved in the pathogenesis of preeclampsia. *J. Cell Physiol.* **2019**, *234*, 5106–5116. [CrossRef] [PubMed]

123. Witvrouwen, I.; Mannaerts, D.; Ratajczak, J.; Boeren, E.; Faes, E.; Van Craenenbroeck, A.H.; Jacquemyn, Y.; Van Craenenbroeck, E.M. MicroRNAs targeting VEGF are related to vascular dysfunction in preeclampsia. *Biosci. Rep.* **2021**, *41*, BSR20210874. [CrossRef] [PubMed]

124. Kim, S.; Park, M.; Kim, J.Y.; Kim, T.; Hwang, J.Y.; Ha, K.S.; Won, M.H.; Ryoo, S.; Kwon, Y.G.; Kim, Y.M. Circulating miRNAs Associated with Dysregulated Vascular and Trophoblast Function as Target-Based Diagnostic Biomarkers for Preeclampsia. *Cells* **2020**, *9*, 2003. [CrossRef] [PubMed]

125. Wang, Z.; Liu, D.; Dai, Y.; Li, R.; Zheng, Y.; Zhao, G.; Wang, J.; Diao, Z.; Cao, C.; Lv, H.; et al. Elevated Placental microRNA-155 Is a Biomarker of a Preeclamptic Subtype. *Hypertension* **2023**, *80*, 370–384. [CrossRef] [PubMed]

126. Motawi, T.M.K.; Sabry, D.; Maurice, N.W.; Rizk, S.M. Role of mesenchymal stem cells exosomes derived microRNAs; miR-136, miR-494 and miR-495 in pre-eclampsia diagnosis and evaluation. *Arch. Biochem. Biophys.* **2018**, *659*, 13–21. [CrossRef] [PubMed]

127. Wang, Y.; Du, X.; Wang, J. Transfer of miR-15a-5p by placental exosomes promotes pre-eclampsia progression by regulating PI3K/AKT signaling pathway via CDK1. *Mol. Immunol.* **2020**, *128*, 277–286. [CrossRef]

128. Ntsethe, A.; Mackraj, I. An Investigation of Exosome Concentration and Exosomal microRNA (miR-155 and miR-222) Expression in Pregnant Women with Gestational Hypertension and Preeclampsia. *Int. J. Womens Health* **2022**, *14*, 1681–1689. [CrossRef]

129. Aharon, A.; Rebibo-Sabbah, A.; Ahmad, R.S.; Dangot, A.; Bar-Lev, T.H.; Brenner, B.; Cohen, A.H.; David, C.B.; Weiner, Z.; Solt, I. Associations of maternal and placental extracellular vesicle miRNA with preeclampsia. *Front. Cell Dev. Biol.* **2023**, *11*, 1080419. [CrossRef]

130. Biro, O.; Fothi, A.; Alasztics, B.; Nagy, B.; Orban, T.I.; Rigo, J., Jr. Circulating exosomal and Argonaute-bound microRNAs in preeclampsia. *Gene* **2019**, *692*, 138–144. [CrossRef]

131. Luque, A.; Farwati, A.; Crovetto, F.; Crispi, F.; Figueras, F.; Gratacos, E.; Aran, J.M. Usefulness of circulating microRNAs for the prediction of early preeclampsia at first-trimester of pregnancy. *Sci. Rep.* **2014**, *4*, 4882. [CrossRef]

132. Sandrim, V.C.; Diniz, S.; Eleuterio, N.M.; Gomes, K.B.; Dusse, L.M.S.; Cavalli, R.C. Higher levels of circulating TIMP-4 in preeclampsia is strongly associated with clinical parameters and microRNA. *Clin. Exp. Hypertens.* **2018**, *40*, 609–612. [CrossRef] [PubMed]

133. Gunel, T.; Kamali, N.; Hosseini, M.K.; Gumusoglu, E.; Benian, A.; Aydinli, K. Regulatory effect of miR-195 in the placental dysfunction of preeclampsia. *J. Matern. Fetal Neonatal Med.* **2020**, *33*, 901–908. [CrossRef] [PubMed]

134. Miura, K.; Higashijima, A.; Murakami, Y.; Tsukamoto, O.; Hasegawa, Y.; Abe, S.; Fuchi, N.; Miura, S.; Kaneuchi, M.; Masuzaki, H. Circulating chromosome 19 miRNA cluster microRNAs in pregnant women with severe pre-eclampsia. *J. Obstet. Gynaecol. Res.* **2015**, *41*, 1526–1532. [CrossRef] [PubMed]

135. Dong, K.; Zhang, X.; Ma, L.; Gao, N.; Tang, H.; Jian, F.; Ma, Y. Downregulations of circulating miR-31 and miR-21 are associated with preeclampsia. *Pregnancy Hypertens.* **2019**, *17*, 59–63. [CrossRef] [PubMed]

136. Whigham, C.A.; MacDonald, T.M.; Walker, S.P.; Pritchard, N.; Hannan, N.J.; Cannon, P.; Nguyen, T.V.; Hastie, R.; Tong, S.; Kaitu'u-Lino, T.J. Circulating GATA2 mRNA is decreased among women destined to develop preeclampsia and may be of endothelial origin. *Sci. Rep.* **2019**, *9*, 235. [CrossRef] [PubMed]

137. Kolkova, Z.; Holubekova, V.; Grendar, M.; Nachajova, M.; Zubor, P.; Pribulova, T.; Loderer, D.; Zigo, I.; Biringer, K.; Hornakova, A. Association of Circulating miRNA Expression with Preeclampsia, Its Onset, and Severity. *Diagnostics* **2021**, *11*, 476. [CrossRef] [PubMed]

138. Whigham, C.A.; MacDonald, T.M.; Walker, S.P.; Hiscock, R.; Hannan, N.J.; Pritchard, N.; Cannon, P.; Nguyen, T.V.; Miranda, M.; Tong, S.; et al. MicroRNAs 363 and 149 are differentially expressed in the maternal circulation preceding a diagnosis of preeclampsia. *Sci. Rep.* **2020**, *10*, 18077. [CrossRef]

139. Akehurst, C.; Small, H.Y.; Sharafetdinova, L.; Forrest, R.; Beattie, W.; Brown, C.E.; Robinson, S.W.; McClure, J.D.; Work, L.M.; Carty, D.M.; et al. Differential expression of microRNA-206 and its target genes in preeclampsia. *J. Hypertens.* **2015**, *33*, 2068–2074. [CrossRef]

140. Pan, M.; Ge, Q.; Li, H.; Yang, Q.; Lu, J.; Zhang, D.; Lu, Z. Sequencing the miRNAs in maternal plasma from women before and after parturition. *J. Nanosci. Nanotechnol.* **2012**, *12*, 4035–4043. [CrossRef]

141. Li, H.; Ge, Q.; Guo, L.; Lu, Z. Maternal plasma miRNAs expression in preeclamptic pregnancies. *Biomed. Res. Int.* **2013**, *2013*, 970265. [CrossRef] [PubMed]

142. Biro, O.; Alasztics, B.; Molvarec, A.; Joo, J.; Nagy, B.; Rigo, J., Jr. Various levels of circulating exosomal total-miRNA and miR-210 hypoxamiR in different forms of pregnancy hypertension. *Pregnancy Hypertens.* **2017**, *10*, 207–212. [CrossRef] [PubMed]

143. Jairajpuri, D.S.; Malalla, Z.H.; Mahmood, N.; Almawi, W.Y. Circulating microRNA expression as predictor of preeclampsia and its severity. *Gene* **2017**, *627*, 543–548. [CrossRef] [PubMed]

144. Bujold, E.; Roberge, S.; Lacasse, Y.; Bureau, M.; Audibert, F.; Marcoux, S.; Forest, J.C.; Giguere, Y. Prevention of preeclampsia and intrauterine growth restriction with aspirin started in early pregnancy: A meta-analysis. *Obstet. Gynecol.* **2010**, *116 2 Pt 1*, 402–414. [CrossRef]

145. Tan, M.Y.; Wright, D.; Syngelaki, A.; Akolekar, R.; Cicero, S.; Janga, D.; Singh, M.; Greco, E.; Wright, A.; Maclagan, K.; et al. Comparison of diagnostic accuracy of early screening for pre-eclampsia by NICE guidelines and a method combining maternal factors and biomarkers: Results of SPREE. *Ultrasound Obstet. Gynecol.* **2018**, *51*, 743–750. [CrossRef]

146. Ura, B.; Feriotto, G.; Monasta, L.; Bilel, S.; Zweyer, M.; Celeghini, C. Potential role of circulating microRNAs as early markers of preeclampsia. *Taiwan. J. Obstet. Gynecol.* **2014**, *53*, 232–234. [CrossRef]

147. Winger, E.E.; Reed, J.L.; Ji, X. First trimester PBMC microRNA predicts adverse pregnancy outcome. *Am. J. Reprod. Immunol.* **2014**, *72*, 515–526. [CrossRef]

148. Hromadnikova, I.; Kotlabova, K.; Doucha, J.; Dlouha, K.; Krofta, L. Absolute and relative quantification of placenta-specific micrornas in maternal circulation with placental insufficiency-related complications. *J. Mol. Diagn.* **2012**, *14*, 160–167. [CrossRef]

149. Winger, E.E.; Reed, J.L.; Ji, X. First-trimester maternal cell microRNA is a superior pregnancy marker to immunological testing for predicting adverse pregnancy outcome. *J. Reprod. Immunol.* **2015**, *110*, 22–35. [CrossRef]

150. Zhang, Y.; Huang, G.; Zhang, Y.; Yang, H.; Long, Y.; Liang, Q.; Zheng, Z. MiR-942 decreased before 20 weeks gestation in women with preeclampsia and was associated with the pathophysiology of preeclampsia in vitro. *Clin. Exp. Hypertens.* **2017**, *39*, 108–113. [CrossRef]

151. Nakashima, A.; Yamanaka-Tatematsu, M.; Fujita, N.; Koizumi, K.; Shima, T.; Yoshida, T.; Nikaido, T.; Okamoto, A.; Yoshimori, T.; Saito, S. Impaired autophagy by soluble endoglin, under physiological hypoxia in early pregnant period, is involved in poor placentation in preeclampsia. *Autophagy* **2013**, *9*, 303–316. [CrossRef] [PubMed]

152. Mavreli, D.; Lykoudi, A.; Lambrou, G.; Papaioannou, G.; Vrachnis, N.; Kalantaridou, S.; Papantoniou, N.; Kolialexi, A. Deep Sequencing Identified Dysregulated Circulating MicroRNAs in Late Onset Preeclampsia. *In Vivo* **2020**, *34*, 2317–2324. [CrossRef] [PubMed]

153. Li, Q.; Han, Y.; Xu, P.; Yin, L.; Si, Y.; Zhang, C.; Meng, Y.; Feng, W.; Pan, Z.; Gao, Z.; et al. Elevated microRNA-125b inhibits cytotrophoblast invasion and impairs endothelial cell function in preeclampsia. *Cell Death Discov.* **2020**, *6*, 35. [CrossRef] [PubMed]

154. Xu, P.; Zhao, Y.; Liu, M.; Wang, Y.; Wang, H.; Li, Y.X.; Zhu, X.; Yao, Y.; Wang, H.; Qiao, J.; et al. Variations of microRNAs in human placentas and plasma from preeclamptic pregnancy. *Hypertension* **2014**, *63*, 1276–1284. [CrossRef] [PubMed]

155. Martinez-Fierro, M.L.; Garza-Veloz, I. Analysis of Circulating microRNA Signatures and Preeclampsia Development. *Cells* **2021**, *10*, 1003. [CrossRef] [PubMed]

156. Hromadnikova, I.; Kotlabova, K.; Ivankova, K.; Krofta, L. First trimester screening of circulating C19MC microRNAs and the evaluation of their potential to predict the onset of preeclampsia and IUGR. *PLoS ONE* **2017**, *12*, e0171756. [CrossRef]

157. Hromadnikova, I.; Dvorakova, L.; Kotlabova, K.; Krofta, L. The Prediction of Gestational Hypertension, Preeclampsia and Fetal Growth Restriction via the First Trimester Screening of Plasma Exosomal C19MC microRNAs. *Int. J. Mol. Sci.* **2019**, *20*, 2972. [CrossRef]

158. Kondracka, A.; Kondracki, B.; Jaszczuk, I.; Staniczek, J.; Kwasniewski, W.; Filip, A.; Kwasniewska, A. Diagnostic potential of microRNAs Mi 517 and Mi 526 as biomarkers in the detection of hypertension and preeclampsia in the first trimester. *Ginekol. Pol.* **2023**, 12. [CrossRef]

159. Martinez-Fierro, M.L.; Garza-Veloz, I.; Gutierrez-Arteaga, C.; Delgado-Enciso, I.; Barbosa-Cisneros, O.Y.; Flores-Morales, V.; Hernandez-Delgadillo, G.P.; Rocha-Pizana, M.R.; Rodriguez-Sanchez, I.P.; Badillo-Almaraz, J.I.; et al. Circulating levels of specific members of chromosome 19 microRNA cluster are associated with preeclampsia development. *Arch. Gynecol. Obstet.* **2018**, *297*, 365–371. [CrossRef]

160. Yin, Y.; Liu, M.; Yu, H.; Zhang, J.; Zhou, R. Circulating microRNAs as biomarkers for diagnosis and prediction of preeclampsia: A systematic review and meta-analysis. *Eur. J. Obstet. Gynecol. Reprod. Biol.* **2020**, *253*, 121–132. [CrossRef]

161. Su, S.; Yang, F.; Zhong, L.; Pang, L. Circulating noncoding RNAs as early predictive biomarkers in preeclampsia: A diagnostic meta-analysis. *Reprod. Biol. Endocrinol.* **2021**, *19*, 177. [CrossRef] [PubMed]

162. Stojanovska, V.; Scherjon, S.A.; Plosch, T. Preeclampsia as Modulator of Offspring Health. *Biol. Reprod.* **2016**, *94*, 53. [CrossRef] [PubMed]

163. Fraser, A.; Nelson, S.M.; Macdonald-Wallis, C.; Sattar, N.; Lawlor, D.A. Hypertensive disorders of pregnancy and cardiometabolic health in adolescent offspring. *Hypertension* **2013**, *62*, 614–620. [CrossRef] [PubMed]

164. Apicella, C.; Ruano, C.S.M.; Mehats, C.; Miralles, F.; Vaiman, D. The Role of Epigenetics in Placental Development and the Etiology of Preeclampsia. *Int. J. Mol. Sci.* **2019**, *20*, 2837. [CrossRef]

165. Ji, L.; Brkic, J.; Liu, M.; Fu, G.; Peng, C.; Wang, Y.L. Placental trophoblast cell differentiation: Physiological regulation and pathological relevance to preeclampsia. *Mol. Aspects Med.* **2013**, *34*, 981–1023. [CrossRef] [PubMed]

166. Chhabra, R. miRNA and methylation: A multifaceted liaison. *Chembiochem* **2015**, *16*, 195–203. [CrossRef]

167. Morales, S.; Monzo, M.; Navarro, A. Epigenetic regulation mechanisms of microRNA expression. *Biomol. Concepts* **2017**, *8*, 203–212. [CrossRef]

168. Wu, D.; Shi, L.; Chen, F.; Lin, Q.; Kong, J. Methylation Status of the miR-141-3p Promoter Regulates miR-141-3p Expression, Inflammasome Formation, and the Invasiveness of HTR-8/SVneo Cells. *Cytogenet. Genome Res.* **2021**, *161*, 501–513. [CrossRef]

169. Brodowski, L.; Schroder-Heurich, B.; von Hardenberg, S.; Richter, K.; von Kaisenberg, C.S.; Dittrich-Breiholz, O.; Meyer, N.; Dork, T.; von Versen-Hoynck, F. MicroRNA Profiles of Maternal and Neonatal Endothelial Progenitor Cells in Preeclampsia. *Int. J. Mol. Sci.* **2021**, *22*, 5320. [CrossRef]

170. Lin, C.; Rajakumar, A.; Plymire, D.A.; Verma, V.; Markovic, N.; Hubel, C.A. Maternal endothelial progenitor colony-forming units with macrophage characteristics are reduced in preeclampsia. *Am. J. Hypertens.* **2009**, *22*, 1014–1019. [CrossRef]

171. Suarez, Y.; Fernandez-Hernando, C.; Pober, J.S.; Sessa, W.C. Dicer dependent microRNAs regulate gene expression and functions in human endothelial cells. *Circ. Res.* **2007**, *100*, 1164–1173. [CrossRef] [PubMed]
172. Suarez, Y.; Fernandez-Hernando, C.; Yu, J.; Gerber, S.A.; Harrison, K.D.; Pober, J.S.; Iruela-Arispe, M.L.; Merkenschlager, M.; Sessa, W.C. Dicer-dependent endothelial microRNAs are necessary for postnatal angiogenesis. *Proc. Natl. Acad. Sci. USA* **2008**, *105*, 14082–14087. [CrossRef] [PubMed]
173. Vitale, S.G.; Fulghesu, A.M.; Mikuš, M.; Watrowski, R.; D'Alterio, M.N.; Lin, L.T.; Shah, M.; Reyes-Muñoz, E.; Sathyapalan, T.; Angioni, S. The Translational Role of miRNA in Polycystic Ovary Syndrome: From Bench to Bedside-A Systematic Literature Review. *Biomedicines* **2022**, *10*, 1816. [CrossRef] [PubMed]

International Journal of
Molecular Sciences

Article

Diagnostic Role of Cell-Free miRNAs in Identifying Placenta Accreta Spectrum during First-Trimester Screening

Angelika V. Timofeeva *, Ivan S. Fedorov , Yuliya V. Suhova, Alla M. Tarasova, Larisa S. Ezhova, Tatyana M. Zabelina, Oksana N. Vasilchenko, Tatyana Y. Ivanets and Gennady T. Sukhikh

Kulakov National Medical Research Center of Obstetrics, Gynecology, and Perinatology, Ministry of Health of Russia, Ac. Oparina 4, 117997 Moscow, Russia; is_fedorov@oparina4.ru (I.S.F.); j_bezzubenko@oparina4.ru (Y.V.S.); am_tarasova@oparina4.ru (A.M.T.); l_ezhova@oparina4.ru (L.S.E.); romashova-1993@bk.ru (T.M.Z.); o_vasilchenko@oparina4.ru (O.N.V.); t_ivanets@oparina4.ru (T.Y.I.); g_sukhikh@oparina4.ru (G.T.S.)
* Correspondence: v_timofeeva@oparina4.ru; Tel.: +7-495-531-4444

Abstract: Placenta accreta spectrum (PAS) is a severe complication of pregnancy associated with excessive invasion of cytotrophoblast cells at the sites of the endometrial–myometrial interface and the myometrium itself in cases of adherent (creta) and invasive (increta and percreta) forms, respectively. This leads to a high risk of massive blood loss, maternal hysterectomy, and preterm birth. Despite advancements in ultrasound protocols and found associations of alpha-fetoprotein, PAPP-A, hCG, PLGF, sFlt-1, IL-8, and IL-33 peripheral blood levels with PAS, there is a high need for an additional non-invasive test to improve the diagnostic accuracy and to select the real PAS from the suspected ones in the first-trimester screening. miRNA signatures of placental tissue, myometrium, and blood plasma from women with PAS in the third trimester of pregnancy, as well as miRNA profiles in exosomes from the blood serum of women in the first trimester with physiologically progressing pregnancy, complicated by PAS or pre-eclampsia, were obtained using deep sequencing. Two logistic regression models were constructed, both featuring statistically significant parameters related to the levels of miR-26a-5p, miR-17-5p, and miR-101-3p, quantified by real-time PCR in native blood serum. These models demonstrated 100% sensitivity in detecting PAS during the first pregnancy screening. These miRNAs were identified as specific markers for PAS, showing significant differences in their blood serum levels during the first trimester in the PAS group compared to those in physiological pregnancies, early- or late-onset pre-eclampsia groups. Furthermore, these miRNAs exhibited differential expression in the PAS placenta and/or myometrium in the third trimester and, according to data from the literature, control angiogenesis. Significant correlations were found between extracellular hsa-miR-101-3p and nuchal translucency thickness, hsa-miR-17-5p and uterine artery pulsatility index, and hsa-miR-26a-5p and hsa-miR-17-5p with PLGF. The developed test system for early non-invasive PAS diagnosis based on the blood serum level of extracellular miR-26a-5p, miR-17-5p, and miR-101-3p can serve as an auxiliary method for first-trimester screening of pregnant women, subject to validation with independent test samples.

Keywords: miRNA; NGS; real-time PCR; peripheral blood; placenta; myometrium; placenta accreta spectrum; first-trimester screening

Citation: Timofeeva, A.V.; Fedorov, I.S.; Suhova, Y.V.; Tarasova, A.M.; Ezhova, L.S.; Zabelina, T.M.; Vasilchenko, O.N.; Ivanets, T.Y.; Sukhikh, G.T. Diagnostic Role of Cell-Free miRNAs in Identifying Placenta Accreta Spectrum during First-Trimester Screening. *Int. J. Mol. Sci.* **2024**, 25, 871. https://doi.org/10.3390/ijms25020871

Academic Editor: Giovanni Tossetta

Received: 9 December 2023
Revised: 6 January 2024
Accepted: 8 January 2024
Published: 10 January 2024

1. Introduction

Placenta accreta spectrum (PAS) is a serious obstetric complication characterized by excessive invasive growth of chorionic villi into adjacent tissue structures. This term encompasses both abnormal adherence (placenta creta, where villi adhere to the myometrium) and abnormal invasion (placenta increta, where villi invade the myometrium; placenta percreta, where villi invade the full thickness of the myometrium) [1]. According to D.A. Carusi, the prevalence of PAS is reported as 1 in 1000 deliveries [2], with an increasing incidence observed over time. For instance, in a tertiary south Italian center, the incidence of PAS tripled from 0.12% to 0.31% between 1970 and 2000 [3]. Other authors report a 100-fold increase in PAS frequency since the 1950s [4,5], which is attributed to the rising rate of cesarean section procedures [6,7]. In turn, the risk of placenta previa significantly increases with the frequency of cesarean sections, representing an additional risk factor for PAS, constituting 50% [8]. This is associated with the increased tropism of trophoblast cells of the blastocyst to the altered scar tissue area, leading to myofiber disarray, inflammatory processes, and dystrophy of elastic and collagen fibers.

Among the etiological factors of PAS, besides changes in scar tissue, curettage, myomectomy, uterine anomalies, endometriosis, and endometritis are noted, all of which may lead to endometrial fibrosis and poor decidualization [8]. Several theories have been proposed to explain the origin of PAS One theory, which involves disorders in the coordinated regulation of extravillous trophoblast differentiation from progenitor cytotrophoblasts, resulting in excessive invasion into the myometrium to remodel the uterine vascular system, causing hypervascularity and vascular dysfunction [9–11]. This trophoblast behavior resembles cancer-like progression [11]. Single-cell transcriptome analysis of PAS and normally detached placenta tissues has revealed close communication between excessive numbers of two cytotrophoblast cell types (LAMB4+ and KRT6A+) and maternal stromal cell subtypes (ADIRF+ and DES+), supporting trophoblast cell migration and invasion, as well as interactions with vascular endothelial cells through FLT1-VEGFA and JAG1-NOTCH2 cell–cell interactions inducing abnormal blood vessels in the myometrium [12]. Another hypothesis suggests that abnormal vascularization with local hypoxia in the uterine scar area impacts decidualization, causing a defect in the regulatory properties of the decidua. This defect allows trophoblast cells to be more aggressive and penetrative at the sites of the endometrial–myometrial interface and the myometrium itself [13,14]. The main complications of PAS include massive blood loss, disseminated intravascular coagulation, hysterectomy, and preterm birth, leading to increased maternal and fetal morbidity and mortality [15]. Despite improvements in ultrasound protocols [1,16,17], the frequency of undiagnosed PAS before delivery is variable [18,19], partly due to the different ultrasound equipment used by ultrasound examinators, the subjective quality of the ultrasound sings of PAS, and the lack of clear evaluation criteria for each of the three grades of PAS. Therefore, the timely and accurate antenatal diagnosis of PAS is essential to formulate the correct patient management algorithm and plan delivery by a multidisciplinary team to reduce the frequency of postpartum complications as also claimed by Pavón-Gomez N. et al. [19]. Therefore, there is a high need for an additional non-invasive test to differentiate real PAS from ultrasound suspected ones antenatally, preferably in the first trimester of pregnancy.

Circulating biomolecules in maternal blood were examined for their potentials use in diagnosing PAS [20–22]. The sensitivity and specificity of maternal serum alpha-fetoprotein in the diagnosis of placenta previa complicated by PAS were only 71% and 46%, respectively [23]. According to a meta-analysis, pregnant women with PAS have a high serum PAPP-A level in the first trimester [24,25], suggesting that this biomarker can be recommended for identifying the risk group for developing PAS. Several studies have shown that, compared to a normal pregnancy, the level of β-hCG in maternal blood serum increases in the first and second trimesters of pregnancy during PAS [23,26]; however, blood serum hCG levels are also associated with miscarriage, ectopic pregnancy, and fetal abnormalities [21]. PlGF levels are significantly higher in subgroups with pathological placental invasion compared to the group with normal placental implantation, while sFlt-1 levels and the sFlt-1/PlGF ratio are lower [27–29]. IL-8 promotes migration and invasion of extravillous trophoblast cells during pregnancy, and its elevation in blood serum may serve as a biomarker for PAS [30]. The level of IL-33 is significantly higher in patients with PAS than in healthy pregnant women [31]. Despite the identified correlations between PAS and the levels of these circulating biomolecules in maternal blood, it is necessary to prove their specificity for PAS and their ability to distinguish it from other pregnancy complications.

Due to the epigenetic regulation of trophoblast differentiation, migration, and invasion [32,33], miRNAs, acting as master regulators of the human genome at the transcriptional and post-transcriptional levels, were analyzed in various biological samples to associate their levels with PAS [21,34–36]. However, miRNA markers for PAS have been identified in the third trimester of pregnancy in studies conducted to date. To individualize the management tactics of pregnant women, preparing for qualified surgical assistance at the time of delivery with the possibility of blood transfusion, it is optimal to conduct screening of women in the first trimester of pregnancy for the content of miRNA markers of PAS in blood serum. Therefore, the aim of this study was to identify extracellular miRNAs circulating in the peripheral blood of women in the first trimester of pregnancy, specific to placental and/or myometrial tissue, and to differentiate PAS from other pregnancy complications, such as pre-eclampsia.

2. Results

To identify circulating cell-free microRNA (miRNA) markers indicative of placental invasion during the first trimester of pregnancy, the study was conducted in four stages:

(I)   Generation of miRNA expression patterns using deep sequencing in both placental and myometrial tissues within the region of pathological trophoblast invasion and beyond at the time of delivery in women with placenta accreta spectrum (PAS) conditions, including creta, increta, and percreta.

(II)   Identification of placenta- and myometrium-specific miRNAs in the blood plasma of pregnant women using deep sequencing and quantitative real-time polymerase chain reaction (qRT-PCR) on the day of delivery, enabling the diagnosis of PAS.

(III)   Qualitative and quantitative analysis of exosomal miRNA composition in the serum of women at 11–14 weeks of pregnancy, facilitating the diagnosis of PAS through deep sequencing with subsequent validation of the obtained data using qRT-PCR.

(IV)   Quantitative analysis of exosomal miRNA markers for PAS in the native serum of women at 11–14 weeks of pregnancy with either a normal course or complications such as the onset of pre-eclampsia symptoms after 20 weeks of gestation or complications involving placental invasion. The goal is to construct a logistic regression model for the accurate diagnosis of PAS.

2.1. Analysis of miRNA Expression Patterns Using Deep Sequencing in Placental and Myometrial Tissues from Patients with Placental Invasion at the Time of Delivery

In the initial stage of the study, deep sequencing of small non-coding RNAs was employed to obtain and compare the expression profiles of microRNAs (miRNAs) in various regions of the placenta and myometrium. Tissues were collected during cesarean section procedures from women in the first cohort (Table 1) with diagnoses of placenta creta (seven patients), placenta increta (six patients), and placenta percreta (four patients).

Placental samples were collected from areas of pathological trophoblast invasion and from regions outside this area. Myometrial samples were obtained from areas adjacent to the placental invasion site. The schematic localization of placental and myometrial sample collection and the designation of corresponding groups are illustrated in Figure 1A.

During the analysis of miRNA expression profiles in placental tissue, two types of comparisons were performed: (1) comparison of placental areas with pathological trophoblast invasion in women with creta (seven samples), increta (six samples), and percreta (four samples): Ppp vs. Ppc, Ppi vs. Ppc; (2) comparison of placental areas outside pathological trophoblast invasion sites in women with creta (three samples), increta (four samples), and percreta (three samples): Pnp vs. Pnc, Pni vs. Pnc.

In the first type of comparison, expression profiles of miRNAs were obtained, significantly differentiating samples of Ppi from Ppc in terms of expression levels for 85 miRNAs (Table S1, Sheet 1), and Ppp from Ppc for 269 miRNAs (Table S2, Sheet 1). Among them, altered expression levels in both types of invasions (increta and percreta) were identified for 78 miRNAs (Figure 1B), constituting 92% of all differentially expressed miRNAs in the case of increta and 29% in the case of percreta. These data indicate similarities in molecular and biological changes in the placenta within the invasion area and the myometrial tissue in cases of increta and percreta, with an exacerbation of these changes in the case of percreta, involving an additional 191 miRNAs.

Table 1. Clinical Characteristics of the First Cohort of Patients.

No	Age	Delivery Time (Weeks)	PAS	Blood Plasma Sample *	Placenta Sample with Pathologic Trophoblast Invasion *	Placenta Sample with Normal Trophoblast Invasion *	Myometrium Sample Adjacent to Placenta with Pathologic Trophoblast Invasion *
1	33	34.4	Absent	✓			
2	29	36.5	Absent	✓			
3	29	38.3	Absent	✓			
4	37	38.2	Absent	✓			
5	30	39	Absent	✓			
6	30	34.1	Creta	✓			
7	42	35.4	Creta	✓	✓		
8	35	36.4	Creta	✓	✓		
9	35	36	Creta	✓	✓		
10	38	36	Creta	✓	✓		
11	40	34.2	Creta	✓	✓	✓	✓
12	42	35.2	Creta	✓	✓	✓	✓
13	28	35	Creta	✓	✓	✓	✓
14	39	37	Increta	✓	✓		
15	34	35.3	Increta	✓	✓		
16	34	34.6	Increta	✓	✓	✓	✓
17	33	35	Increta	✓	✓	✓	✓
18	35	34.2	Increta	✓	✓	✓	✓
19	32	36.1	Increta	✓	✓	✓	✓
20	37	35.2	Percreta	✓			
21	37	36	Percreta	✓	✓		
22	37	33	Percreta	✓	✓	✓	✓
23	32	35.1	Percreta	✓	✓	✓	✓
24	37	35.4	Percreta	✓	✓	✓	✓

* ✓ Checkmark indicates the analysis of miRNA levels in the sample using deep sequencing.

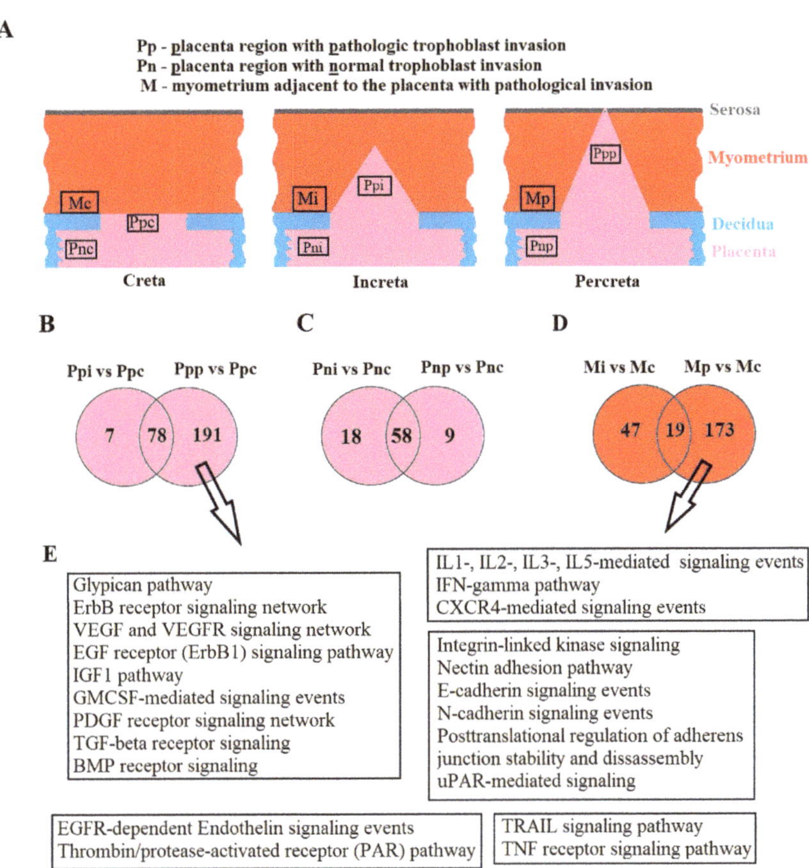

Figure 1. Analysis of placental and myometrial tissues on the day of delivery in women with placenta creta, increta, and percreta using deep sequencing of miRNA. Schematic representation of sample collection locations for placenta and myometrium (**A**); Venn–Euler diagrams of differentially expressed miRNAs in the placenta within the pathological trophoblast invasion site for percreta and increta relative to creta (**B**), in the placenta outside areas of pathological trophoblast invasion for percreta and increta relative to creta (**C**), in myometrial tissues adjacent to areas of pathological trophoblast invasion for percreta and increta relative to creta (**D**), common signaling pathways regulated by differentially expressed miRNAs in the pathological trophoblast invasion site for percreta (Ppp vs. Ppc) and in adjacent myometrial tissue (Mp vs. Mc), according to MirTarbase and Funrich (**E**).

In the second type of comparison, expression profiles of miRNAs significantly differentiating samples of Pni from Pnc in terms of expression levels for 76 miRNAs (Table S3, Sheet 1), and Pnp from Pnc for 67 miRNAs (Table S4, Sheet 1) were obtained. Among them, altered expression levels in both types of invasions (increta and percreta) were identified for 58 miRNAs (Figure 1C), constituting 76% of all differentially expressed miRNAs in the case of increta and 87% in the case of percreta. These data suggest commonalities in molecular and biological changes throughout the entire placenta in women with increta and percreta, possibly formed during the embryo implantation stage due to the interaction of trophoblasts with a pathologically altered deciduous layer of the endometrium. Cells of the cytotrophoblast in certain areas of such a placenta acquire a phenotype with excessive invasive activity under predisposing conditions, such as structural changes in the myometrium.

Therefore, deep-sequencing analysis was performed on miRNA expression levels in myometrial tissue adjacent to the site of pathological trophoblast invasion for percreta (three samples) and increta (four samples) compared to creta (three samples). Expression profiles of miRNAs significantly differentiating samples Mi from Mc (Table S5, Sheet 1) and Mp from Mc (Table S6, Sheet 1) were obtained. Among them, altered expression levels in myometrial tissue for both types of PAS (increta and percreta) were identified for only 19 miRNAs (Figure 1D), constituting 29% of all differentially expressed miRNAs in the case of increta and 10% in the case of percreta. It is noteworthy that more pronounced changes in the qualitative and quantitative composition of miRNAs in myometrial tissue adjacent to placenta percreta were observed compared to that adjacent to placenta increta, possibly contributing to a greater depth of trophoblast cell invasion in the case of placenta percreta.

For each miRNA profile in a specific type of comparison (see above), the use of the MirWalk program (http://mirwalk.umm.uni-heidelberg.de/search_mirnas/, last accessed on 15 October 2023) allowed the identification of experimentally proven target genes according to the MiRTarBase algorithm (Sheet 2 in Tables S1, S2, S5, and S6). Additionally, using the FunRich program (http://www.funrich.org/, last accessed on 15 October 2023), signaling pathways regulated by differentially expressed miRNAs in placental and myometrial tissues in different types of PAS (Sheet 3 in Tables S1, S2, S5, and S6) were revealed. Since the most significant molecular and biological changes were identified in the placenta within the pathological trophoblast invasion site for percreta and in adjacent myometrial tissue compared to other types of PAS; for clarity, in Figure 1E we have presented common signaling pathways regulated by differentially expressed miRNAs in the comparison of Ppp vs. Ppc and Mp vs. Mc. Imbalances in the activity of these signaling pathways may account for local inflammatory processes, structural rearrangements in the extracellular matrix, neovascularization, apoptosis, and changes in the proliferative and invasive properties of cells, such as cytotrophoblast cells.

2.2. Identification of Placenta- and Myometrium-Specific miRNAs Circulating in the Blood Plasma of Pregnant Women on the Day of Delivery Using Deep Sequencing and Real-Time Quantitative PCR

With the aim of developing a non-invasive diagnostic test for PAS in the third trimester of pregnancy, expression profiles of miRNAs were obtained in the peripheral blood plasma of 24 women from the first cohort (Table 1) with various types of PAS using deep sequencing. Lists of differentially expressed (DE) miRNAs were generated for creta (Table S7, Sheet 1), increta (Table S7, Sheet 2), and percreta (Table S7, Sheet 3) relative to the group with a normal pregnancy. These lists of DE miRNAs were compared by constructing a Venn–Euler diagram (Figure 2A), indicating that most miRNAs differentially expressed in the blood plasma of women with creta and increta had significant altered expression levels in the blood plasma of women with percreta. As expression profiles of miRNAs in placental, myometrial, and blood plasma samples were analyzed in the same cohort of patients using deep sequencing, the tissue specificity of circulating blood miRNA markers for PAS was evaluated.

The common DE list of 165 miRNAs in the blood plasma (BPi/p, Figure 2A) for increta and percreta was compared with the lists of DE miRNAs in Ppi (Table S1, Sheet 1), Ppp (Table S2, Sheet 1), Mi (Table S5, Sheet 1), and Mp (Table S6, Sheet 1). Intersections were found between BPi/p and Ppi for 38 miRNAs, BPi/p and Ppp for 128 miRNAs, BPi/p and Mi for 33 miRNAs, BPi/p and Mp for 90 miRNAs (Figure 2B). For further validation of the obtained data using real-time quantitative PCR, 40 miRNAs were selected that were differentially expressed in the blood plasma of women with invasive types of PAS (increta and percreta), as well as in placenta and/or myometrium in both invasive PAS types. The list of the selected 40 miRNAs is presented in Figure 2C.

The peripheral blood plasma samples from the second cohort of patients ($n = 46$, Table 2) were used to validate the sequencing data.

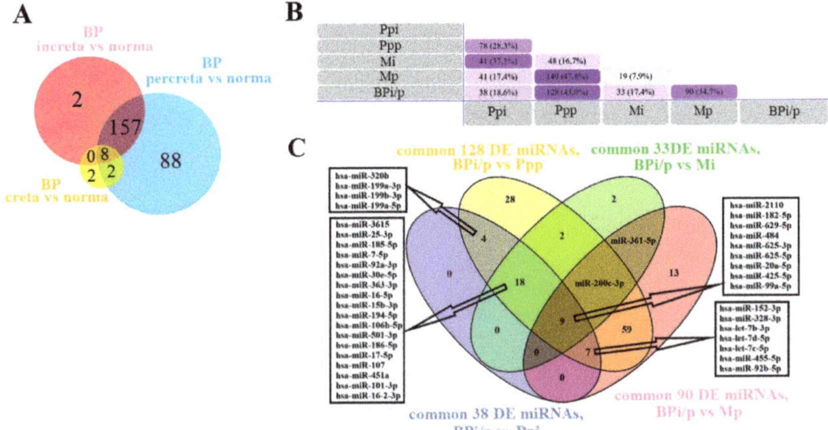

Figure 2. Comparison of miRNA Expression Profiles in placental tissue, adjacent myometrium, and peripheral blood plasma of Patients in Cohort 1. (**A**) Venn–Euler diagram for comparative analysis of lists of differentially expressed (DE) miRNAs in the blood plasma of women with creta, increta, and percreta relative to the group with a normal pregnancy using FunRich v. 3.1.3 (http://www.funrich.org/, last accessed on 15 October 2023) and Venny 2.1.0 (https://bioinfogp.cnb.csic.es/tools/venny/, last accessed on 15 October 2023). (**B**) Venn–Euler diagram for comparative analysis of the 165 DE miRNAs list in the blood plasma common for placenta increta and percreta cases (BPi/p, from (**A**)) with the DE miRNAs lists in placenta and myometrium for increta (Ppi and Mi, respectively) and percreta (Ppp and Mp, respectively) cases using FunRich v. 3.1.3 (http://www.funrich.org/, last accessed on 15 October 2023). (**C**) Venn–Euler diagram for comparative analysis of the DE miRNA list common to BPi/p and Ppi, BPi/p and Ppp, BPi/p and Mi, BPi/p and Mp (all obtained from (**B**)), constructed using Venny 2.1.0 (https://bioinfogp.cnb.csic.es/tools/venny/, last accessed on 15 October 2023).

Table 2. Clinical Characteristics of the Second Cohort of Patients.

Group Name	Group Size	Scar on the Uterus	Placenta Previa	Delivery Time (Weeks) *
Norma (pregnancy without complications)	14	yes	no	37.5 (34.3; 39.2)
Creta	9	yes	yes	36.4 (35; 37.4)
Increta	16	yes	yes	34.6 (33.1; 36.3)
Percreta	7	yes	yes	33 (32; 33)

* Data are presented as the median (Me) and quartiles Q1 and Q3 in the format: Me (Q1; Q3).

The $-\Delta Ct$ values were calculated based on the difference between the Ct values of the analyzed 40 miRNAs and the Ct value of the exogenous RNA UniSp6 (see Section 4) in each sample. Using the partial least squares regression (PLS-A) method, the greatest contribution to the clustering of blood plasma samples from women with different types of PAS was made by hsa-miR-92a-3p, hsa-miR-25-3p, hsa-miR-629-5p, hsa-miR-320b, hsa-let-7d-5p, hsa-miR-17-5p, hsa-miR-16-5p, hsa-miR-106b-5p (Figure 3).

From Figure 3, it is evident that all samples of peripheral blood plasma from women with PAS differ from control samples in the expression profile of eight miRNAs, forming a distinct cluster. For this reason, it was decided to combine the creta, increta, and percreta groups to develop logistic regression models for creating a non-invasive diagnostic test for PAS based on the quantitative assessment of miRNAs in blood plasma in the third trimester of pregnancy. Using RStudio, optimal combinations of miRNAs associated with the presence of PAS in pregnant women were found step by step, considering their contribution to building logistic regression models. In these models, the dependent variable (response variable) was the presence or absence of PAS in pregnant women (0—absence of PAS; 1—presence of PAS). The selected models, presented in Figure 4, included statistically significant independent variables. The parameters of the models in Figure 4 are listed in Table 3. Formulas 1, 2, and 3, describing the models 1, 2, and 5 in Figure 4, respectively, are provided below.

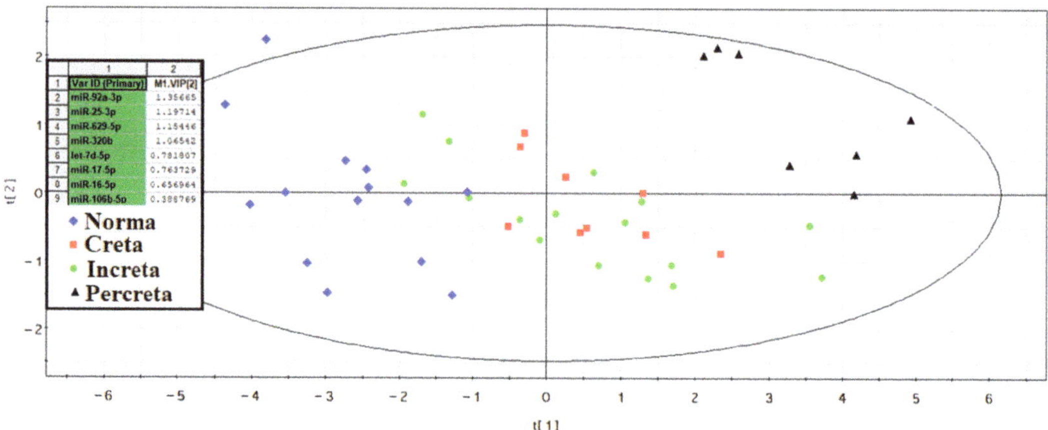

Figure 3. PLS-A analysis of quantitative RT-PCR data on the expression level of miRNAs in the peripheral blood plasma of pregnant women with physiological pregnancy and PAS.

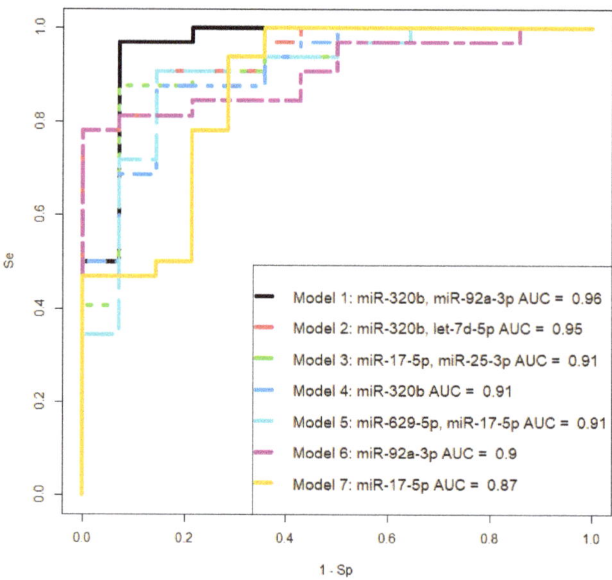

Figure 4. Receiver operating characteristic (ROC) curves of the logistic regression models based on real-time quantitative PCR data when comparing the combined groups "creta, increta, percreta" with the "norma" group for the content of miRNAs in the blood plasma of pregnant women in the third trimester. Se—sensitivity, Sp—specificity.

Table 3. Parameters of the logistic regression models in Figure 4.

	Model	Estimate (95% CI)	Wald	*p*-Value	OR (95% CI)	Se	Sp
	(Intercept)	12.4342 (5.8431; 25.1735)	2.6862	0.0072	251,251.6477 (344.8485; 85,646,274,941.2615)		
1	miR-320b	−2.6129 (−5.2514; −0.9477)	−2.504	0.0123	0.0733 (0.0052; 0.3876)	0.97	0.93
	miR-92a-3p	−2.1527 (−4.8586; −0.5747)	−2.0715	0.0383	0.1162 (0.0078; 0.5629)		
	(Intercept)	9.7122 (4.807; 18.8667)	2.8577	0.0043	16,518.5771 (122.3664; 156,204,305.6105)		
2	miR-320b	−4.0078 (−7.9435; −1.83)	−2.6992	0.007	0.0182 (0.0004; 0.1604)	0.78	1
	let-7d-5p	−0.6938 (−1.4087; −0.1879)	−2.3393	0.0193	0.4996 (0.2445; 0.8287)		
	(Intercept)	11.9713 (5.7515; 21.3935)	3.0688	0.0021	158,149.4506 (314.6611; 1,954,609,961.5401)		
3	miR-17-5p	−0.7119 (−1.3684; −0.2466)	−2.5573	0.0105	0.4907 (0.2545; 0.7815)	0.88	0.93
	miR-25-3p	−0.5773 (−1.0826; −0.2314)	−2.7482	0.006	0.5614 (0.3387; 0.7934)		
	(Intercept)	5.4891 (2.9334; 9.5854)	3.3713	<0.001	242.042 (18.7919; 14,550.8746)		
4	miR-320b	−2.8885 (−5.1368; −1.4507)	−3.1781	0.0015	0.0557 (0.0059; 0.2344)	0.88	0.86
	(Intercept)	9.5615 (4.6005; 16.6831)	3.1352	0.0017	14,206.4468 (99.5329; 17,593,816.9241)		
5	miR-629-5p	−0.4622 (−0.8576; −0.1463)	−2.61	0.0091	0.6299 (0.4242; 0.8639)	0.78	1
	miR-17-5p	−0.5663 (−1.1491; −0.1557)	−2.2973	0.0216	0.5676 (0.3169; 0.8558)		
	(Intercept)	8.6533 (4.3535; 15.0105)	3.2633	0.0011	5,729.094 (77.7512; 3,303,537.1587)		
6	miR-92a-3p	−2.2867 (−3.9974; −1.1017)	−3.1624	0.0016	0.1016 (0.0184; 0.3323)	0.91	0.86
	(Intercept)	5.891 (2.3393; 11.7884)	2.5906	0.0096	361.7674 (10.3736; 131,711.3987)		
7	miR-17-5p	−0.5333 (−1.0924; −0.1761)	−2.4108	0.0159	0.5866 (0.3354; 0.8385)	0.94	0.71

(1) Model 1 formula:

$$\frac{1}{1 + e^{-12.43 + 2.6x_1 + 2.15x_2}} \tag{1}$$

wherre x_1—«−ΔCt» value for hsa-miR-320b, x_2—«−ΔCt» value for hsa-miR-92a-3p;

(2) Model 2 formula:

$$\frac{1}{1 + e^{-9.7 + 4x_1 + 0.69x_2}} \tag{2}$$

where x_1—«−ΔCt» value for hsa-miR-320b, x_2—«−ΔCt» value for hsa-let-7d-5p;

(3) Model 5 formula:

$$\frac{1}{1 + e^{-9.56 + 0.46x_1 + 0.57x_2}} \tag{3}$$

where x_1—«−ΔCt» value for hsa-miR-629-5p, x_2—«−ΔCt» value for hsa-miR-17-5p.

Model 1 demonstrates the highest sensitivity (96.9%) compared to Models 2 (Se = 78.1%) and 5 (Se = 78.1%), indicating a superior diagnostic value in identifying PAS in the third trimester of pregnancy. On the other hand, Models 2 and 5 exhibit 100% specificity, unlike Model 1 (Sp = 93%), ensuring 100% accuracy in detecting pregnancies without PAS. Simultaneous utilization of all three models on an extended training and testing dataset is essential for understanding their diagnostic value and selecting one for further consideration.

2.3. Analysis of Qualitative and Quantitative Composition of Exosome microRNA (miRNA) in the Blood Serum of Women at 11–14 Weeks of Pregnancy

One of the means of intercellular communication is the directed delivery of molecules, including miRNA, contained within exosomes. We hypothesized that morphofunctionally altered tissues of the placenta and myometrium in the area of pathological trophoblast invasion secrete exosomes with a modified miRNA profile into the blood. This profile could potentially diagnose the presence of PAS in early pregnancy stages, such as the 11–14 weeks of gestation.

A retrospective study of exosomes from 48 women's peripheral blood serum at 11–14 weeks of pregnancy was performed, and four main groups were formed according to the diagnose at delivery (Table 4): normal pregnancy (N, 10 women); high risk of developing pre-eclampsia according to the Astraia program without clinical manifestations after 20 weeks of pregnancy (Nhr, 7 women); development of clinical symptoms of early- or late-onset pre-eclampsia after 20 weeks of pregnancy (PE, 21 women); placenta creta, increta, or percreta in pregnant women (PAS, 10 women). miRNA expression profiles were obtained using deep sequencing as described in Section 4.

Table 4. Clinical characteristics of the third cohort of patients.

| Sample ID | Diagnose at Delivery, Group Name | Biochemical Data | | | | | | Ultrasound Data | | | |
		1st Trimester Screening, Peripheral Blood Sampling, GW	PAPP-A (0.7–6.0 IU/L)	PAPP-A (0.5–2.0 MoM)	b-hCG (50.0–55.0 IU/mL)	b-hCG (0.5–2.0 MoM)	PLGF (21.9–71.2 пг/мл)	Crown Rump Length, CRL (43.0–84.0 mm)	Nuchal Translucency Thickness, NT (1.6–1.7 mm)	Uterine Artery Pulsatility Index, UA (PI), MoM	Uterine Artery Pulsatility Index, UA (Pi), 0.9–2.6 (5th and 95th Percentiles)
1		12.1	1.72	0.65	51.1	0.94	23.8	57.4	1.3	0.75	1.27
6	late-onset pre-eclampsia, PE	12.2	1	0.5	53.6	1.64	15.8	58	1.7	0.94	1.6
7		12.1	4.3	1.7	27.9	0.57	15.79	60.5	1.9	1.21	2.06
11		12.5	1.92	0.64	40.5	0.81	20.9	63.2	1.91	1.4	2.35
12		12	4.27	1.21	28.1	0.46	8.6	58.7	1.5	0.73	1.28
14		12.5	3.13	0.65	27.8	0.45	24.01	63.3	1.4	1.33	2.32
16		12.5	2.96	1.03	35.6	0.79	15.6	64	2	0.4	0.65
17		12.1	0.61	0.4	28.3	0.76	15.8	58.6	1.5	1.45	2.28
18		11.6	1.63	0.55	42.9	0.69	9.34	55.1	1.5	1.04	1.86
19		12.4	5.01	2.65	28.1	0.51	26.5	57.6	1.3	0.81	1.41
21		12.1	1.86	0.9	54	1.09		57	1.6	2.07	3.5
22		12.1	0.8	0.46	36	0.6		50.6	1.76	1.18	2.09
23	early-onset pre-eclampsia, PE	12.2	3.36	1.17	45.9	0.92	11.7	60.2	2.9	1.22	2.04
24		11.2	1.96	1.04	94.3	1.66		50	1.07	1.2	2.12
25		11.6	6.2	2.96	43.8	0.76	10	56	1.45	1.03	1.83
26		12.1	2.06	0.79	48.8	1.05	9.2	60.3	1.7	1.56	2.6
27		12.3	0.86	0.45	15.6	0.36		59	1.9	1.2	2
28		11.6	2.47	2.27	34.3	0.98	12.84	57	1.3	1.01	1.59
29		12.4	2.9	1	56.8	1.1	18	62	1.3	0.79	1.31
30		12.4	1.71	0.64	31.3	0.7	6.3	62	1.5	0.98	1.68
31		12	2.09	0.83	16.9	0.33	16	57.7	1.7	0.33	1.686
33		12.5	2.28	0.86	89.7	2.17	19.7	63.9	2.2	0.7	1.12
34	physiological pregnancy, N	12.1	2	0.71	89.8	2.25	16.39	66	1.6	1.09	1.7
35		13.4	2.57	0.83	62.2	1.81	32.7	74.7	1.8	1.19	1.77
39		12	3.02	1.52	66	1.32	16.11	54	1.1	1.028	1.75
41		13	3.1	0.88	61	1.26	26.68	67.5	1.6	0.28	0.44
43		12.2	1.62	0.52	86.7	1.77	14.7	58	1.2	1.14	1.95
44		13	3.42	1.3	56.6	1.44	15.97	66	1.21	1.02	1.6
46		12	3.011	1.23	64.2	1.15	15.49	54.5	1.2	1.002	1.77
47		12.6	2.9	0.61	52.3	1.06	32.52	65	1.17	1.3	2.15
49		12	6.9	3.15	58.8	1.105	26.2	55	1.3	0.88	1.49

Table 4. *Cont.*

Sample ID	Diagnose at Delivery, Group Name	1st Trimester Screening, Peripheral Blood Sampling, GW	PAPP-A (0.7–6.0 IU/L)	PAPP-A (0.5–2.0 MoM)	b-hCG (50.0–55.0 IU/mL)	b-hCG (0.5–2.0 MoM)	PLGF (21.9–71.2 пг/мл)	Crown Rump Length, CRL (43.0–84.0 mm)	Nuchal Translucency Thickness, NT (1.6–1.7 mm)	Uterine Artery Pulsatility Index, UA (PI), MoM	Uterine Artery Pulsatility Index, UA (Pi), 0.9–2.6 (5th and 95th Percentiles)
					Biochemical Data				**Ultrasound Data**		
51	high risk of pre-eclampsia according to Astraia but no pre-eclampsia at delivery, Nhr	11.2	1.05	0.41	25.5	0.36		50	1	1.33	2.48
52		13	3.38	0.68	39.8	0.74		68	2	1.15	1.92
53		11.6	1.55	1.32	23.7	0.717	12.7	59.7	1.9	1.4	2.17
54		12.1	2.1	1.07	89.8	2.5		56	1.1	1.02	1.68
55		12.2	2.25	0.97	23.2	0.48	19.99	57.6	1.6	1.19	2.02
56		11.6	2.27	1.23	114.7	2.3		53	1.29	1.05	1.8
58		12.1	4.18	1.8	51.7	1.02		57	1.9	0.81	1.36
110	Placenta percreta, PAS	12.2	13.7	4.6	101.21	2.047	37.7	61.9	1.7	1.069	1.75
111	Placenta increta, PAS	12.5	5.117	1.983	47	0.872	25.6	61	1.3	0.49	0.855
112	Placenta creta, PAS	12	4.46	2.145	38.14	0.69	22.8	53	1.2	0.981	1.715
113	Placenta increta, PAS	12.3	1.891	0.995	35	0.86		59.6	1.8	0.845	1.365
114	Placenta percreta, PAS	12.6	3.257	1.043	65.17	1.428	23.3	65	1.9	0.552	0.89
115	Placenta percreta, PAS	13.1	2.5	1.151	28.06	0.9	56.2	70	2	0.999	1.475
116	Placenta increta, PAS	12.0	5.84	2.021	223.6	3.97		57.9	1.2	1.164	1.995
117	Placenta increta, PAS	13.0	8.6	2.8	53	1.309	34.8	68	1.5	0.485	0.755
118	Placenta increta, PAS	12.1	3.838	1.562	45.84	0.865		56.8	1.6	1.312	2.21
119	Placenta percreta, PAS	12.4	2.252	0.989	86.85	2.06		61.7	1.9	0.753	1.22

The comparison of read numbers in the "PAS" group with those in the combined "N + PE" group is presented in Table S8. The partial least squares regression (PLS) method was used to assess the contribution of each identified miRNA to the separation of samples from the analyzed groups (Figure 5).

The distribution of samples in Figure 5 shows that blood serum samples from women with PAS are distant from all other samples and form a separate cluster. The molecules hsa-miR-92a-3p, hsa-miR-320a, hsa-miR-101-3p, hsa-miR-26a-5p, hsa-miR-148a-3p, hsa-miR-1307-3p, hsa-miR-16-5p, and hsa-miR-17-5p contributed most to this distribution. The levels of these molecules were evaluated by quantitative PCR in samples from the N and PAS groups (Figure 6). The "$-\Delta Ct$" values were calculated based on the difference between the Ct value of the analyzed miRNA and the Ct value of the endogenous hsa-let-7a-5p, given its stable expression in all analyzed samples by deep sequencing (coefficient of variation was equal to 0.165, see Table S8). The statistical significance of differences between the compared groups is presented in Table 5.

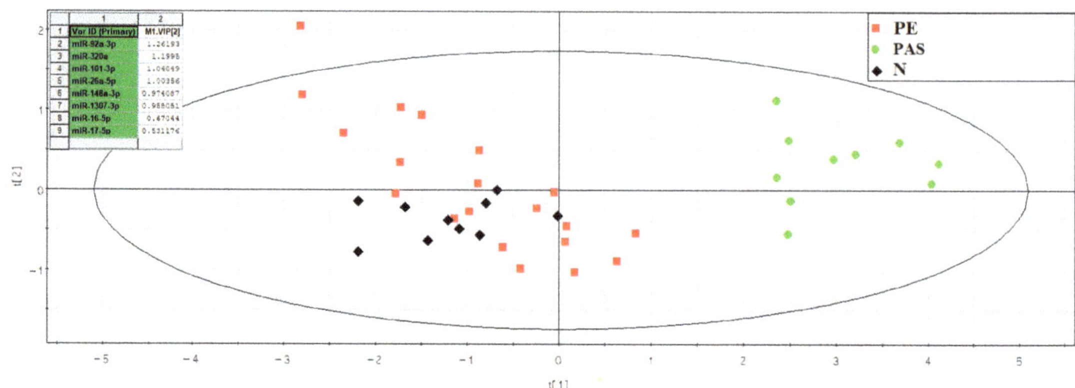

Figure 5. PLS analysis of deep-sequencing data of miRNAs in women's peripheral blood serum at 11–14 weeks of pregnancy with physiological course, early- or late-onset pre-eclampsia, and PAS.

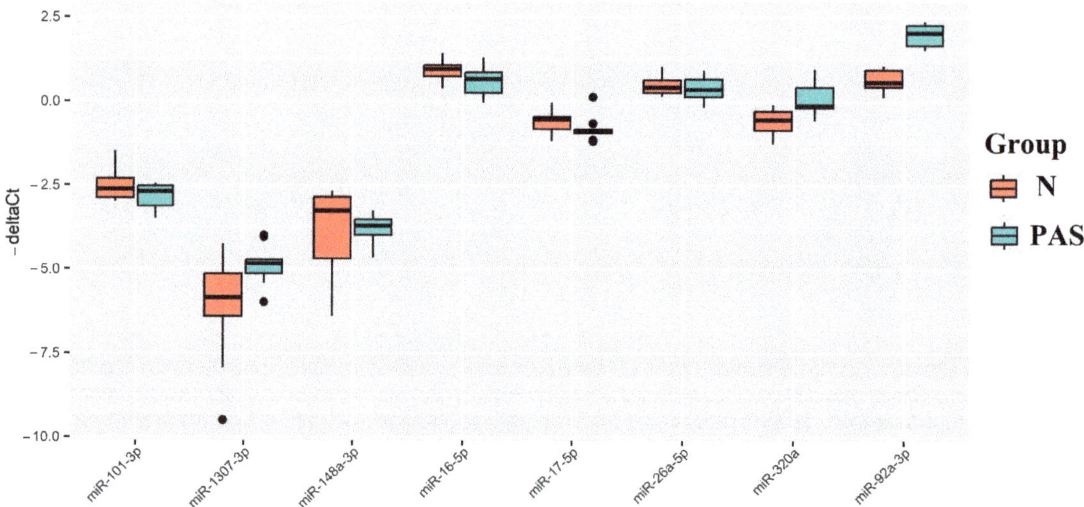

Figure 6. Box plot of miRNA content in exosomes of blood serum from women at 11–14 weeks of pregnancy with a physiological course and PAS. The position of the median inside the box in the form of a horizontal line and outliers in the form of points are indicated.

From Figure 6 and Table 5, it can be inferred that the PAS group statistically significantly differed from the N group only in the elevated levels of hsa-miR-320a, hsa-miR-92a-3p, and hsa-miR-1307-3p among all the analyzed miRNAs. As miRNAs are secreted by cells not only as part of exosomes but also as part of lipoproteins [37], we analyzed the levels of hsa-miR-92a-3p, hsa-miR-320a, hsa-miR-101-3p, hsa-miR-26a-5p, hsa-miR-148a-3p, hsa-miR-1307-3p, hsa-miR-16-5p, and hsa-miR-17-5p in the native blood serum sample of the third patient cohort (Table 4). Two types of normalization were used to calculate the "$-\Delta Ct$" values: normalization to the exogenous RNA UniSp6 (Figure 7A, Table 6) and endogenous hsa-let-7a-5p (Figure 7B, Table 6).

Table 5. Comparison of groups "N" and "PAS" by the miRNA "$-\Delta Ct$" value plotted as a box diagram in Figure 6.

miRNA	Group	Me	Q1	Q3	*p*-Value
miR-101-3p	N	−2.62	−2.89	−2.31	0.25
	PAS	−2.69	−3.13	−2.52	
miR-16-5p	N	0.96	0.75	1.08	0.09
	PAS	0.67	0.26	0.86	
miR-17-5p	N	−0.53	−0.82	−0.44	0.08
	PAS	−0.89	−0.93	−0.82	
miR-26a-5p	N	0.43	0.26	0.64	0.53
	PAS	0.37	0.15	0.67	
miR-320a	N	−0.54	−0.85	−0.28	0.006
	PAS	−0.13	−0.18	0.43	
miR-92a-3p	N	0.58	0.45	0.95	<0.001
	PAS	2.05	1.68	2.28	
miR-1307-3p	N	−5.85	−6.41	−5.15	0.04
	PAS	−4.84	−5.14	−4.72	
miR-148a-3p	N	−3.26	−4.69	−2.85	0.44
	PAS	−3.71	−3.99	−3.52	

Table 6. Pairwise comparison of third patient cohort groups listed in Table 4 by miRNA "$-\Delta Ct$" value, presented as a box diagram in Figure 7.

Normalization to UniSp6								
					\ *p*-Value			
miRNA	Group	Me	Q1	Q3	N	Nhr	PE	PAS
miR-101-3p	N	−16.39	−16.72	−15.75	-	0.53	0.54	<0.001
	Nhr	−17	−19.25	−15.75	0.53	-	0.43	<0.001
	PE	−16.08	−17.07	−15.53	0.54	0.43	-	<0.001
	PAS	−14.81	−15.26	−13.48	<0.001	<0.001	<0.001	-
miR-16-5p	N	−11.26	−11.85	−11.04	-	0.06	0.98	0.02
	Nhr	−11.74	−12.29	−11.64	0.06	-	0.08	<0.001
	PE	−11.27	−11.85	−10.96	0.98	0.08	-	0.01
	PAS	−10.87	−11.03	−9.71	0.02	<0.001	0.01	-
miR-17-5p	N	−14.63	−15.24	−14.23	-	0.07	0.85	<0.001
	Nhr	−15.35	−16.6	−15.03	0.07	-	0.17	<0.001
	PE	−14.93	−16.05	−13.95	0.85	0.17	-	<0.001
	PAS	−13.44	−13.5	−12.75	<0.001	<0.001	<0.001	-
miR-26a-5p	N	−14.82	−15	−14.51	-	0.02	0.88	0.002
	Nhr	−15.6	−16.98	−15.38	0.02	-	0.08	<0.001
	PE	−14.93	−16.02	−13.98	0.88	0.08	-	<0.001
	PAS	−12.92	−14.01	−12.46	0.002	<0.001	<0.001	-
miR-320a-3p	N	−12.79	−13.11	−12.31	-	0.88	0.39	0.02
	Nhr	−12.78	−13.57	−12.58	0.88	-	0.09	0.01
	PE	−12.39	−13.01	−12.01	0.39	0.09	-	0.07
	PAS	−11.86	−12.56	−11.2	0.02	0.01	0.07	-

Table 6. *Cont.*

					Normalization to UniSp6			
						p-Value		
miRNA	Group	Me	Q1	Q3	N	Nhr	PE	PAS
miR-92a-3p	N	−9.46	−10.13	−9.03	-	0.13	0.12	0.05
	Nhr	−10.23	−10.46	−9.98	0.13	-	0.001	0.009
	PE	−8.84	−9.59	−8.46	0.12	0.001	-	0.24
	PAS	−8.56	−9.23	−7.72	0.05	0.009	0.24	-
miR-1307-3p	N	−21.19	−21.25	−21.05	-	0.31	0.34	0.31
	Nhr	−18.41	−21.15	−18.06	0.31	-	0.14	0.60
	PE	−21.24	−21.35	−21.13	0.34	0.14	-	0.09
	PAS	−18.92	−21.2	−16.2	0.31	0.60	0.09	-
miR-148a-3p	N	−14.55	−15.44	−13.87	-	0.07	0.63	0.73
	Nhr	−15.88	−16.4	−15.48	0.07	-	0.10	0.003
	PE	−14.52	−15.93	−13.99	0.63	0.10	-	0.30
	PAS	−14.42	−14.86	−13.81	0.73	0.003	0.30	-

					Normalization to Hsa-Let-7a-5p			
						p-Value		
miRNA	Group	Me	Q1	Q3	N	Nhr	PE	PAS
miR-101-3p	N	−3.68	−4.24	−3.46	-	0.36	0.12	<0.001
	Nhr	−2.59	−4.93	−2.14	0.36	-	0.95	0.05
	PE	−3.09	−3.78	−2.39	0.12	0.95	-	0.01
	PAS	−1.9	−2.09	−1.22	<0.001	0.05	0.01	-
miR-16-5p	N	1.13	0.96	1.98	-	0.16	0.28	0.02
	Nhr	1.91	1.74	2.09	0.16	-	0.83	0.41
	PE	1.96	1.03	2.38	0.28	0.83	-	0.17
	PAS	2.12	1.62	3.34	0.02	0.41	0.17	-
miR-17-5p	N	−2.46	−2.84	−1.72	-	0.41	0.21	<0.001
	Nhr	−1.58	−2.51	−0.95	0.41	-	0.53	0.009
	PE	−1.85	−2.62	−1.37	0.21	0.53	-	<0.001
	PAS	−0.51	−0.89	−0.31	<0.001	0.009	<0.001	-
miR-26a-5p	N	−2.33	−2.56	−1.55	-	0.73	0.23	<0.001
	Nhr	−1.35	−2.83	−1.15	0.73	-	0.91	<0.001
	PE	−1.67	−2.4	−1.24	0.23	0.91	-	<0.001
	PAS	−0.71	−0.92	−0.35	<0.001	<0.001	<0.001	-
miR-320a-3p	N	−0.04	−0.78	0.4	-	0.05	0.11	0.05
	Nhr	1.35	0.24	1.52	0.05	-	0.29	0.60
	PE	0.9	−0.1	1.3	0.11	0.29	-	0.88
	PAS	0.65	0.43	1.05	0.05	0.60	0.88	-
miR-92a-3p	N	2.99	2.37	3.64	-	0.10	0.03	0.03
	Nhr	3.6	3.44	3.99	0.10	-	0.46	0.53
	PE	4.25	3.55	4.93	0.03	0.46	-	1
	PAS	3.97	3.43	4.72	0.03	0.53	1	-

Table 6. *Cont.*

| | | | | | Normalization to Hsa-Let-7a-5p | | | |
| | | | | | *p*-Value | | | |
miRNA	Group	Me	Q1	Q3	N	Nhr	PE	PAS
miR-1307-3p	N	−7.78	−9.1	−7.14	-	0.04	0.41	0.19
	Nhr	−5.17	−6.71	−3.98	0.04	-	0.03	0.60
	PE	−7.74	−8.34	−6.15	0.41	0.03	-	0.30
	PAS	−5.91	−8.02	−3.87	0.19	0.60	0.30	-
miR-148a-3p	N	−1.9	−2.42	−1.47	-	0.96	0.95	0.48
	Nhr	−2.07	−2.17	−1.86	0.96	-	1	0.36
	PE	−2.08	−2.76	−1.33	0.95	1	-	0.41
	PAS	−1.54	−2.44	−1.14	0.48	0.36	0.41	-

In comparison with the results of quantitative assessment of miRNA in the exosomal fraction of blood serum from women at 11–14 weeks of pregnancy, we observed a statistically significant increase in the levels of hsa-miR-101-3p, hsa-miR-26a-5p, hsa-miR-16-5p, and hsa-miR-17-5p in the native blood serum of women in the PAS group relative to the N group (Table 6). We conclude that such increase in the levels of these miRNAs occurs in the non-exosomal fraction of blood serum. Notably, only hsa-miR-101-3p, hsa-miR-26a-5p, and hsa-miR-17-5p showed statistical significance in distinguishing the PAS group from all other groups (N, Nhr, PE) using two types of normalization: on UniSp6 or hsa-let-7a-5p. Hence, these miRNAs can be considered specific markers for PAS. The other three miRNAs—hsa-miR-92a-3p, hsa-miR-320a, and hsa-miR-16-5p—did not show statistically significant differences between the PAS group and the PE and/or Nhr groups (Table 6), indicating that they cannot be considered specific markers for PAS.

The quantitative analysis of hsa-miR-92a-3p, hsa-miR-320a, hsa-miR-101-3p, hsa-miR-26a-5p, hsa-miR-148a-3p, hsa-miR-1307-3p, hsa-miR-16-5p, and hsa-miR-17-5p in the native blood serum of women at 11–14 weeks of pregnancy, using endogenous hsa-let-7a-5p as a normalizing RNA, was employed to develop logistic regression models for creating a non-invasive diagnostic system for PAS. In RStudio, optimal combinations of miRNAs associated with the presence of PAS in pregnant women were identified via stepwise inclusion and exclusion of each molecule in logistic regression models, where the dependent variable (response variable) was the presence or absence of PAS in pregnant women (0—combined groups N, Nhr, PE; 1—PAS group). The models presented in Figure 8 were selected, where all independent variables were statistically significant. The parameters of the models in Figure 8 are indicated in Table 7. Formulas 4 and 5, describing Models 2 and 3 in Figure 8, respectively, are provided below.

From Table 7, it can be inferred that logistic regression models 1, 2, and 3 have 100% sensitivity in detecting PAS in women during the first-trimester screening based on the quantitative analysis of the corresponding miRNAs in native blood serum. Since model 1 includes miR-92a-3p, the expression level of which was statistically significantly changed in both PE and PAS compared to normal pregnancy (Table 6), it cannot be considered as a specific marker for PAS, and therefore Model 1 was excluded from consideration.

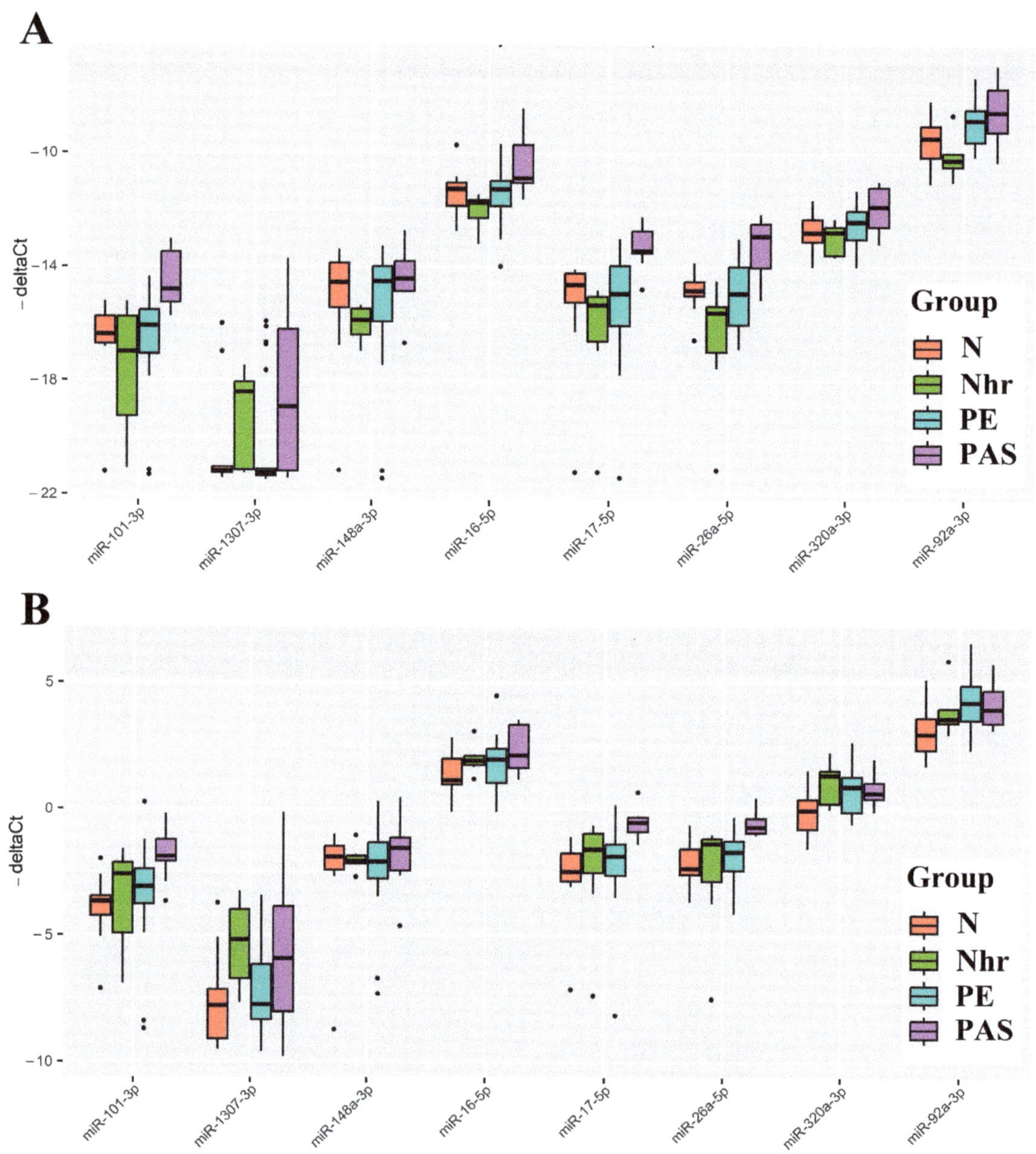

Figure 7. Box plot of miRNA content in native blood serum of women at 11–14 weeks of pregnancy in groups "N," "Nhr," "PE," and "PAS." "−ΔCt" Values were calculated using exogenous RNA UniSp6 (**A**). "−ΔCt" values were calculated relative to the content of endogenous miRNA hsa-let-7a-5p (**B**). The position of the median inside the box in the form of a horizontal line and outliers in the form of points are indicated.

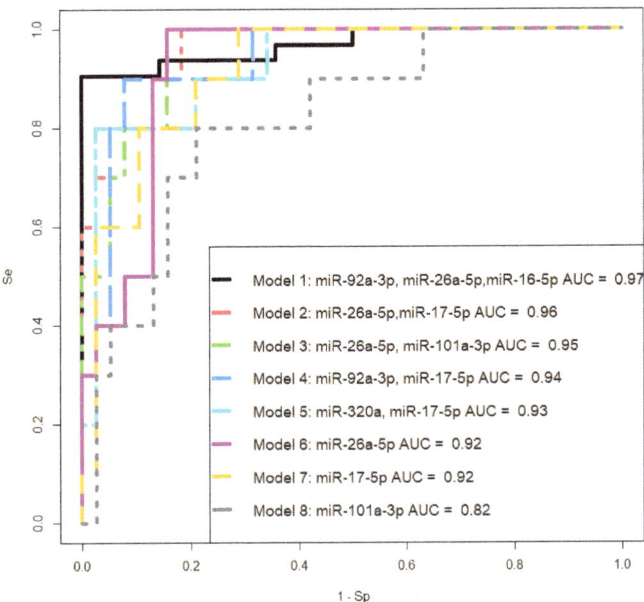

Figure 8. ROC curves of logistic regression models based on real-time PCR data comparing the PAS group and combined groups "N + Nhr + PE" for the content of miRNA in native blood serum of women at 11–14 weeks of pregnancy. Se—sensitivity, Sp—specificity.

Table 7. Parameters of logistic regression models in Figure 8.

Models	Estimate (95% CI)	Wald	p-Value	OR(95% CI)	Se	Sp
(Intercept)	12.264 (3.532; 27.813)	2.149	0.032	211,919.348 (34.208; 1.2×10^{12})	**Model 1**	
miR-92a-3p	−4.955 (−10.911; −1.566)	−2.246	0.025	0.007 (0.00001; 0.209)		
miR-26a-5p	5.093 (2.068; 10.91)	2.355	0.019	162.828 (7.905; 54,728.702)	1	0.84
miR-16-5p	5.584 (1.896; 12.356)	2.267	0.023	266.031 (6.658; 232,425.187)		
(Intercept)	5.661 (1.863; 12.559)	2.157	0.031	287.563 (6.445; 284,717.908)	**Model 2**	
miR-26a-5p	4.189 (1.259; 9.276)	2.157	0.031	65.939 (3.523; 10,674.696)	1	0.82
miR-17-5p	2.921 (0.646; 6.887)	1.916	0.055	18.564 (1.908; 979.604)		
(Intercept)	8.049 (2.747; 17.86)	2.193	0.028	3129.605 (15.589; 5.7×10^7)	**Model 3**	
miR-26a-5p	5.538 (2.059; 12.068)	2.274	0.023	254.057 (7.835; 174,165.021)	1	0.84
miR-101-3p	1.989 (0.502; 4.585)	2.012	0.044	7.308 (1.652; 98.023)		
(Intercept)	9.303 (2.248; 19.251)	2.21	0.027	10,967.88 (9.473; 2.3×10^8)	**Model 4**	
miR-92a-3p	−1.451 (−3.193; −0.173)	−1.941	0.052	0.234 (0.041; 0.841)	0.9	0.9211
miR-17-5p	4.333 (1.956; 8.034)	2.872	0.004	76.151 (7.071; 3084.714)		
(Intercept)	4.152 (1.217; 8.272)	2.386	0.017	63.567 (3.378; 3911.935)	**Model 5**	
miR-320a-3p	−1.477 (−3.168; −0.148)	−1.973	0.048	0.228 (0.042; 0.862)	0.8	0.9737
miR-17-5p	4.108 (1.878; 7.607)	2.899	0.004	60.854 (6.539; 2013.226)		
(Intercept)	2.256 (0.321; 4.915)	1.974	0.048	9.544 (1.378; 136.287)	**Model 6**	
miR-26a-5p	3.389 (1.483; 6.373)	2.773	0.006	29.638 (4.405; 585.998)	1	0.8421
(Intercept)	1.628 (−0.009; 3.742)	1.742	0.082	5.093 (0.991; 42.181)	**Model 7**	
miR-17-5p	2.533 (1.127; 4.672)	2.877	0.004	12.592 (3.087; 106.945)	1	0.7105
(Intercept)	1.136 (−0.568; 3.188)	1.218	0.223	3.115 (0.566; 24.245)	**Model 8**	
miR-101-3p	0.985 (0.33; 1.887)	2.533	0.011	2.679 (1.391; 6.598)	0.8	0.7895

(4) Formula for Model 2:

$$\frac{1}{1 + e^{-5.66 - 4.19x_1 - 2.92x_2}} \tag{4}$$

where x_1—«$-\Delta$Ct» value for hsa-miR-26a-5p, x_2—«$-\Delta$Ct» value for hsa-miR-17-5p;
(5) Formula for Model 3:

$$\frac{1}{1 + e^{-8.05 - 5.54x_1 - 1.99x_2}} \tag{5}$$

where x_1—«$-\Delta$Ct» value for hsa-miR-26a-5p, x_2—«$-\Delta$Ct» value for hsa-miR-101-3p.

Considering all 48 blood serum samples (Table 4), a reliable correlation of miRNA level with biochemical and instrumental analysis data at 11–14 GW were found using the non-parametric Spearman rank correlation method: hsa-miR-101-3p with NT (r = 0.34, p = 0.0188), hsa-miR-17-5p with UA(PI)MoM (r = -0.3, p = 0.038), hsa-miR-26a-5p with PLGF (r = 0.49, p = 0.025), hsa-miR-17-5p with PLGF (r = 0.36, p = 0.0327). At the same time, statistically significant correlations were found for UA(PI)MoM with PAPP-A (r = -0.33, p = 0.0242) and with PAPP-A(MoM) (r = -0.29, p = 0.043), for PLGF with PAPP-A (r = 0.37, p = 0.0264).

3. Discussion

In this study, the miRNA signature in placental tissues was analyzed both within the invasive region and outside this area, in the adjacent myometrial tissues, and in the blood plasma from women using deep sequencing. The obtained data were validated through quantitative real-time PCR to construct logistic regression models as a non-invasive diagnostic approach for differentiating various types of placenta accreta spectrum (PAS) in the third trimester of pregnancy. It was observed that the quantitative assessment of hsa-miR-320b, hsa-let-7d-5p, hsa-miR-629-5p, and hsa-miR-17-5p in the blood plasma of women allows for the statistically significant identification of PAS cases with high specificity (93–100%) and sensitivity (78–97%). These molecules were found to be differentially expressed in both placental and myometrial tissues of women with different types of PAS. The constructed logistic regression models can be considered as an additional diagnostic method alongside commonly used instrumental diagnostic approaches such as ultrasound (US) and magnetic resonance imaging (MRI).

To formulate an individual management strategy for patients with PAS, involving the referral to a specialized hospital with a multidisciplinary team of surgeons, intensivists, neonatologists, and preparedness for blood transfusion in case of hemorrhage, it is imperative to diagnose this pregnancy complication in the first trimester. Due to the absence of precise biochemical and instrumental tests for detecting PAS during this gestational period, we conducted a retrospective study using deep sequencing and quantitative real-time PCR of cell-free miRNAs in the blood serum of women who underwent first-trimester screening and continued examination until the parturition with a clear diagnosis (physiological pregnancy or pre-eclampsia or PAS) at the Kulakov National Medical Research Center of Obstetrics, Gynecology, and Perinatology. The search for miRNA markers of placental invasion specifically in the exosomal fraction of blood serum was motivated by our previous findings [38] indicating a decrease in the concentration of identified miRNA markers for pre-eclampsia upon repeated cycles of freezing/thawing of the analyzed blood serum sample. The question of the miRNA stability in the body's biological fluids has been thoroughly investigated by other researchers such as Coenen-Stass et al. [39]. Exosomes, being membrane-containing structures, protect the encapsulated miRNAs from degradation by extracellular RNases and facilitate the targeted delivery of miRNAs to specific cells and tissues [40]. We discovered that the most significant contributors to the differentiation of exosome fractions in the blood serum of women at 11–14 weeks of pregnancy with various types of PAS from combined groups of women with physiological pregnancy and pre-eclampsia, using principal component analysis, were miR-92a-3p, miR-

320a, miR-101-3p, miR-26a-5p, miR-148a-3p, miR-1307-3p, miR-16-5p, and miR-17-5p based on deep-sequencing data. Among these, statistically significant differences between the PAS group and all other comparable groups were identified only for miR-320a, miR-92a-3p, and miR-1307-3p based on quantitative real-time PCR data.

However, the isolation of exosomes for identifying cases of placenta accreta spectrum (PAS) using marker mRNA is an additional time-consuming and costly method, which may not be feasible given the high patient volume during screening studies. Therefore, we decided to analyze exosome mRNA markers for PAS in the native blood serum of pregnant women at 11–14 weeks of gestation using quantitative real-time PCR. In comparison with the analysis of the exosome fraction of blood serum, we observed a significant increase in the levels of miR-101-3p, miR-26a-5p, miR-16-5p, and miR-17-5p, in addition to changes in miR-92a-3p and miR-320a, in the native blood serum of women from the PAS group compared to the group of women with physiological pregnancies. This increase in the levels of these mRNA markers occurs in the non-exosome fraction, possibly as part of very-low-density lipoproteins (VLDL), low-density lipoproteins (LDL), or high-density lipoproteins (HDL). For example, other researchers demonstrated the presence of some of the analyzed here miRNAs in blood lipoproteins, namely miR-16-5p in VLDL and HDL, miR-17-5p and miR-26a-5p in HDL [37], dependent on miRNA sequence motifs. Mechanistically, it was found that HDL-miRNAs have roles in metabolic homeostasis and angiogenesis, whereas targets for LDL-miRNAs were enriched in pathways related to inflammation, immune system function, and different cardiomyopathies.

When calculating "$-\Delta Ct$" values for each analyzed miRNA, any of the RNA species was used as a reference: either UniSp6, introduced into the sample during the reverse transcription stage according to the Qiagen's recommendation, or the endogenous let-7a-5p, exhibiting a stable high expression level in the exosomal fraction of blood serum and native blood serum of women at 11–14 weeks of pregnancy according to our deep-sequencing data. The use of UniSp6 allows for accounting for variations in reverse transcription and PCR efficiency but does not account for possible mRNA degradation due to extracellular RNases during repeated sample freezing/thawing cycles, unlike the use of endogenous let-7a-5p. The presence of let-7a-5p in exosomes, as identified in the current study, and in VLDL, as reported by Guido Rossi-Herring [37], when used as an endogenous reference RNA, would consider detrimental processes affecting the concentration of analyzed miRNA in exosomes and the non-exosomal fraction of blood serum ex vivo.

It is important to note that miR-101-3p, miR-26a-5p, and miR-17-5p were found to be the most specific markers for PAS, as statistically significant differences in their expression levels were identified in the PAS group compared to all other comparison groups (physiological pregnancy, high risk of developing pre-eclampsia according to Astraia during the first pregnancy screening without clinical manifestations of pre-eclampsia after 20 weeks of pregnancy, development of early and late pre-eclampsia) using two types of normalization: on UniSp6 or let-7a-5p. Meanwhile, other miRNA associated with PAS (miR-92a-3p, miR-320a, miR-16-5p) did not significantly differentiate the PAS group from the pre-eclampsia and/or high-risk group for developing pre-eclampsia without clinical manifestations after 20 weeks of pregnancy. Therefore, the latter cannot be considered specific markers for PAS.

The potential role of miR-92a-3p, miR-320a, and miR-16-5p in the pathogenesis of pre-eclampsia (PE) has been studied in several works, revealing that (i) intravascular inflammation occurs in PE as a sequence of Th1 polarization [41,42] through the targeting of GATA3 by upregulated miR-92a-3p, contained in vesicles of activated NK cells [43]; (ii) miR-320a overexpression observed in PE inhibits trophoblast cell invasion and causes anomalous placentation by targeting estrogen-related receptor-gamma [44], IL-4 [45], and IGF-1R [46]; (iii) in a PE rat model, the upregulation of miR-16-5p directly downregulates IGF-2 and provides inhibition of trophoblast cell viability and migration [47]. Thus, the involvement of the same miRNA molecules in different diseases necessitates the search for a unique combination of marker miRNAs that can differentiate one pregnancy complication from another. Therefore, when selecting logistic regression models for diagnosing PAS at 11–14 weeks of gestation, we relied on the following criteria: the combination of molecules should include any or all the miRNAs miR-101-3p, miR-26a-5p, and miR-17-5p; all model parameters must be significant, and the model should have high specificity and sensitivity. These criteria are met by two models we developed: the combination of miR-26a-5p and miR-17-5p, and the combination of miR-26a-5p and miR-101-3p, both of which exhibit 100% sensitivity in detecting PAS in women during the first pregnancy screening through their quantitative analysis in native blood serum using real-time PCR.

It is important to note that the circulating miRNA markers for PAS identified in this study during the first trimester of pregnancy are differentially expressed in the placenta within the area of invasion and in the adjacent myometrium, specifically in the case of invasive forms (increta and percreta) relative to the adherent form (creta) in the third trimester of pregnancy at the time of delivery, as per deep-sequencing data. Comparing areas of the placenta outside the invasive region led us to hypothesize that for placental invasion into the myometrium, molecular and biological changes in the myometrial tissue are necessary. This assumption is based on the observation that the intersection of differentially expressed miRNA lists in areas of the placenta outside the invasion site, in the case of increta and percreta relative to creta, is found for 70–80% of all miRNAs. And more pronounced quantitative and qualitative changes in the miRNA signature are observed in the area of trophoblast invasion in both placental tissue and adjacent myometrium from women with placenta percreta. In other words, for the invasive type of PAS, there are fundamental changes in the miRNA signature in the placenta, similar between increta and percreta but distinct from PAS adherent form (creta); the invasion of such altered placenta into the myometrial tissue occurs only if there are changes in the myometrium itself, most pronounced in the case of percreta. These findings are consistent with those of other researchers who adhere to the concept of a primary deciduomyometrium defect in PAS that impacts the formation of the migratory and invasive phenotype of interstitial and endovascular extravillous trophoblast cells [13,14].

When analyzing experimentally validated target genes of identified here miRNAs differentially expressed in the placenta and myometrium within the PAS site, signaling pathways involving growth factors, glypicans, cell adhesion proteins, integrins, interleukins, and chemokines were identified. These pathways are responsible for processes such as cell adhesion, proliferation, migration, angiogenesis, inflammation, and apoptosis. These data align with discussions in published articles on pathways that stimulate trophoblast invasion [12,48].

A distinctive feature of invasive PAS, as observed through instrumental research methods and macroscopic examination of the uterine surface, is uteroplacental vascular changes in the accreta area resulting from both neovascularization and/or increased infiltration of deep uterine vessels (radial and even the arcuate arteries) by extravillous trophoblasts (EVT) [1]. The role in angiogenesis of the identified here miRNAs as the markers of PAS in the first trimester of pregnancy has been demonstrated by numerous scientific teams. Hypoxia-responsive hsa-miR-101-3p, known as angiomiR, regulate angiogenesis by targeting cullin 3 thereby promoting Nrf2 nuclear accumulation and causing heme oxygenase-1 induction, VEGF expression, and nitric oxide production [49], or by targeting c-Met [50]—a receptor for hepatocyte growth factor (HGF), which is one of the key molecules that stimulate endothelial cells to proliferate and migrate via the upregulation of VEGF and its receptor KDR [51] as well as metalloproteinases [52] to degrade extracellular matrix for vascular growth. In addition, the inverse correlation between the expression of the miR-320a and the HGF gene was found [50]. Different studies demonstrated the suppressor function of miR-320a in cell invasion and angiogenesis in ovary cancer [53], hepatocellular carcinoma [54,55], and endometrial cancer [56]. The anti-inflammatory effect and promotion of the angiogenesis in the skeletal muscle injury model were found for miR-320a and miR-26a via reduction of the protein expression of their target genes—PTEN and TLR3, respectively [57]. Enriched by miR-17-5p exosomes from endothelial progenitor cell decrease cell apoptosis, increase microvessel density and capillary angiogenesis as well as promote muscle structural integrity in a diabetic hind-limb ischemia mode through increasing the levels of PI3K and phosphorylated Akt [58]. The participation of miR-92a-3p in exosome-mediated angiogenesis was found in retinoblastoma by targeting transcription factor KLF2 [59] which is able to modulate tumor proliferation and metastasis [60]. The significant correlations of hsa-miR-17-5p expression level with uterine artery pulsatility index, and hsa-miR-26a-5p and hsa-miR-17-5p with PLGF revealed in the present study prove the important role of these PAS miRNA markers in angiogenesis.

4. Materials and Methods

4.1. Patients

All patients (Tables 1, 2 and 4) enrolled to investigation were admitted to the National Medical Research Center for Obstetrics, Gynecology, and Perinatology, named after the Academician V.I. Kulakov of Ministry of Healthcare of the Russian Federation for management of pregnancy and delivery, and signed informed consent to participate in the study; the study was approved by the Ethics Committee of the Center. Clinical and biochemical blood tests, ultrasound examination of the pelvic and fetal organs, fetoplacental blood flow dopplerometry, cardiotocography, blood pressure measurement, the determination of protein levels in urine, and concentrations of PLGF, sFlt-1, PAPP-A, and β-HCG in blood serum using diagnostic test systems were carried out for each patient. The criteria for non-inclusion in the study were the onset of pregnancy via assisted reproductive technologies, multiple pregnancy, and fetal aneuploidy.

4.2. RNA Isolation from Blood Plasma or Serum

A total of 200 µL of blood plasma or serum, purified from cells and cell debris via stepwise centrifugations at $300\times g$ for 20 min and at $16{,}000\times g$ for 10 min, were used for RNA isolation using an miRNeasy Serum/Plasma kit (Qiagen, Hilden, Germany) according to the manufacturer's protocol.

4.3. RNA Isolation from Placenta or Myometrium Tissue

Placenta or myometrium tissue samples were taken for research no later than 10 min after delivery as shown in Figure 1A, and immediately frozen in liquid nitrogen for subsequent storage at −80 °C. Total RNA was extracted from 20 to 40 mg of tissue using the miRNeasy Micro Kit (Qiagen, Hilden, Germany) followed by the RNeasy MinElute Cleanup Kit (Qiagen, Hilden, Germany) according to the manufacturer's protocol. The RNA concentration was measured using a Qubit fluorimeter 3.0 (Life Technologies, Petaling Jaya, Malaysia). The sample quality of the total RNA was examined using the Agilent Bioanalyzer 2100 (Agilent, Waldbronn, Germany) and the RNA 6000 Nano Kit (Agilent Technologies, Santa Clara, CA, USA). Total RNA samples with a 28S/18S ribosomal RNA ratio equal to 1.5–1.8 were used for further studies.

4.4. RNA Isolation from Blood Serum Exosomes

A total of 200 µL of blood serum, purified from cells and cell debris via stepwise centrifugations at $300\times g$ for 20 min and at $16{,}000\times g$ for 10 min, were used for RNA isolation using an exoRNeasy Midi Kit (Qiagen, Hilden, Germany) according to the manufacturer's protocol.

4.5. miRNA Deep Sequencing

cDNA libraries were synthesized using 6 µL of total RNA column eluate (miRNeasy Serum/Plasma Kit) extracted from 200 µL of blood plasma, 6 µL of total RNA column eluate (exoRNeasy Midi Kit) extracted from 200 µL of blood serum, and 500 ng of total RNA from placenta or myometrium tissue, using the NEBNext® Multiplex Small RNA Library Prep Set for Illumina® (Set11 and Set2, New England Biolab®, Frankfurt am Main, Germany, cat. nos. E7300S and E7580S), amplified for 19 cycles in case of blood plasma samples, 21 PCR cycles in case of blood serum exosomes, and 14 PCR cycles for placenta or myometrium samples, purified by QIAQuick PCR Purification Kit (Qiagen, Hilden, Germany), and 6% polyacrylamide gel electrophoresis for isolation of 136–150 bp bands corresponding to the adapter-ligated miRNAs, and sequenced using the NextSeq 500 platform (Illumina, San Diego, CA, USA, cat. no. SY-415-1001). The adapters were removed using Cutadapt. All trimmed reads shorter than 16 bp and longer than 55 bp were filtered, and only reads with a mean quality higher than 15 were retained. The remaining reads were mapped to the GRCh38.p15 human genome and miRBase v21 using the bowtie aligner [61]. Aligned reads were counted using the featureCount tool from the Subread package [62] and the fracOverlap 0.9 option; thus, the whole read was forced to have a 90% intersection with sncRNA features. Differential expression analysis of the sncRNA count data was performed using the DESeq2 package [63].

4.6. Reverse Transcription and Quantitative Real-Time PCR

Two microliters of total RNA column eluate (miRNeasy Serum/Plasma Kit, Qiagen, Hilden, Germany) or total RNA column eluate (exoRNeasy Midi Kit) was converted into cDNA in accordance with the miRCURY LNA RT Kit (Qiagen, Hilden, Germany) protocol; then, the sample volume was diluted 1:60, and cDNA was amplified in accordance with the miRCURY LNA miRNA SYBR Green PCR (Qiagen, Hilden, Germany) protocol using the miRCURY LNA miRNA PCR Assay (cat. no. 339306) in a StepOnePlus^{TM} thermocycler (Applied Biosystems, Waltham, MA, USA). The relative expression of miRNA in the sample was determined by the ΔCt method, using UniSp6 or hsa-let-7a-5p as the reference RNA.

4.7. Statistical Analysis of the Obtained Data

For statistical processing, scripts written in R language [62] and RStudio [64] were used. The correspondence of the analyzed parameters to the normal distribution law was assessed via the Shapiro–Wilk test. When the distribution of data was different from normal, the Mann–Whitney test for paired comparison was used. Since both quantitative and qualitative characteristics were analyzed, a rank-order correlation analysis was performed using Spearman's non-parametric correlation test. The 95% confidence interval for the correlation coefficient was determined using Fisher transformation. The value of the threshold significance level (p) was taken as being equal to 0.05.

5. Conclusions

Two logistic regression models have been developed for diagnosing PAS in women at 11–14 weeks of pregnancy by quantifying cell-free miRNAs hsa-miR-101-3p, hsa-miR-26a-5p, and hsa-miR-17-5p circulating in the peripheral blood, which are crucial for improving maternal and perinatal outcomes. These models demonstrated 100% sensitivity in detecting PAS during the first pregnancy screening. Hsa-miR-101-3p, hsa-miR-26a-5p, and hsa-miR-17-5p were identified as specific markers for PAS but not for other pregnancy complications such as early- and late-onset pre-eclampsia, and significantly correlated with nuchal translucency thickness (in the case of hsa-miR-101-3p), uterine artery pulsatility index (in the case of hsa-miR-17-5p), and PLGF (in the case of hsa-miR-26a-5p and hsa-miR-17-5p). A limitation of the developed diagnostic method is the small number of training samples due to the lack of a complete examination of the majority of pregnant women within one clinical center, namely the National Medical Research Center for Obstetrics, Gynecology, and Perinatology of Ministry of Healthcare of the Russian Federation, and the admission of pregnant women with PAS to this center only for delivery without the ability to analyze biological material (peripheral blood serum) donated at another clinical center during the first-trimester screening and vice versa. In connection with the emerging new screening method in the context of this study, pregnant women with an identified high risk of PAS by this method in the National Medical Research Center for Obstetrics, Gynecology, and Perinatology of Ministry of Healthcare of the Russian Federation will be offered to undergo further management of pregnancy and delivery in the same Center, which will expand the training set followed by validation of method efficiency on an independent test set with the integration of clinical and biochemical PAS markers to improve the accuracy of timely PAS diagnosis.

Supplementary Materials: The supporting information can be downloaded at: https://www.mdpi.com/article/10.3390/ijms25020871/s1.

Author Contributions: Conceptualization, A.V.T.; methodology, A.V.T., I.S.F. and T.Y.I.; software, I.S.F.; validation, A.V.T., I.S.F., T.M.Z. and T.Y.I.; investigation, A.V.T., I.S.F., Y.V.S., A.M.T., L.S.E. and T.Y.I.; data curation, A.V.T., O.N.V. and T.Y.I.; writing—original draft preparation, A.V.T.; writing—review and editing, A.V.T.; visualization, A.V.T. and I.S.F.; project administration, G.T.S.; and funding acquisition, A.V.T. and G.T.S. All authors have read and agreed to the published version of the manuscript.

Funding: This study was funded by the Russian Science Foundation under grant no. 22-15-00363 "Epigenetic and biochemical aspects of the pathology of pregnancy in violations of the invasive properties of the trophoblast: from early diagnosis to the prevention of maternal and perinatal morbidity".

Institutional Review Board Statement: The study was conducted in accordance with the Declaration of Helsinki and was approved by the Institutional Ethics Committee of National Medical Research Center for Obstetrics, Gynecology, and Perinatology, named after the Academician V.I. Kulakov of Ministry of Healthcare of the Russian Federation (protocol No. 2, approval date: 9 March 2017).

Informed Consent Statement: Written informed consent has been obtained from the patients to publish this paper.

Data Availability Statement: Data are contained within the article.

Conflicts of Interest: The authors declare no conflicts of interest.

References

1. Jauniaux, E.; Collins, S.; Burton, G.J. Placenta accreta spectrum: Pathophysiology and evidence-based anatomy for prenatal ultrasound imaging. *Am. J. Obstet. Gynecol.* **2018**, *218*, 75–87. [CrossRef]
2. Carusi, D.A. The Placenta Accreta Spectrum: Epidemiology and Risk Factors. *Clin. Obstet. Gynecol.* **2018**, *61*, 733–742. [CrossRef]
3. Morlando, M.; Sarno, L.; Napolitano, R.; Capone, A.; Tessitore, G.; Maruotti, G.M.; Martinelli, P. Placenta accreta: Incidence and risk factors in an area with a particularly high rate of cesarean section. *Acta Obstet. Gynecol. Scand.* **2013**, *92*, 457–460. [CrossRef]
4. Matsuzaki, S.; Mandelbaum, R.S.; Sangara, R.N.; McCarthy, L.E.; Vestal, N.L.; Klar, M.; Matsushima, K.; Amaya, R.; Ouzounian, J.G.; Matsuo, K. Trends, characteristics, and outcomes of placenta accreta spectrum: A national study in the United States. *Am. J. Obstet. Gynecol.* **2021**, *225*, 534.e1–534.e38. [CrossRef] [PubMed]
5. Wu, S.; Kocherginsky, M.; Hibbard, J.U. Abnormal placentation: Twenty-year analysis. *Am. J. Obstet. Gynecol.* **2005**, *192*, 1458–1461. [CrossRef]
6. De Mucio, B.; Serruya, S.; Alemán, A.; Castellano, G.; Sosa, C.G. A systematic review and meta-analysis of cesarean delivery and other uterine surgery as risk factors for placenta accreta. *Int. J. Gynaecol. Obstet. Off. Organ Int. Fed. Gynaecol. Obstet.* **2019**, *147*, 281–291. [CrossRef]
7. Thurn, L.; Lindqvist, P.G.; Jakobsson, M.; Colmorn, L.B.; Klungsoyr, K.; Bjarnadóttir, R.I.; Tapper, A.M.; Børdahl, P.E.; Gottvall, K.; Petersen, K.B.; et al. Abnormally invasive placenta-prevalence, risk factors and antenatal suspicion: Results from a large population-based pregnancy cohort study in the Nordic countries. *BJOG* **2016**, *123*, 1348–1355. [CrossRef] [PubMed]
8. Jauniaux, E.; Chantraine, F.; Silver, R.M.; Langhoff-Roos, J. FIGO consensus guidelines on placenta accreta spectrum disorders: Epidemiology. *Int. J. Gynaecol. Obstet.* **2018**, *140*, 265–273. [CrossRef] [PubMed]
9. Jauniaux, E.; Jurkovic, D. Placenta accreta: Pathogenesis of a 20th century iatrogenic uterine disease. *Placenta* **2012**, *33*, 244–251. [CrossRef] [PubMed]
10. Goh, W.A.; Zalud, I. Placenta accreta: Diagnosis, management and the molecular biology of the morbidly adherent placenta. *J. Matern. Fetal Neonatal Med.* **2016**, *29*, 1795–1800. [CrossRef]
11. Jauniaux, E.; Burton, G.J. Pathophysiology of placenta accreta spectrum disorders: A review of current findings. *Clin. Obstet. Gynecol.* **2018**, *61*, 743–754. [CrossRef] [PubMed]
12. Ma, J.; Liu, Y.; Guo, Z.; Sun, R.; Yang, X.; Zheng, W.; Ma, Y.; Rong, Y.; Wang, H.; Yang, H.; et al. The diversity of trophoblast cells and niches of placenta accreta spectrum disorders revealed by single-cell RNA sequencing. *Front. Cell Dev. Biol.* **2022**, *10*, 1044198. [CrossRef] [PubMed]
13. Wehrum, M.J.; Buhimschi, I.A.; Salafia, C.; Thung, S.; Bahtiyar, M.O.; Werner, E.F.; Campbell, K.H.; Laky, C.; Sfakianaki, A.K.; Zhao, G.; et al. Accreta complicating complete placenta previa is characterized by reduced systemic levels of vascular endothelial

growth factor and by epithelial-to-mesenchymal transition of the invasive trophoblast. *Am. J. Obstet. Gynecol.* **2011**, *204*, 411.e1–411.e11. [CrossRef] [PubMed]

14. Krstic, J.; Deutsch, A.; Fuchs, J.; Gauster, M.; Gorsek Sparovec, T.; Hiden, U.; Krappinger, J.C.; Moser, G.; Pansy, K.; Szmyra, M.; et al. (Dis)similarities between the Decidual and Tumor Microenvironment. *Biomedicines* **2022**, *10*, 1065. [CrossRef] [PubMed]

15. Silver, R.M.; Barbour, K.D. Placenta accreta spectrum: Accreta, increta, and percreta. *Obstet. Gynecol. Clin. N. Am.* **2015**, *42*, 381–402. [CrossRef]

16. Adu-Bredu, T.K.; Owusu, Y.G.; Owusu-Bempah, A.; Collins, S.L. Absence of abnormal vascular changes on prenatal imaging aids in differentiating simple uterine scar dehiscence from placenta accreta spectrum: A case series. *Front. Reprod. Health* **2023**, *5*, 1068377. [CrossRef]

17. Jauniaux, E.; Collins, S.L.; Jurkovic, D.; Burton, G.J. Accreta placentation: A systematic review of prenatal ultrasound imaging and grading of villous invasiveness. *Am. J. Obstet. Gynecol.* **2016**, *215*, 712–721. [CrossRef]

18. Silveira, C.; Kirby, A.; Melov, S.J.; Nayyar, R. Placenta accreta spectrum: We can do better. *Aust. N. Z. J. Obstet. Gynaecol.* **2022**, *62*, 376–382. [CrossRef]

19. Pavón-Gomez, N.; López, R.; Altamirano, L.; Cabrera, S.B.; Rosales, G.P.; Chamorro, S.; González, K.; Morales, A.; Maya, J.; Sinisterra, S.; et al. Relationship between the Prenatal Diagnosis of Placenta Acreta Spectrum and Lower Use of Blood Components. *Rev. Bras. Ginecol. Obstet. Rev. Fed. Bras. Soc. Ginecol. Obstet.* **2022**, *44*, 1090–1093. [CrossRef]

20. Wu, X.; Yang, H.; Yu, X.; Zeng, J.; Qiao, J.; Qi, H.; Xu, H. The prenatal diagnostic indicators of placenta accreta spectrum disorders. *Heliyon* **2023**, *9*, e16241. [CrossRef]

21. Zhang, T.; Wang, S. Potential Serum Biomarkers in Prenatal Diagnosis of Placenta Accreta Spectrum. *Front. Med.* **2022**, *9*, 860186. [CrossRef] [PubMed]

22. Penzhoyan, G.A.; Makukhina, T.B. Significance of the routine first-trimester antenatal screening program for aneuploidy in the assessment of the risk of placenta accreta spectrum disorders. *J. Perinat. Med.* **2019**, *48*, 21–26. [CrossRef] [PubMed]

23. Berezowsky, A.; Pardo, J.; Ben-Zion, M.; Wiznitzer, A.; Aviram, A. Second Trimester Biochemical Markers as Possible Predictors of Pathological Placentation: A Retrospective Case-Control Study. *Fetal Diagn. Ther.* **2019**, *46*, 187–192. [CrossRef]

24. Wang, F.; Chen, S.; Wang, J.; Wang, Y.; Ruan, F.; Shu, H.; Zhu, L.; Man, D. First trimester serum PAPP-A is associated with placenta accreta: A retrospective study. *Arch. Gynecol. Obstet.* **2021**, *303*, 645–652. [CrossRef] [PubMed]

25. Li, Y.; Meng, Y.; Chi, Y.; Li, P.; He, J. Meta-analysis for the relationship between circulating pregnancy-associated plasma protein A and placenta accreta spectrum. *Medicine* **2023**, *102*, e34473. [CrossRef]

26. Pekar-Zlotin, M.; Melcer, Y.; Maymon, R.; Jauniaux, E. Second-trimester levels of fetoplacental hormones among women with placenta accreta spectrum disorders. *Int. J. Gynaecol. Obstet. Off. Organ Int. Fed. Gynaecol. Obstet.* **2018**, *140*, 377–378. [CrossRef] [PubMed]

27. Wang, F.; Zhang, L.; Zhang, F.; Wang, J.; Wang, Y.; Man, D. First trimester serum PIGF is associated with placenta accreta. *Placenta* **2020**, *101*, 39–44. [CrossRef] [PubMed]

28. Faraji, A.; Akbarzadeh-Jahromi, M.; Bahrami, S.; Gharamani, S.; Raeisi Shahraki, H.; Kasraeian, M.; Vafaei, H.; Zare, M.; Asadi, N. Predictive value of vascular endothelial growth factor and placenta growth factor for placenta accreta spectrum. *J. Obstet. Gynaecol. J. Inst. Obstet. Gynaecol.* **2022**, *42*, 900–905. [CrossRef]

29. Zhang, F.; Gu, M.; Chen, P.; Wan, S.; Zhou, Q.; Lu, Y.; Li, L. Distinguishing placenta accreta from placenta previa via maternal plasma levels of sFlt-1 and PLGF and the sFlt-1/PLGF ratio. *Placenta* **2022**, *124*, 48–54. [CrossRef]

30. Al-Khan, A.; Youssef, Y.H.; Feldman, K.M.; Illsley, N.P.; Remache, Y.; Alvarez-Perez, J.; Mannion, C.; Alvarez, M.; Zamudio, S. Biomarkers of abnormally invasive placenta. *Placenta* **2020**, *91*, 37–42. [CrossRef]

31. Ozler, S.; Oztas, E.; Guler, B.G.; Caglar, A.T. Increased levels of serum IL-33 is associated with adverse maternal outcomes in placenta previa accreta. *J. Matern. Fetal Neonatal Med.* **2021**, *34*, 3192–3199. [CrossRef] [PubMed]

32. Kohan-Ghadr, H.-R.; Kadam, L.; Jain, C.; Armant, D.R.; Drewlo, S. Potential role of epigenetic mechanisms in regulation of trophoblast differentiation, migration, and invasion in the human placenta. *Cell Adh. Migr.* **2016**, *10*, 126–135. [CrossRef] [PubMed]

33. Kannampuzha, S.; Ravichandran, M.; Mukherjee, A.G.; Wanjari, U.R.; Renu, K.; Vellingiri, B.; Iyer, M.; Dey, A.; George, A.; Gopalakrishnan, A.V. The mechanism of action of non-coding RNAs in placental disorders. *Biomed. Pharmacother.* **2022**, *156*, 113964. [CrossRef]

34. Murrieta-Coxca, J.M.; Barth, E.; Fuentes-Zacarias, P.; Gutiérrez-Samudio, R.N.; Groten, T.; Gellhaus, A.; Köninger, A.; Marz, M.; Markert, U.R.; Morales-Prieto, D.M. Identification of altered miRNAs and their targets in placenta accreta. *Front. Endocrinol.* **2023**, *14*, 1021640. [CrossRef] [PubMed]

35. Chen, S.; Pang, D.; Li, Y.; Zhou, J.; Liu, Y.; Yang, S.; Liang, K.; Yu, B. Serum miRNA biomarker discovery for placenta accreta spectrum. *Placenta* **2020**, *101*, 215–220. [CrossRef]

36. Yang, T.; Li, N.; Hou, R.; Qiao, C.; Liu, C. Development and validation of a four-microRNA signature for placenta accreta spectrum: An integrated competing endogenous RNA network analysis. *Ann. Transl. Med.* **2020**, *8*, 919. [CrossRef] [PubMed]

37. Rossi-Herring, G.; Belmonte, T.; Rivas-Urbina, A.; Benítez, S.; Rotllan, N.; Crespo, J.; Llorente-Cortés, V.; Sánchez-Quesada, J.L.; de Gonzalo-Calvo, D. Circulating lipoprotein-carried miRNome analysis reveals novel VLDL-enriched microRNAs that strongly correlate with the HDL-microRNA profile. *Biomed. Pharmacother.* **2023**, *162*, 114623. [CrossRef]

38. Timofeeva, A.V.; Fedorov, I.S.; Sukhova, Y.V.; Ivanets, T.Y.; Sukhikh, G.T. Prediction of Early- and Late-Onset Pre-Eclampsia in the Preclinical Stage via Placenta-Specific Extracellular miRNA Profiling. *Int. J. Mol. Sci.* **2023**, *24*, 8006. [CrossRef]

39. Coenen-Stass, A.M.L.; Pauwels, M.J.; Hanson, B.; Martin Perez, C.; Conceição, M.; Wood, M.J.A.; Mäger, I.; Roberts, T.C. Extracellular microRNAs exhibit sequence-dependent stability and cellular release kinetics. *RNA Biol.* **2019**, *16*, 696–706. [CrossRef]

40. Boon, R.A.; Vickers, K.C. Intercellular transport of microRNAs. *Arterioscler. Thromb. Vasc. Biol.* **2013**, *33*, 186–192. [CrossRef]

41. Kanninen, T.; Jung, E.; Gallo, D.M.; Diaz-Primera, R.; Romero, R.; Gotsch, F.; Suksai, M.; Bosco, M.; Chaiworapongsa, T. Soluble suppression of tumorigenicity-2 in pregnancy with a small-for-gestational-age fetus and with preeclampsia. *J. Matern. Fetal Neonatal Med.* **2023**, *36*, 2153034. [CrossRef] [PubMed]

42. Wei, X.; Yang, X. The central role of natural killer cells in preeclampsia. *Front. Immunol.* **2023**, *14*, 1009867. [CrossRef] [PubMed]

43. Dosil, S.G.; Lopez-Cobo, S.; Rodriguez-Galan, A.; Fernandez-Delgado, I.; Ramirez-Huesca, M.; Milan-Rois, P.; Castellanos, M.; Somoza, A.; Gómez, M.J.; Reyburn, H.T.; et al. Natural killer (NK) cell-derived extracellular-vesicle shuttled microRNAs control T cell responses. *Elife* **2022**, *11*, e76319. [CrossRef]

44. Gao, T.; Deng, M.; Wang, Q. MiRNA-320a inhibits trophoblast cell invasion by targeting estrogen-related receptor-gamma. *J. Obstet. Gynaecol. Res.* **2018**, *44*, 756–763. [CrossRef]

45. Xie, N.; Jia, Z.; Li, L. miR-320a upregulation contributes to the development of preeclampsia by inhibiting the growth and invasion of trophoblast cells by targeting interleukin 4. *Mol. Med. Rep.* **2019**, *20*, 3256–3264. [CrossRef] [PubMed]

46. Liao, G.; Cheng, D.; Li, J.; Hu, S. Clinical significance of microRNA-320a and insulin-like growth factor-1 receptor in early-onset preeclampsia patients. *Eur. J. Obstet. Gynecol. Reprod. Biol.* **2021**, *263*, 164–170. [CrossRef]

47. Yuan, Y.; Zhao, L.; Wang, X.; Lian, F.; Cai, Y. Ligustrazine-induced microRNA-16-5p inhibition alleviates preeclampsia through IGF-2. *Reproduction* **2020**, *160*, 905–917. [CrossRef]

48. Rekowska, A.K.; Obuchowska, K.; Bartosik, M.; Kimber-Trojnar, Ż.; Słodzińska, M.; Wierzchowska-Opoka, M.; Leszczyńska-Gorzelak, B. Biomolecules Involved in Both Metastasis and Placenta Accreta Spectrum-Does the Common Pathophysiological Pathway Exist? *Cancers* **2023**, *15*, 2618. [CrossRef]

49. Kim, J.-H.; Lee, K.-S.; Lee, D.-K.; Kim, J.; Kwak, S.-N.; Ha, K.-S.; Choe, J.; Won, M.-H.; Cho, B.-R.; Jeoung, D.; et al. Hypoxia-responsive microRNA-101 promotes angiogenesis via heme oxygenase-1/vascular endothelial growth factor axis by targeting cullin 3. *Antioxid. Redox Signal.* **2014**, *21*, 2469–2482. [CrossRef]

50. Tokarski, M.; Cierzniak, A.; Baczynska, D. Role of hypoxia on microRNA-dependant regulation of HGFA—HGF—c-Met signalling pathway in human progenitor and mature endothelial cells. *Int. J. Biochem. Cell Biol.* **2022**, *152*, 106310. [CrossRef]

51. Gerritsen, M.E. HGF and VEGF: A dynamic duo. *Circ. Res.* **2005**, *96*, 272–273. [CrossRef] [PubMed]

52. Morishita, R.; Aoki, M.; Hashiya, N.; Yamasaki, K.; Kurinami, H.; Shimizu, S.; Makino, H.; Takesya, Y.; Azuma, J.; Ogihara, T. Therapeutic angiogenesis using hepatocyte growth factor (HGF). *Curr. Gene Ther.* **2004**, *4*, 199–206. [CrossRef]

53. Huang, Y.; Xu, M.; Jing, C.; Wu, X.; Chen, X.; Zhang, W. Extracellular vesicle-derived miR-320a targets ZC3H12B to inhibit tumorigenesis, invasion, and angiogenesis in ovarian cancer. *Discov. Oncol.* **2021**, *12*, 51. [CrossRef]

54. Zhang, Z.; Li, X.; Sun, W.; Yue, S.; Yang, J.; Li, J.; Ma, B.; Wang, J.; Yang, X.; Pu, M.; et al. Loss of exosomal miR-320a from cancer-associated fibroblasts contributes to HCC proliferation and metastasis. *Cancer Lett.* **2017**, *397*, 33–42. [CrossRef] [PubMed]

55. Hao, X.; Xin, R.; Dong, W. Decreased serum exosomal miR-320a expression is an unfavorable prognostic factor in patients with hepatocellular carcinoma. *J. Int. Med. Res.* **2020**, *48*, 300060519896144. [CrossRef] [PubMed]

56. Zhang, N.; Wang, Y.; Liu, H.; Shen, W. Extracellular vesicle encapsulated microRNA-320a inhibits endometrial cancer by suppression of the HIF1α/VEGFA axis. *Exp. Cell Res.* **2020**, *394*, 112113. [CrossRef]

57. Jiang, X.; Yang, J.; Lin, Y.; Liu, F.; Tao, J.; Zhang, W.; Xu, J.; Zhang, M. Extracellular vesicles derived from human ESC-MSCs target macrophage and promote anti-inflammation process, angiogenesis, and functional recovery in ACS-induced severe skeletal muscle injury. *Stem Cell Res. Ther.* **2023**, *14*, 331. [CrossRef] [PubMed]

58. Pan, Q.; Xu, X.; He, W.; Wang, Y.; Xiang, Z.; Jin, X.; Tang, Q.; Zhao, T.; Ma, X. Enrichment of miR-17-5p enhances the protective effects of EPC-EXs on vascular and skeletal muscle injury in a diabetic hind limb ischemia model. *Biol. Res.* **2023**, *56*, 16. [CrossRef]

59. Chen, S.; Chen, X.; Luo, Q.; Liu, X.; Wang, X.; Cui, Z.; He, A.; He, S.; Jiang, Z.; Wu, N.; et al. Retinoblastoma cell-derived exosomes promote angiogenesis of human vesicle endothelial cells through microRNA-92a-3p. *Cell Death Dis.* **2021**, *12*, 695. [CrossRef]

60. Tetreault, M.-P.; Yang, Y.; Katz, J.P. Krüppel-like factors in cancer. *Nat. Rev. Cancer* **2013**, *13*, 701–713. [CrossRef]

61. Langmead, B.; Trapnell, C.; Pop, M.; Salzberg, S.L. Ultrafast and memory-efficient alignment of short DNA sequences to the human genome. *Genome Biol.* **2009**, *10*, R25. [CrossRef] [PubMed]

62. Team, R.C. A Language and Environment for Statistical Computing. R Foundation for Statistical Computing, Vienna, Austria. Available online: https://www.r-project.org (accessed on 10 March 2021).

63. Love, M.I.; Huber, W.; Anders, S. Moderated estimation of fold change and dispersion for RNA-seq data with DESeq2. *Genome Biol.* **2014**, *15*, 550. [CrossRef] [PubMed]

64. RS Team. RStudio: Integrated Development for R. RStudio. Available online: http://www.rstudio.com/ (accessed on 23 March 2021).

International Journal of
Molecular Sciences

Review

Molecular Determinants of Uterine Receptivity: Comparison of Successful Implantation, Recurrent Miscarriage, and Recurrent Implantation Failure

Veronika Günther [1,2], Leila Allahqoli [3], Anupama Deenadayal-Mettler [2], Nicolai Maass [1], Liselotte Mettler [2], Georgios Gitas [4], Kristin Andresen [1], Melanie Schubert [1], Johannes Ackermann [1], Sören von Otte [2] and Ibrahim Alkatout [1,*]

1   Department of Obstetrics and Gynecology, University Hospitals Schleswig-Holstein, Campus Kiel, Arnold-Heller-Strasse 3 (House C), 24105 Kiel, Germany; veronika.guenther@uksh.de (V.G.)
2   University Fertility Center, Ambulanzzentrum of University Hospitals Schleswig-Holstein, Campus Kiel, Arnold-Heller-Strasse 3 (House C), 24105 Kiel, Germany
3   School of Public Health, Iran University of Medical Sciences (IUMS), Tehran 14535, Iran
4   Private Gynecologic Practice, Chrisostomou Smirnis 11B, 54622 Thessaloniki, Greece
*   Correspondence: ibrahim.alkatout@uksh.de; Tel.: +49-(0)431-500-21401

Abstract: Embryo implantation is one of the most remarkable phenomena in human reproduction and is not yet fully understood. Proper endometrial function as well as a dynamic interaction between the endometrium itself and the blastocyst—the so-called embryo–maternal dialog—are necessary for successful implantation. Several physiological and molecular processes are involved in the success of implantation. This review describes estrogen, progesterone and their receptors, as well as the role of the cytokines interleukin (IL)-6, IL-8, leukemia inhibitory factor (LIF), IL-11, IL-1, and the glycoprotein glycodelin in successful implantation, in cases of recurrent implantation failure (RIF) and in cases of recurrent pregnancy loss (RPL). Are there differences at the molecular level underlying RIF or RPL? Since implantation has already taken place in the case of RPL, it is conceivable that different molecular biological baseline situations underlie the respective problems.

Keywords: implantation; uterine receptivity; cytokines; recurrent implantation failure; recurrent pregnancy loss

Citation: Günther, V.; Allahqoli, L.; Deenadayal-Mettler, A.; Maass, N.; Mettler, L.; Gitas, G.; Andresen, K.; Schubert, M.; Ackermann, J.; von Otte, S.; et al. Molecular Determinants of Uterine Receptivity: Comparison of Successful Implantation, Recurrent Miscarriage, and Recurrent Implantation Failure. *Int. J. Mol. Sci.* **2023**, *24*, 17616. https://doi.org/10.3390/ijms242417616

Academic Editor: Giovanni Tossetta

Received: 1 November 2023
Revised: 6 December 2023
Accepted: 6 December 2023
Published: 18 December 2023

1. Introduction

Embryo implantation is one of the most remarkable phenomena in human reproduction and is not yet fully understood. Proper endometrial function as well as a dynamic interaction between the endometrium and the blastocyst—the so-called embryo–maternal dialog—is necessary for successful implantation. Several physiological and molecular processes are responsible for this success. The time frame is the so-called window of implantation (WOI), which is the specific time slot during which the endometrium is of an appropriate structure and receptive for implantation. This period is usually between the 20th and the 24th days of the menstrual cycle, during the secretory phase [1]. The adhesion of the blastocyst occurs during the implantation phase. The adhesion molecules are the generic term for a whole family of molecules that fulfill functions in various areas of the reproductive system [2]. Cadherins, integrins, trophinin, and selectin are just a few examples of adhesion molecules, which are responsible for the adhesion and the required physical interaction between the endometrium and the blastocyst [2,3]. There might be cell–cell and cell–extracellular matrix interactions leading to a successful implantation [3]. The adhesion is followed by the invasion of trophoblast cells surrounded by an immune-modulated environment, tolerating the embryo even though half of the genes are foreign to the maternal organism because they are of paternal origin [4,5]. Only a downregulation of the maternal immune system, which affects different cell populations such as the natural killer cells or T helper cells, makes successful implantation possible [5]. Pinopodes are small microvilli on the apical surface of the epithelial cells that appear on the 20th and 21st days of the cycle (may vary by 5 days) and characterize endometrial receptivity [1]. However, this receptivity is complex; it is a result of several physiological and molecular mechanisms [6].

In the following, implantation and its molecular processes, and especially the involved cytokines, will be described in detail in three different situations: successful implantation, recurrent implantation failure (RIF), and recurrent pregnancy loss (RPL). RIF is defined as the failure to achieve a pregnancy after three or more consecutive in vitro attempts, with the transfer of at least four high-quality embryos in three fresh or frozen cycles [4,7]. RPL is defined as three consecutive pregnancy losses prior to 20 weeks of gestation. The American Society for Reproductive Medicine defines RPL as the failure to achieve pregnancy even after two or more pregnancy losses with clinical (ultrasonography or histopathology) evidence of pregnancy [4,8]. There are a number of causes that are discussed to be associated with RIF or RPL, such as anatomical factors, chromosomal causes, or thrombophilia. Table 1 provides an overview of these possible factors.

Table 1. Possible causes for the occurrence of RIF or RPL [9–11].

Genetic factors	Chromosomal aberrations De novo chromosomal abnormalities of the embryo or certain gene polymorphisms
Anatomical factors	Uterine malformation Adhesions Polyp Submucosal fibroid
Microbiological factors	Bacterial vaginosis
Chronic endometritis	Factor V Leiden mutation Antithrombin deficiency Prothrombin mutation Protein C deficiency Protein S deficiency
Thrombophilia	Antiphospholipid syndrome
Endocrine disorders	Hyperthyroidism Hypothyroidism Diabetes mellitus Hyperandrogenemia Hyperprolactinemia PCOS Luteal phase defects
Immunological causes	Cytokine levels Natural killer cells (uterine and peripheral) T-helper cell type 1/type 2 quotient KIR receptors HLA antibodies
Lifestyle	Overweight Underweight Increased stress Alcohol consumption Nicotine consumption
Idiopathic	

Concerning the implantation process, there are several cytokines involved, such as: IL-6, IL-8, leukemia inhibitory factor, IL-11, IL-1, and the glycoprotein glycodelin. Accordingly, an imbalance could occur in the event of RIF or RPL. Are there differences at the molecular level underlying RIF or RPL? Since implantation has already taken place in the case of RPL, it is conceivable that different baseline states of molecular biology underlie the respective problems. In the following, we will first describe the molecular factors of successful implantation and early pregnancy development, and then explain the respective molecular situation in RIF and RPL. By accurately analyzing the particular pathomechanism at the molecular level, it may be possible to develop strategies that will help, in the clinical setting, to devise a suitable therapeutic approach for the individual patient with RIF or RPL and achieve a successful pregnancy. Furthermore, this review will focus on estrogen, progesterone, and their receptors but will not take the ovarian function, including FSH, LH, and AMH and their relation to pregnancy, into account.

2. Estrogen, Progesterone, and Their Receptors

Estrogen and progesterone and their respective receptors are among the first factors that influence the uterus and prepare the uterine environment for implantation. During the menstrual cycle, the estrogen level rises with the growing follicle, reaches its peak during ovulation, and then decreases. Alongside the falling estrogen level, progesterone rises and is the leading hormone during the luteal phase, inducing decidualization and opening the window of implantation (WOI) [12]. If pregnancy occurs, the progesterone level remains elevated. The WIO is characterized by balanced estrogen and progesterone levels, leading to a proper proliferated and transformed endometrium for blastocyst invasion. Any disruptions of this well-balanced system results in a failed implantation of the blastocyst.

However, not only estrogen and progesterone levels are important for implantation; their receptors also appear to play an important role in this process. Estrogen receptors exist in two forms: ERα and ERβ. Both are expressed in the endometrium but have different functions. Studies on knockout mice have addressed these respective functions. ERα knockout mice show endometrial hypoplasia and are infertile. Thus, ERα appears to be essential for implantation [13]. In contrast, ERβ knockout mice have a normal endometrium and are fertile, suggesting that ERβ is involved in other aspects of endometrial function [14]. During the proliferative phase, estrogen (through ERα) causes the progesterone receptor (PR) in endometrial cells to induce progesterone responsiveness during the luteal phase. As a negative feedback, progesterone inhibits ERα expression for correct endometrial function. The effects of progesterone in endometrial cells are mediated by the progesterone receptor, which exists in two isoform: PR-A and PR-B. Progesterone effects mediated by PR-A appear to be responsible for correct implantation, pregnancy, and parturition. This has been shown in PR-B knockout mice. The effect of progesterone is mediated by PR-B in PR-A knockout mice, and, here, scientists have found endometrial hyperplasia, inflammation, and the absence of decidualization of the endometrium [15]. Knockout mice for both PR-A and PR-B are infertile, showing severely reduced or no further ovulation, uterine hyperplasia, the absence of decidualization, severely limited mammary gland development, and impaired sexual behavior [16]. These study data can be extrapolated to humans, demonstrating the importance of progesterone receptors and their different functions in fertility.

2.1. Estrogen and RPL

Rising estrogen levels during follicle growth are responsible for endometrial proliferation, myometrium thickness, and increased blood supply: all of these are factors for successful implantation. Estradiol levels during early pregnancy can reflect the quality of the dominant follicle and the function of the corpus luteum as well as help in maintaining the corpus luteum [17]. Furthermore, estradiol appears to be an important factor in preserving early pregnancy [18]. In the 4 to 8 weeks of pregnancy, a positive correlation was found between serum estradiol level and gestational age. Serum estradiol levels were significantly lower in pregnant women with abortion than in those with normal pregnancy [17]. Deng et al. analyzed serum levels of estradiol, progesterone, and human chorionic gonadotropin (hCG) in 165 women during 9 weeks of gestation. A total of 71 women had a miscarriage, whereas 94 women had a normal pregnancy. Low levels of estradiol, progesterone, and hCG were associated with a miscarriage in the first trimester. The authors concluded that estradiol and progesterone or estradiol alone at 7–9 weeks and hCG or progesterone in combination with estradiol at 5–6 weeks of gestation can be used to predict miscarriage [19].

2.2. Progesterone and RPL

A deficiency of progesterone itself and an aberrant PR-mediated signaling may play a role in RPL. As mentioned above, progesterone induces decidualization, opens the WOI, and maintains pregnancy. A progesterone deficiency and a shortened luteal phase may result in suboptimal endometrial development, which has been associated with RPL [20]. On the other hand, it has been stated that a progesterone deficiency is rather an expression of a previously insufficient follicular phase, which should be optimized in advance. While the effect of progesterone through PRs plays an important role in implantation and the maintenance of pregnancy, progesterone supplementation in patients with sporadic miscarriage does not appear to improve the outcome of pregnancy [20]. In a randomized controlled trial in the UK, the authors analyzed 4153 women with vaginal bleeding during early pregnancy, who were receiving either progesterone or placebo. The progesterone therapy did not result in a significantly higher rate of live births among women with threatened miscarriage overall, but women with early-pregnancy bleeding and previous miscarriages in the past had a higher number of live births [21]. The authors of a meta-analysis conducted in 2021 reached similar conclusions: again, vaginal micronized progesterone resulted in a higher live birth rate in female patients with a history of vaginal bleeding and miscarriage(s) in the past [22].

2.3. Estrogen, Progesterone, and RIF

Estrogen plays a crucial role in endometrial receptivity due to the initiation of paracrine or autocrine signaling [23]. Higher or lower estrogen levels in the periconceptional period lead to lower pregnancy rates, as well as in natural cycles and also in assisted reproductive technology (ART) cycles. In naturally conceived pregnancies, low estrogen concentrations are associated with non-conception cycles [23]. On the other hand, excessive supraphysiologic estrogen levels at the time of the luteinizing hormone (LH) peak correlate with lower live birth rates and a higher risk of pregnancy complications [23]. In very early pregnancy, estrogen plays an important role in placentation by modulating the angiogenic factor expression and causing an immune-tolerant environment by influencing the concentration of a uterine natural killer and T helper cells, which are important for implantation [23].

In women with RIF undergoing ART treatment, hormonal stimulation leads to supraphysiological estrogen levels and possibly to a premature increase in progesterone. The endometrium is exposed to the premature impact of progesterone, which causes asynchronicity between the embryo and the endometrium, and leads to suboptimal implantation conditions [1]. A premature rise in progesterone before ovulation can be detected by a laboratory analysis. Consequently, no embryo transfer but a freezing of all fertilized eggs should take place.

Furthermore, a decrease in ER and a polymorphism in estrogen receptor 1 was noticed in patients with a recurrent implantation failure [24,25].

3. Molecular and Cellular Events Involved in Successful Embryo Implantation

Interleukins, interferons, chemokines, and numerous other mediators can be summarized as cytokines which are produced by different effector cells and play an important role in the human innate and adaptive immune systems [26]. Based on their properties, cytokines may be pro- or anti-inflammatory, as shown in Table 2. The fragile balance and complicated relationship of cytokines are caused by their overlapping biological activities; any alteration of one cytokine is likely to affect the others.

Table 2. Pro- and anti-inflammatory cytokines.

Pro-inflammatory cytokines	IFN-γ IL-2 TNF-α IL-1β IL-6 IL-8 IL-17 IL-12
Anti-inflammatory cytokines	IL-4 IL-5 IL-9 IL-10 IL-11 IL-13 TGF-β1 LIF

Abbreviations: INF-γ: interferon-γ, TNF-α: tumor necrosis factor-α, TGF-β1: transforming growth factor-β1.

The interleukin (IL)-6 family is a class of cytokines consisting of IL-6, IL-11, ciliary neurotrophic factor, leukemia inhibitor factor (LIF), oncostatin M, cardiotrophin 1, cardiotrophin-like cytokine, and IL-27. They belong to one family because the receptor complex of each cytokine contains two (IL-6 and IL-11) or one molecule (all other cytokines) of the signaling receptor subunit gp130 [27]. In this review, we will focus on IL-6, LIF, and IL-11 because they play key roles in the implantation of the embryo [1]. Furthermore, we will analyze IL-8, IL-1, glycodelin, and their roles in successful implantation, as well as RPL and RIF.

Figure 1 summarizes the above-mentioned molecular determinants and their functions in regular implantation and pregnancy. These components will be described in the following.

IL-6 is a pleiotropic cytokine and is responsible for many physiological processes. Its role in the immune response, inflammation, and metabolic regulation is known. However, it is also indispensable for the development of early pregnancy [28]. IL-6 is synthesized by macrophages, fibroblasts, epithelial cells, and placental trophoblasts, and it is found in high concentrations in the luteal phase, especially in the receptive window. It is responsible for the development of the placenta and pregnancy itself by regulation of trophoblast invasion and spiral artery remodeling [29]. IL-6 interacts by binding to its receptors and subsequently activating the Janus kinase/signal transducer and activator of the transcription (JAK/STAT) pathway [28]. The classical signal transduction of IL-6 is induced by the binding of IL-6 to its specific membrane IL-6α-receptor (IL-6R). This pathway is confined to a few tissues because of the restricted expression of IL-6R [30].

In the first trimester of pregnancy, different cell populations of the uteroplacental tissues (decidual stromal cells and different populations of immune cells of the decidua and syncytiotrophoblast, extravillous trophoblast, and cytotrophoblast cells of the placenta) express IL-6 [1]. IL-6 expression increases with gestational age [1].

IL-8 is a proinflammatory chemokine produced by immune cells and other cells under inflammatory conditions, and it is responsible for the attraction of neutrophils in cases of inflammation, monocyte-macrophage growth and their differentiation, endothelial cell survival, proliferation, and angiogenesis [31–33]. Furthermore, IL-8 contributes to endometrial receptivity and maintains the dialog between the embryo and the human endometrium during implantation [34]. The upregulation of endometrial IL-8 mRNA occurs in the receptive phase of the menstrual cycle and also in the presence of an embryo [35]. IL-8 appears to stimulate progesterone secretion in order to maintain pregnancy [36].

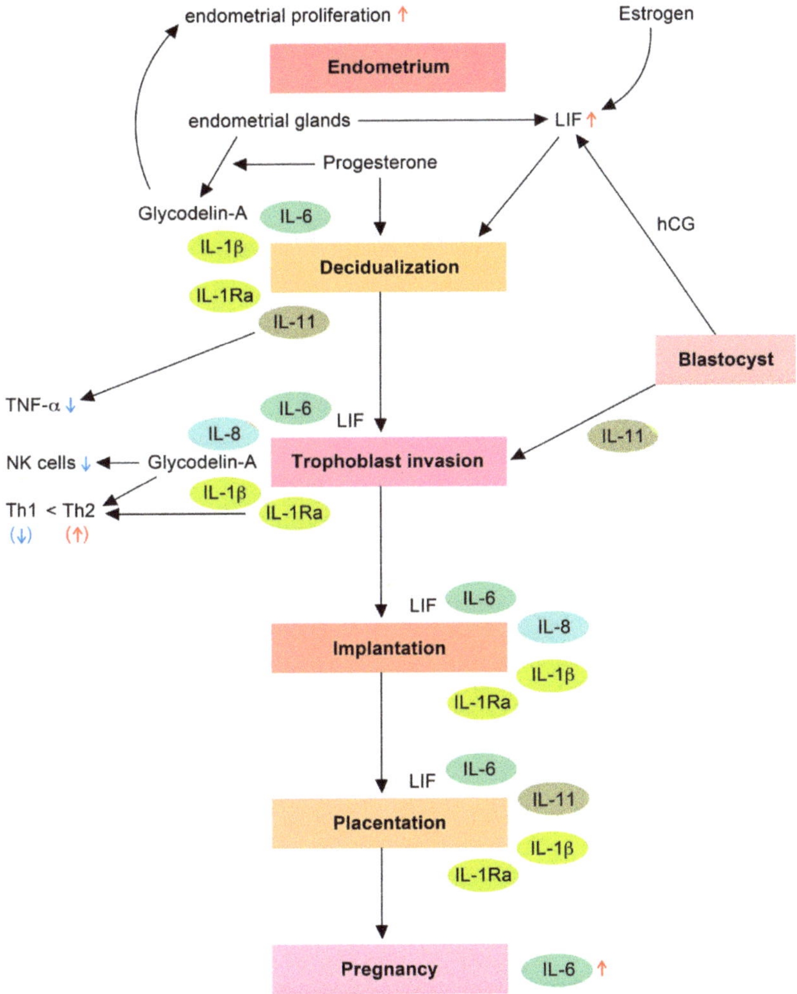

Figure 1. Function of hormones and molecular determinants in implantation and early pregnancy. Progesterone, LIF, IL-6, IL-11, IL-1β, and IL-1Ra are involved in regulating decidualization. LIF is increased under the influence of estrogen and hCG. LIF, IL-6, IL-8, glycodelin-A, IL-1β, and IL-1Ra are responsible for trophoblast invasion. As regards shifts in the immune system, IL-11 reduces TNF-α, glycodelin-A lowers the levels of NK cells, while glycodelin-A, IL-1β, and IL-1Ra are responsible for the shift within Th cells. LIF, IL-6, IL-8, IL-1β, and IL-1Ra are associated with embryo implantation. Furthermore, LIF, IL-6, IL-11, IL-1β, and IL-1Ra participate in placentation. Abbreviations: LIF (leukemia inhibitory factor), TNF-α (tumor necrosis factor-α), NK cells (natural killer cells), Th1/2 (type 1 T helper cells and type 2 T helper cells), IL-1Ra (IL-1 receptor antagonist), ↓ (down-regulated), ↑ (up-regulated).

3.1. IL-6, IL-8, and RPL

Several studies have shown a higher risk of pregnancy loss in cases of inadequate expression of IL-6 and IL-8 in the feto–maternal interface [37–39]. Interestingly, IL-6/IL-8 levels can be too low or too high, which could be responsible for sporadic abortion or recurrent pregnancy loss, respectively. In cases of sporadic abortion, decidual macrophages and decidual natural killer cells produce less IL-6 and IL-8 compared to the corresponding cells in normal pregnancy, which leads to suboptimal IL-6 and IL-8 levels and results in inadequate trophoblast invasion and spiral artery remodeling, and, thus, a higher risk of early pregnancy loss [37].

On the other hand, increased levels of IL-6 and IL-8 appear to be related to RPL [39]. Increased IL-6 and IL-8 levels in decidual tissues indicate a pro-inflammatory state at the feto–maternal interface, which could be disturbing to the implanted embryo and lead to an abortion. Previous findings have indicated that both insufficient and excessive levels of IL-6/IL-8 disturb the inflammatory network at the feto–maternal interface, which may compromise the pregnancy. Furthermore, differences in the expression profile of IL-6 and IL-8 in reproductive tissues between spontaneous pregnancy loss and RPL support the hypothesis that these complications may have a completely different pathogenetic background [28,40].

3.2. IL-6 and RIF

A small number of studies with limited case numbers have been focused on the relationship between cytokines and RIF [41–43].

Liang et al. analyzed the balance state of pro- and anti-inflammatory cytokines in the plasma of patients with RIF [41]. A total of 34 patients with RIF and 25 women with successful pregnancies were included. The interferon (IFN)-γ, IL-1β, IL-6, and IL-4 concentrations were higher, whereas the transforming growth factor (TGF)-β1 concentration was lower in the RIF group than in the control group [41].

Ozgu-Erdinc et al. reported different data regarding IL-6 and pregnancy outcome. In 129 patients, the authors correlated high-sensitivity C-reactive protein and IL-6 with pregnancy outcomes after in vitro fertilization (IVF)/intracytoplasmic sperm injection (ICSI). Serum levels were measured at the beginning of the IVF/ICSI cycle. No differences in C-reactive protein or IL-6 concentrations were seen on comparing implantation/no implantation, clinical pregnancy, miscarriage, and live birth [43].

3.3. IL-6, IL-8, and RIF

A study group in Beijing analyzed inflammatory cytokine levels in serum samples of women undergoing IVF [42]. A total of 84 women were included, of whom 46 conceived and 38 did not. Serum samples were taken on the second day of the menstrual cycle before the treatment cycle, and the inflammatory cytokines (interleukin-1β, interleukin-6, interleukin-8, and monocyte chemotactic protein-1) were measured. Women who did not become pregnant had significantly higher serum IL-8 levels than those who achieved a pregnancy [42]. Furthermore, a dose–response correlation between serum IL-8 levels and the risk of IVF-ET failure was noted, especially when the IL-8 concentration was >11.2 pg/mL [42].

Leukemia inhibitory factor (LIF) also belongs to the IL-6 family and is a pleiotropic cytokine. IL-6 in the uterus activates the Janus kinase (JAK)-signal transducer and activator of the transcription protein (STAT) pathway, and, therefore, phosphorylates STAT3, whose activation is necessary for implantation [15,44,45]. LIF is produced by the endometrial glands in order to make the endometrium receptive for blastocyst attachment [46]. The highest concentrations are registered during blastocyst formation immediately before implantation [47]. The rising estrogen levels during the menstrual cycle may be a stimulator of LIF expression [48]. LIF secretion is influenced by human chorionic gonadotropin as well as by the male seminal fluid [49,50]. Additionally, LIF is responsible for the decidualization of the stroma for implantation and, finally, for placenta development. During embryo implantation, LIF maintains the development of pinopodes and is, thus, responsible for embryo attachment, which is followed by invasion [51]. Furthermore, LIF is important for the formation of maternal decidua and trophoblast giant cells, as well as for the interaction of feto–maternal blood vessels [46].

3.4. LIF and RPL

Again, a small number of studies with limited case numbers have addressed the relationship between LIF and RPL. A case–control study comprised 30 fertile women and 30 suffering from RPL. Endometrium samples were taken in order to investigate the expression of LIF. The authors found a statistically higher mRNA expression of LIF in women with RPL [52].

Xu et al. came to a different conclusion [53]: the authors analyzed endometrial biopsy samples from women with RPL and compared the expression of LIF, integrin-ß3, and mucin-1, as well as the pinopode morphology, with those of normal fertile women. No differences were seen between the two groups with regard to any of the studied parameters [53].

A further recent study from Iran analyzed the role of LIF and inflammatory cytokines in women with RPL compared to a control group of fertile women [54]. Gene expression levels of LIF, tumor necrosis factor-alpha (TNF-α), and interleukin-17 were measured in peripheral blood. The mRNA levels of LIF were significantly lower in women with RPL compared to the fertile control group ($p = 0.003$) [54]. Regarding the other cytokines, no significant difference was seen between the two groups ($p \geq 0.05$) [54].

LIF is regarded as a characteristic biomarker of endometrial receptivity. High expression of LIF appears to be an indicator of a receptive endometrium in fertile women. Nevertheless, the three studies mentioned above obtained different results concerning LIF concentrations in women with RPL. Higher, lower, and no differences in LIF expression were reported. Further studies with larger case numbers will be needed to make conclusive statements.

3.5. LIF and RIF

One of the first studies analyzing the correlation between LIF and RIF was performed in 1998 by Hambartsoumian [55]. Women with unexplained infertility and RIF were compared with fertile women as a control group. Endometrial samples were taken in the proliferative and secretory phases. Fertile women showed an elevation of LIF secretion in the secretory phase compared to the proliferative phase (2.2 times higher) [55]. In contrast, infertile women had no elevation of cytokine production. Furthermore, in infertile women with RIF, the LIF level fell in the secretory phase [55]. When the quantity of cytokine secretion was compared on the same day of the cycle between the two groups of women, LIF production in fertile women on days 18–21 of the menstrual cycle was 3.5 times higher than that in infertile women with RIF and 2.2 times higher than LIF production in women without RIF ($p < 0.01$ and $p < 0.05$, respectively) [55]. On days 8–11 of the cycle, the levels of LIF in these groups did not differ significantly. However, the distribution of cytokines varied in infertile women: the highest amplitude of variations was seen in patients with RIF. The author concluded that the majority of infertile women (especially those with RIF) have a dysregulation of LIF production in the endometrium during both the proliferative and the secretory phases of the cycle [55].

Wu et al. confirmed the above-mentioned data by analyzing the expression of LIF in endometrial samples of fertile women and women with RIF [56]. Fertile women had a moderate expression of LIF in the proliferative phase and a high expression of LIF in the secretory phase. In addition to the previous knowledge that LIF is decreased in the secretory phase, women with RIF had lower LIF levels in the proliferative phase compared to fertile women [56]. The data suggest that low LIF in the proliferation phase may also be partly responsible for repeated implantation failure.

Another interesting approach concerning the molecular causes of LIF downregulation in patients with RIF was described by a Chinese study group [57]. Women with a history of RIF showed abnormally increased levels of Krüppel-like factor 12 (KLF12), which in turn leads to reduced LIF expression with negative effects on endometrial receptivity, embryo adhesion, and implantation [57]. After analyzing the LIF gene sequence, the authors concluded that KL12 can directly suppress LIF expression, as it can bind to two KLF12 binding sites in the promoter of the LIF gene [57]. Interestingly, the authors also showed that KLF12 expression could be significantly reduced in vitro when Ishikawa cells overexpressing

KLF12 are treated with medroxyprogesterone acetate (MPA). Consequently, progesterone may be a novel therapeutic approach for patients with RIF, as it acts upstream of both LIF and KLF12, inhibiting KLF12 expression and, thus, promoting LIF secretion [57,58].

Interleukin-11 (IL-11) is a pleiotropic cytokine with anti-inflammatory functions [59]. It is produced by stromal and epithelial cells and maintains adequate decidualization [60]. The function of IL-11 is mediated by the IL-11 receptor (IL-11Rα), which is expressed by the luminal and glandular epithelia [60]. Unlike IL-11, which is at its highest concentration during decidualization, the receptor is not subject to cyclic variation. The different concentrations of IL-11 depend on its production, influenced by steroid hormones, and local factors such as relaxin and PGE_2 [60,61]. Interestingly, there are differences between the stromal and epithelia cells: both express IL-11 but stromal cells produce the highest concentrations during decidualization. In contrast, epithelial cells express IL-11 primarily during the early secretory phase of the cycle. This suggests that epithelial and stromal IL-11 may play different roles in endometrial function [60,62].

In vitro studies have shown that IL-11 leads to a dose-dependent decrease in the production of the proinflammatory cytokine TNF-α. TNF-α production is highest in the secretory phase but then diminishes in early pregnancy. This reduction may be caused by a higher concentration of IL-11 [60]. In contrast, LIF and IL-6 were not found to affect TNF-α production [60].

IL-11Rα knockout mice are infertile, probably as a result of inadequate or absent decidualization and possibly over-invasiveness of the trophoblast [61]. This phenomenon can be explained by the downregulation of a metalloproteinase inhibitor [63].

The embryo itself produces IL-11 during trophoblast invasion, probably in order to influence the endometrium [59]. Interestingly, no receptors for IL-11 have been found so far in the human embryo [61].

3.6. IL-11 and RPL

A study group from the UK analyzed the endometrial peri-implantation biopsies of fertile women and women with RPL [64]. Both groups showed increased expression of IL-11 and IL-11Rα in epithelial cells compared to stromal cells [64]. Furthermore, there was a significant reduction in epithelial cell IL-11, but not stromal cell IL-11, expression in the endometrium of RPL women compared to fertile women. The expression of IL-11Rα protein did not differ between the two groups [64].

3.7. IL-11 and RIF

Karpovich et al. analyzed endometrial stroma cells and their IL-11 expression in women with primary infertility and compared them with a fertile control group [65]. The authors showed that the decidualization of endometrial stromal cells derived from women with primary infertility is defective, and that the production of IL-11 is compromised in these cells [65]. No differences were seen between the two groups in the expression of IL-11Rα in decidualized stromal cells, maintaining the hypothesis that the modulation of IL-11 expression, and not its receptor, is important in controlling the process of decidualization [65]. Thus, defects in endometrial IL-11 signaling may be associated with RIF.

Different results were reported by Sabry et al. The authors obtained endometrial tissue samples of women undergoing IVF and correlated IL-11 expression with pregnancy rates. No differences in gene expression levels of IL-11 and IL-11Rα were seen in women with implantation vs. non-implantation after IVF [66].

The **interleukin-1 (IL-1)** system consists of different components, including IL-1α, IL-1β, IL-1 receptor type 1 (IL-1R1), IL-1 receptor type 2 (decoy receptor, IL-1R2), IL-1 receptor accessory protein, and IL-1 receptor antagonist (IL-1Ra) [67]. IL-1α and IL-1β have proinflammatory properties and appear to play the most important roles in the development of pregnancy [59]. Unlike IL-1α, which is constitutively expressed, IL-1β is inducible [67]. The activity of IL-1β is regulated by IL-1Ra [59]. IL-1β is also an immune system modulator and is produced by dendritic cells, blood monocytes, T cells, and tissue macrophages [67]. With high affinity, IL-1Ra binds to IL-1R1 and, thus, inhibits IL-1α or IL-1β activity [68]. The extent of inflammation is influenced by the levels of circulating IL-1β, IL-1Ra, and IL-1R2 [69]. The expression of the components of the IL-1 system differs during the menstrual cycle, implantation, and early pregnancy [67]. Several studies have shown that the IL-1 system—especially IL-1β and IL-1—are responsible for the embryo–maternal dialog and the immunological shift that is necessary for the successful implantation of the blastocyst. The tissue of the fallopian tubes and myometrial smooth muscle synthesize IL-1β during pregnancy [70]. IL-1Ra has been found in the maternal, fetal, and amniotic fluid compartments, especially the decidua, and increases with rising gestational age in fetal and maternal plasma [71]. IL-1β plays an important role in the decidualization of stromal cells and in successful blastocyst implantation. IL-1β is produced by cytotrophoblast cells from the placenta of the first trimester, while lower expressions are seen in cultures from second- and third-trimester placenta [67].

In addition to its function in pregnancy, IL-1β influences the coagulation system by increasing the expression of several proteins that in turn increase platelet stability and raise the risk of thrombosis. This pro-thrombotic state with a higher risk of microthrombi during early pregnancy may be the reason for miscarriages [72].

3.8. IL-1 and RPL

The immune system plays an important role in implantation and early pregnancy [4]. The T helper (Th) cell populations and their corresponding cytokines have different, even contradictory, functions: whereas type 1 T helper cells (Th1) (including interferon-γ, IL-2, TNF-α) seem to disrupt pregnancy, type 2 T helper cells (Th2) (including IL-4, IL-5, IL-6, IL-10, IL-13) support and maintain the development of early pregnancy [4]. Women with RPL had increased levels of Th1 and low concentrations of Th2 [73]. IL-1β can induce Th2 responses in humans [67,74]. Furthermore, low levels of IL-1β and IL-6 mRNA have been found in the endometrium of women with RPL [67,75].

Löb et al. analyzed IL-1β levels in the decidua and obtained different results. Patients with spontaneous and recurrent miscarriages had a significantly higher expression of IL-1β than the control group with healthy pregnancies [76].

In addition, the IL-1 system appears to influence the outcome of pregnancy in women with the antiphospholipid syndrome. The latter is an autoimmune disease with, inter alia, a hypercoagulable state leading to thrombosis. The patients suffer from RPL and have elevated antiphospholipid antibodies [77]. The activation of inflammasome as well as secretion of IL-1β seem to increase inflammation, which has been proposed as one of the causes of RPL in the antiphospholipid syndrome [78].

3.9. IL-1 and RIF

Several studies have analyzed the relationship between IL-1 concentrations and pregnancy outcomes after IVF in order to describe RIF.

A retrospective investigation performed by Kreines et al. addressed IL-1β and IL-1Ra in maternal serum after IVF [79]. In healthy pregnancies, increasing IL-1β levels were registered over the duration of pregnancy. In cases of a negative pregnancy test or subsequent pregnancy loss, no increase was detected. Interestingly, women who initially had a positive pregnancy test and an abortion a few weeks later had undetectable IL-1β levels. In contrast, no significant data were registered in the serum IL1-Ra analysis [79].

A group from The Netherlands studied endometrial–embryonic interactions by taking endometrial secretions aspirated prior to embryo transfer from 210 women undergoing IVF [80]. Seventeen soluble cytokines, including IL-1β, were correlated with pregnancy outcomes using a multivariate logistic regression analysis. A significant negative association was registered between IL-1β levels and pregnancy outcome: clinical pregnancy was associated with lower levels of IL-1β. The predictive value of IL-1β (and TNF-alpha) for pregnancy was equivalent to, and in addition to, that of embryo quality [80].

These data are consistent with a previous report comparing women with RIF after IVF and controls, which demonstrated significantly higher IL-1β levels in endometrial flushings from women with implantation failure [81].

As mentioned earlier, women with recurrent miscarriage had significantly lower levels of IL-1β mRNA in the endometrium [75]. It is postulated that women with habitual abortion may have an 'over-receptive' endometrium that allows for the implantation of aneuploid embryos instead of leading to implantation failure, resulting in recurrent miscarriages of these embryos [82]. IL-1β levels appear to be related to achieving a clinical pregnancy rather than embryo implantation. This suggests that this cytokine may have a more important role in later stages of implantation rather than initial apposition and adhesion of the embryo [80].

3.10. Glycodelin

Glycodelin is a secretory protein responsible for different processes such as immuno-suppression, fertilization, and implantation. Analogous to its functions, it has a temporospa-tial pattern of expression, primarily in the reproductive tract, where glycodelin is expressed in the mid-secretory phase. The peak occurs 10 days after ovulation and is regulated by progesterone. In addition to the secretory and decidualized endometrium, glycodelin is found in mammary glands as well as in bone marrow, ovaries, fallopian tubes, and seminal vesicles [83].

Depending on the respective attached glycosylation form, glycodelin may be divided into four glycoforms (glycodelin-A, -S, -F, and -C). The respective capital letter describes where each glycoform occurs. In the male reproductive tract, glycodelin-S is secreted by seminal vesicles to the seminal fluid. In the female genital tract, glycodelin-A is mainly expressed in endometrial epithelial cells and secreted into the uterine fluid and amniotic fluid. Granulosa cells secrete glycodelin-F into the follicular fluid and glycodelin-C is found in cumulus cells [83].

Glycodelin-A appears to play an important role in changing the immune system during early pregnancy by suppressing the cytotoxicity of natural killer cells, as well as shifting the concentration of Th-1 and Th-2 cells towards a Th-2-dominated area which is necessary for a successful implantation [84]. Glycodelin-A inhibits T cell proliferation, stimulates T cell apoptosis, and controls the immune response from B cells. Furthermore, glycodelin-A stimulates endometrial proliferation and maintains the attachment of the trophoblast as well as its hCG production [1].

3.11. Glycodelin and RPL

Toth et al. analyzed the expression of glycodelin protein and mRNA in the en-dometrium and compared patients with a normal pregnancy to those who had an abortion and those who had a molar pregnancy [85]. Patients with an abortion had a significantly lower glycodelin expression than those with a healthy pregnancy. In contrast, women with a molar pregnancy showed an upregulation of glycodelin. Reduced glycodelin concentra-tions may lead to an overactive immune system that rejects the developing embryo and causes abortion. Furthermore, hCG appears to influence glycodelin expression because there are equal proportions between hCG and glycodelin in cases of molar pregnancy and in cases of abortion [85].

A German study group achieved similar results [86]. Glycodelin and the pregnancy zone protein in the decidua of patients with spontaneous and recurrent miscarriages were compared to the expression of these entities in the placental tissue of term infants. The spontaneous and the recurrent miscarriages groups had significantly lower glycodelin levels than controls [86].

3.12. Glycodelin and RIF

In order to investigate the correlation between glycodelin and RIF, endometrial cells and blood samples were taken during the window of implantation in women with RIF and from a control group of fertile women [87]. A significantly lower expression of glycodelin was measured in the blood and endometrial tissue of women with RIF compared with controls [87].

These data were confirmed by a Polish study group. The authors analyzed glycodelin expression in uterine fluid, as well as in serum and endometrium samples of women with RPL or RIF, and compared these with fertile controls [88]. In general, glycodelin levels in the serum were three times lower than those in the uterine fluid, and there were no differences between the three study groups [88]. Concerning the uterine fluid, the lowest glycodelin expression was noted in women with idiopathic infertility and infertile women with endometriosis. Compared with the fertile control group, these results were statistically significant ($p < 0.0001$) [88]. Women suffering from RPL had lower glycodelin concentrations than controls, but the difference did not achieve statistical significance [88].

In summary, glycodelin is an endometrial receptivity marker. The expression of glycodelin in the endometrium is mirrored in the uterine fluid and serum.

4. Conclusions

Successful pregnancy needs the correct interaction between (a) the individual hormones and their respective receptors. Furthermore, (b) adhesion molecules are responsible for the required physical interaction between the endometrium and the blastocyst, and, finally, (c) immune cells and cytokine signaling pathways act as mediators for the so-called embryo–maternal-dialogue. A dysregulation or inappropriate expression in one of these three sections may lead to an implantation failure or a pregnancy loss [3,89]. In the present review, we focused on the hormones estrogen and progesterone; the cytokines IL-6, IL-8, LIF, IL-11, and IL-1; and the glycoprotein glycodelin. We analyzed their respective functions during implantation and early pregnancy, and we compared the situations of normal pregnancy vs. recurrent pregnancy loss vs. recurrent implantation failure. The intention was to determine differences at the molecular level as to how cytokines behave at the time of implantation or in early pregnancy in order to draw conclusions about RIF and RPL. It would be conceivable that a cytokine is upregulated in the case of RIF and downregulated in the case of RPL.

As shown in Table 3, studies on this topic are very inhomogeneous. Even the investigation of RIF or RPL alone reveals diverse results regarding cytokine levels. Accordingly, no clear statement can be made as to how the individual cytokines behave in RIF or RPL.

Table 3. Molecular determinants, their function in normal pregnancy, and changes in RPL and RIF.

Hormone/Cytokine/Glycoprotein	Normal Pregnancy	RPL	RIF
Estrogen	Proliferation of endometrium, ERα: essential for implantation, ERβ: no restriction of fertility	Estrogen ↓ [17,19]	Estrogen ↓ [23,24] Estrogen ↑ → premature progesterone influence [23]
Progesterone	Inducing decidualization, opening the WOI PR-A (PR-B knock-out): implantation, pregnancy, parturation PR-B (PR-A knock-out): endometrial hyperplasia, inflammation, absence of decidualization	Progesterone ↓ [19]	Estrogen ↑ → premature progesterone influence, asynchronicity between endometrium and embryo [23]
IL-6	Luteal phase: high concentration, regulation of trophoblast invasion and spiral artery remodeling	Sporadic abortion: IL-6 ↓ [37] RPL: IL-6 ↑ [39]	IL-6 ↑ [41] IL-6 = [43] IL-6 = [42]
IL-8	Luteal phase and during embryo presentation: high concentration; stimulates progesterone secretion	Sporadic abortion: IL-8 ↓ [37] RPL: IL-8 ↑ [39]	IL-8 ↑ [42]

Table 3. *Cont.*

Hormone/Cytokine/Glycoprotein	Normal Pregnancy	RPL	RIF
LIF	Receptivity of endometrium, decidualization, maintaining pinopodes, embryo attachment, implantation, placenta development	LIF ↑ (endometrial samples) [52] LIF = (endometrial samples) [53] LIF ↓ (serum) [54]	LIF ↓ (secretory phase), high amplitude (proliferative phase) [55] LIF ↓ (proliferative phase) [56] KLF12 ↑ → LIF ↓ [57]
IL-11	Produced by stromal and epithelial cells, during early secretory phase (epithelial cells), during decidualization (stromal cells), TNF-α ↓	IL-11 (epithelial cells) ↓ [64]	IL-11 (stromal cells) ↓ [65] IL-11 = [66]
IL-1	IL-1β and IL-1Ra are responsible for the embryo–maternal dialog and the immunological shift. The tissue of the fallopian tubes and the myometrial smooth muscle synthesize IL-1β during pregnancy, IL-1β ↑ during pregnancy, responsible for prothrombotic state	IL-1β ↓ [67,75] IL-1β ↑ [76] IL-1β ↑ and antiphospholipid syndrome [78]	IL-1β= no increase in contrast to healthy pregnancies, IL1-Ra = [79], IL-1 β ↑ [41,80,81]
Glycodelin	Glycodelin-A: shifting the immune system, endometrial proliferation, fertilization and implantation	Glycodelin-A ↓ [85,86]	Glycodelin-A ↓ [87,88]

Abbreviations: ER: estrogen receptor, PR: progesterone receptor, WOI: window of implantation, IL-1Ra: IL-1 receptor antagonist, KLF12: Krüppel-like factor 12, ↑: higher expression compared to regular implantation, =: no changes concerning the expression, ↓: lower expression compared to regular implantation, →: leading to.

This can be explained by the different populations and, in part, small sample sizes, the different timings of sample collection, and whether the cytokines were measured in the plasma, a whole tissue, or a specific cell population. The complicated relationship between cytokines and their overlapping biological activities means that the alteration of one cytokine is likely to affect others; this hinders the study of their roles in implantation, RIF, and RPL. Furthermore, large-scale investigations should address the role of cytokines during implantation and in different types of reproduction failure.

It is difficult to derive a direct clinical consequence or recommendation for action from the above-mentioned cytokine changes. In theory, it would be conceivable to use glucocorticoids or i.v.-immunoglobulins to reduce cytokines in general. However, this would be an undirected therapy and would affect all cytokines, including those that are already low or not elevated at all.

Targeted approaches to lower, for example, IL-6 levels, e.g., using tocilizumab, an IL-6 antagonist, are contraindicated in pregnancy and should be discontinued at least 3 months before conception [90].

Immunomodulatory therapy approaches to the treatment of RIF or RPL have already been analyzed in detail and are not included in the guideline due to their inhomogeneous data situation and their sometimes-severe side effects [91].

Author Contributions: Conceptualization, V.G., L.A., A.D.-M., L.M., G.G., S.v.O., N.M. and I.A.; Project administration, V.G., I.A., K.A., M.S. and J.A.; Supervision, N.M. and I.A.; Visualization, S.v.O., L.A. and J.A.; Writing—original draft, V.G.; Writing—review and editing, I.A., S.v.O. and L.A. All authors have read and agreed to the published version of the manuscript.

Funding: This research received no external funding.

Institutional Review Board Statement: All procedures involving human participants were performed in accordance with the ethical standards of the institutional and/or national research committee, as well as the 1964 Helsinki declaration and its later amendments or comparable ethical standards.

Informed Consent Statement: Not applicable.

Data Availability Statement: The data presented in this study are available on request from the corresponding author.

Acknowledgments: We would like to thank Markus Voll, Medical Illustrations, 81245 Munich, for the support with the graphical representation.

Conflicts of Interest: The authors declare no conflict of interest.

Abbreviations

IL	interleukin
LIF	leukemia inhibitory factor
RIF	recurrent implantation failure
RPL	recurrent pregnancy loss
WOI	window of implantation
ER	estrogen receptor
PR	progesterone receptor
hCG	human chorionic gonadotropin
ART	assisted reproductive technology
LH	luteinizing hormone
JAK	Janus kinase
STAT	signal transducer and activator of transcription
IL-6R	IL-6α-receptor
INF-γ	interferon-γ
TGF	transforming growth factor
IVF	in vitro fertilization
ICSI	intracytoplasmic sperm injection
TNF	tumor necrosis factor
KLF12	Krüppel-like factor 12
IL-11Rα	IL-11 receptor
PGE2	prostaglandin E2
IL-1R1	IL-1 receptor type 1
IL-1R2	IL-1 receptor type 2
IL-1Ra	IL-1 receptor antagonist
Th1/2	Type 1 T helper cells and type 2 T helper cells

References

1. Mrozikiewicz, A.E.; Ozarowski, M.; Jedrzejczak, P. Biomolecular Markers of Recurrent Implantation Failure—A Review. *Int. J. Mol. Sci.* **2021**, *22*, 82. [CrossRef] [PubMed]
2. Bowen, J.A.; Hunt, J.S. The role of integrins in reproduction. *Proc. Soc. Exp. Biol. Med.* **2000**, *223*, 331–343. [CrossRef] [PubMed]
3. D'Occhio, M.J.; Campanile, G.; Zicarelli, L.; Visintin, J.A.; Baruselli, P.S. Adhesion molecules in gamete transport, fertilization, early embryonic development, and implantation-role in establishing a pregnancy in cattle: A review. *Mol. Reprod. Dev.* **2020**, *87*, 206–222. [CrossRef] [PubMed]
4. Gunther, V.; Alkatout, I.; Meyerholz, L.; Maass, N.; Gorg, S.; von Otte, S.; Ziemann, M. Live Birth Rates after Active Immunization with Partner Lymphocytes. *Biomedicines* **2021**, *9*, 1350. [CrossRef] [PubMed]
5. Gunther, V.; Alkatout, I.; Junkers, W.; Maass, N.; Ziemann, M.; Gorg, S.; von Otte, S. Active Immunisation with Partner Lymphocytes in Female Patients Who Want to Become Pregnant—Current Status. *Geburtsh. Frauenheilkd.* **2018**, *78*, 260–273. [CrossRef] [PubMed]
6. Achache, H.; Revel, A. Endometrial receptivity markers, the journey to successful embryo implantation. *Hum. Reprod. Update* **2006**, *12*, 731–746. [CrossRef]
7. Sun, Y.; Zhang, Y.; Ma, X.; Jia, W.; Su, Y. Determining Diagnostic Criteria of Unexplained Recurrent Implantation Failure: A Retrospective Study of Two vs. Three or More Implantation Failure. *Front. Endocrinol.* **2021**, *12*, 619437. [CrossRef]
8. Practice Committee of the American Society for Reproductive Medicine. Definitions of infertility and recurrent pregnancy loss: A committee opinion. *Fertil. Steril.* **2020**, *113*, 533–535. [CrossRef]
9. Bashiri, A.; Halper, K.I.; Orvieto, R. Recurrent Implantation Failure-update overview on etiology, diagnosis, treatment and future directions. *Reprod. Biol. Endocrinol. RBE* **2018**, *16*, 121. [CrossRef]
10. Deng, T.; Liao, X.; Zhu, S. Recent Advances in Treatment of Recurrent Spontaneous Abortion. *Obstet. Gynecol. Surv.* **2022**, *77*, 355–366. [CrossRef]

11. La, X.; Wang, W.; Zhang, M.; Liang, L. Definition and Multiple Factors of Recurrent Spontaneous Abortion. *Adv. Exp. Med. Biol.* **2021**, *1300*, 231–257. [CrossRef] [PubMed]

12. Paulson, R.J. Hormonal induction of endometrial receptivity. *Fertil. Steril.* **2011**, *96*, 530–535. [CrossRef] [PubMed]

13. Chen, M.; Wolfe, A.; Wang, X.; Chang, C.; Yeh, S.; Radovick, S. Generation and characterization of a complete null estrogen receptor alpha mouse using Cre/LoxP technology. *Mol. Cell. Biochem.* **2009**, *321*, 145–153. [CrossRef] [PubMed]

14. Lee, H.R.; Kim, T.H.; Choi, K.C. Functions and physiological roles of two types of estrogen receptors, ERalpha and ERbeta, identified by estrogen receptor knockout mouse. *Lab. Anim. Res.* **2012**, *28*, 71–76. [CrossRef] [PubMed]

15. Massimiani, M.; Lacconi, V.; La Civita, F.; Ticconi, C.; Rago, R.; Campagnolo, L. Molecular Signaling Regulating Endometrium-Blastocyst Crosstalk. *Int. J. Mol. Sci.* **2019**, *21*, 23. [CrossRef] [PubMed]

16. Lydon, J.P.; DeMayo, F.J.; Funk, C.R.; Mani, S.K.; Hughes, A.R.; Montgomery, C.A., Jr.; Shyamala, G.; Conneely, O.M.; O'Malley, B.W. Mice lacking progesterone receptor exhibit pleiotropic reproductive abnormalities. *Genes Dev.* **1995**, *9*, 2266–2278. [CrossRef] [PubMed]

17. Gao, Q.W.M.; Rong, T. Serum estradiol level in pregnant women of four to eight weeks and its relationship with threatened abortion. *J. Shandong Univ. (Health Sci.)* **2008**, *46*, 884–886.

18. Zhang, W.; Tian, Y.; Xie, D.; Miao, Y.; Liu, J.; Wang, X. The impact of peak estradiol during controlled ovarian stimulation on the cumulative live birth rate of IVF/ICSI in non-PCOS patients. *J. Assist. Reprod. Genet.* **2019**, *36*, 2333–2344. [CrossRef]

19. Deng, W.; Sun, R.; Du, J.; Wu, X.; Ma, L.; Wang, M.; Lv, Q. Prediction of miscarriage in first trimester by serum estradiol, progesterone and beta-human chorionic gonadotropin within 9 weeks of gestation. *BMC Pregnancy Childbirth* **2022**, *22*, 112. [CrossRef]

20. Patel, B.; Elguero, S.; Thakore, S.; Dahoud, W.; Bedaiwy, M.; Mesiano, S. Role of nuclear progesterone receptor isoforms in uterine pathophysiology. *Hum. Reprod. Update* **2015**, *21*, 155–173. [CrossRef]

21. Coomarasamy, A.; Harb, H.M.; Devall, A.J.; Cheed, V.; Roberts, T.E.; Goranitis, I.; Ogwulu, C.B.; Williams, H.M.; Gallos, I.D.; Eapen, A.; et al. Progesterone to prevent miscarriage in women with early pregnancy bleeding: The PRISM RCT. *Health Technol. Assess.* **2020**, *24*, 1–70. [CrossRef]

22. Devall, A.J.; Papadopoulou, A.; Podesek, M.; Haas, D.M.; Price, M.J.; Coomarasamy, A.; Gallos, I.D. Progestogens for preventing miscarriage: A network meta-analysis. *Cochrane Database Syst. Rev.* **2021**, *4*, CD013792. [CrossRef]

23. Parisi, F.; Fenizia, C.; Introini, A.; Zavatta, A.; Scaccabarozzi, C.; Biasin, M.; Savasi, V. The pathophysiological role of estrogens in the initial stages of pregnancy: Molecular mechanisms and clinical implications for pregnancy outcome from the periconceptional period to end of the first trimester. *Hum. Reprod. Update* **2023**, *29*, 699–720. [CrossRef]

24. Long, N.; Liu, N.; Liu, X.L.; Li, J.; Cai, B.Y.; Cai, X. Endometrial expression of telomerase, progesterone, and estrogen receptors during the implantation window in patients with recurrent implantation failure. *Genet. Mol. Res. GMR* **2016**, *15*, 15027849. [CrossRef]

25. Vagnini, L.D.; Renzi, A.; Petersen, B.; Canas, M.; Petersen, C.G.; Mauri, A.L.; Mattila, M.C.; Ricci, J.; Dieamant, F.; Oliveira, J.B.A.; et al. Association between estrogen receptor 1 (ESR1) and leukemia inhibitory factor (LIF) polymorphisms can help in the prediction of recurrent implantation failure. *Fertil. Steril.* **2019**, *111*, 527–534. [CrossRef] [PubMed]

26. Jarczak, D.; Nierhaus, A. Cytokine Storm-Definition, Causes, and Implications. *Int. J. Mol. Sci.* **2022**, *23*, 11740. [CrossRef] [PubMed]

27. Rose-John, S. Interleukin-6 Family Cytokines. *Cold Spring Harb. Perspect. Biol.* **2018**, *10*, a028415. [CrossRef] [PubMed]

28. Vilotic, A.; Nacka-Aleksic, M.; Pirkovic, A.; Bojic-Trbojevic, Z.; Dekanski, D.; Jovanovic Krivokuca, M. IL-6 and IL-8: An Overview of Their Roles in Healthy and Pathological Pregnancies. *Int. J. Mol. Sci.* **2022**, *23*, 14574. [CrossRef]

29. Perrier d'Hauterive, S.; Charlet-Renard, C.; Berndt, S.; Dubois, M.; Munaut, C.; Goffin, F.; Hagelstein, M.T.; Noel, A.; Hazout, A.; Foidart, J.M.; et al. Human chorionic gonadotropin and growth factors at the embryonic-endometrial interface control leukemia inhibitory factor (LIF) and interleukin 6 (IL-6) secretion by human endometrial epithelium. *Hum. Reprod.* **2004**, *19*, 2633–2643. [CrossRef]

30. Jones, S.A.; Jenkins, B.J. Recent insights into targeting the IL-6 cytokine family in inflammatory diseases and cancer. *Nat. Rev. Immunol.* **2018**, *18*, 773–789. [CrossRef]

31. Matsushima, K.; Yang, D.; Oppenheim, J.J. Interleukin-8: An evolving chemokine. *Cytokine* **2022**, *153*, 155828. [CrossRef] [PubMed]

32. Corre, I.; Pineau, D.; Hermouet, S. Interleukin-8: An autocrine/paracrine growth factor for human hematopoietic progenitors acting in synergy with colony stimulating factor-1 to promote monocyte-macrophage growth and differentiation. *Exp. Hematol.* **1999**, *27*, 28–36. [CrossRef] [PubMed]

33. Li, A.; Dubey, S.; Varney, M.L.; Dave, B.J.; Singh, R.K. IL-8 directly enhanced endothelial cell survival, proliferation, and matrix metalloproteinases production and regulated angiogenesis. *J. Immunol.* **2003**, *170*, 3369–3376. [CrossRef] [PubMed]

34. Caballero-Campo, P.; Dominguez, F.; Coloma, J.; Meseguer, M.; Remohi, J.; Pellicer, A.; Simon, C. Hormonal and embryonic regulation of chemokines IL-8, MCP-1 and RANTES in the human endometrium during the window of implantation. *Mol. Hum. Reprod.* **2002**, *8*, 375–384. [CrossRef]

35. Dominguez, F.; Galan, A.; Martin, J.J.; Remohi, J.; Pellicer, A.; Simon, C. Hormonal and embryonic regulation of chemokine receptors CXCR1, CXCR4, CCR5 and CCR2B in the human endometrium and the human blastocyst. *Mol. Hum. Reprod.* **2003**, *9*, 189–198. [CrossRef] [PubMed]

36. Tsui, K.H.; Chen, L.Y.; Shieh, M.L.; Chang, S.P.; Yuan, C.C.; Li, H.Y. Interleukin-8 can stimulate progesterone secretion from a human trophoblast cell line, BeWo. *In Vitro Cell. Dev. Biol. Anim.* **2004**, *40*, 331–336. [CrossRef] [PubMed]

37. Pitman, H.; Innes, B.A.; Robson, S.C.; Bulmer, J.N.; Lash, G.E. Altered expression of interleukin-6, interleukin-8 and their receptors in decidua of women with sporadic miscarriage. *Hum. Reprod.* **2013**, *28*, 2075–2086. [CrossRef]

38. Chen, P.; Zhou, L.; Chen, J.; Lu, Y.; Cao, C.; Lv, S.; Wei, Z.; Wang, L.; Chen, J.; Hu, X.; et al. The Immune Atlas of Human Deciduas with Unexplained Recurrent Pregnancy Loss. *Front. Immunol.* **2021**, *12*, 689019. [CrossRef]

39. Zhao, L.; Han, L.; Hei, G.; Wei, R.; Zhang, Z.; Zhu, X.; Guo, Q.; Chu, C.; Fu, X.; Xu, K.; et al. Diminished miR-374c-5p negatively regulates IL (interleukin)-6 in unexplained recurrent spontaneous abortion. *J. Mol. Med.* **2022**, *100*, 1043–1056. [CrossRef]

40. Laisk, T.; Soares, A.L.G.; Ferreira, T.; Painter, J.N.; Censin, J.C.; Laber, S.; Bacelis, J.; Chen, C.Y.; Lepamets, M.; Lin, K.; et al. The genetic architecture of sporadic and multiple consecutive miscarriage. *Nat. Commun.* **2020**, *11*, 5980. [CrossRef]

41. Liang, P.Y.; Diao, L.H.; Huang, C.Y.; Lian, R.C.; Chen, X.; Li, G.G.; Zhao, J.; Li, Y.Y.; He, X.B.; Zeng, Y. The pro-inflammatory and anti-inflammatory cytokine profile in peripheral blood of women with recurrent implantation failure. *Reprod. Biomed. Online* **2015**, *31*, 823–826. [CrossRef] [PubMed]

42. Xie, J.; Yan, L.; Cheng, Z.; Qiang, L.; Yan, J.; Liu, Y.; Liang, R.; Zhang, J.; Li, Z.; Zhuang, L.; et al. Potential effect of inflammation on the failure risk of in vitro fertilization and embryo transfer among infertile women. *Hum. Fertil.* **2020**, *23*, 214–222. [CrossRef] [PubMed]

43. Ozgu-Erdinc, A.S.; Gozukara, I.; Kahyaoglu, S.; Yilmaz, S.; Yumusak, O.H.; Yilmaz, N.; Erkaya, S.; Engin-Ustun, Y. Is there any role of interleukin-6 and high sensitive C-reactive protein in predicting IVF/ICSI success? A prospective cohort study. *Horm. Mol. Biol. Clin. Investig.* **2021**, *43*, 35–40. [CrossRef] [PubMed]

44. Cheng, J.G.; Chen, J.R.; Hernandez, L.; Alvord, W.G.; Stewart, C.L. Dual control of LIF expression and LIF receptor function regulate Stat3 activation at the onset of uterine receptivity and embryo implantation. *Proc. Natl. Acad. Sci. USA* **2001**, *98*, 8680–8685. [CrossRef] [PubMed]

45. Catalano, R.D.; Johnson, M.H.; Campbell, E.A.; Charnock-Jones, D.S.; Smith, S.K.; Sharkey, A.M. Inhibition of Stat3 activation in the endometrium prevents implantation: A nonsteroidal approach to contraception. *Proc. Natl. Acad. Sci. USA* **2005**, *102*, 8585–8590. [CrossRef] [PubMed]

46. Nicola, N.A.; Babon, J.J. Leukemia inhibitory factor (LIF). *Cytokine Growth Factor Rev.* **2015**, *26*, 533–544. [CrossRef] [PubMed]

47. Charnock-Jones, D.S.; Sharkey, A.M.; Fenwick, P.; Smith, S.K. Leukaemia inhibitory factor mRNA concentration peaks in human endometrium at the time of implantation and the blastocyst contains mRNA for the receptor at this time. *J. Reprod. Fertil.* **1994**, *101*, 421–426. [CrossRef] [PubMed]

48. Chen, J.R.; Cheng, J.G.; Shatzer, T.; Sewell, L.; Hernandez, L.; Stewart, C.L. Leukemia inhibitory factor can substitute for nidatory estrogen and is essential to inducing a receptive uterus for implantation but is not essential for subsequent embryogenesis. *Endocrinology* **2000**, *141*, 4365–4372. [CrossRef]

49. Licht, P.; Losch, A.; Dittrich, R.; Neuwinger, J.; Siebzehnrubl, E.; Wildt, L. Novel insights into human endometrial paracrinology and embryo-maternal communication by intrauterine microdialysis. *Hum. Reprod. Update* **1998**, *4*, 532–538. [CrossRef]

50. Gutsche, S.; von Wolff, M.; Strowitzki, T.; Thaler, C.J. Seminal plasma induces mRNA expression of IL-1beta, IL-6 and LIF in endometrial epithelial cells in vitro. *Mol. Hum. Reprod.* **2003**, *9*, 785–791. [CrossRef]

51. Yue, X.; Wu, L.; Hu, W. The regulation of leukemia inhibitory factor. *Cancer Cell Microenviron.* **2015**, *2*, 877. [CrossRef]

52. Karaer, A.; Cigremis, Y.; Celik, E.; Urhan Gonullu, R. Prokineticin 1 and leukemia inhibitory factor mRNA expression in the endometrium of women with idiopathic recurrent pregnancy loss. *Fertil. Steril.* **2014**, *102*, 1091–1095.e1. [CrossRef] [PubMed]

53. Xu, B.; Sun, X.; Li, L.; Wu, L.; Zhang, A.; Feng, Y. Pinopodes, leukemia inhibitory factor, integrin-beta3, and mucin-1 expression in the peri-implantation endometrium of women with unexplained recurrent pregnancy loss. *Fertil. Steril.* **2012**, *98*, 389–395. [CrossRef] [PubMed]

54. Miresmaeili, S.M.; Fesahat, F.; Kazemi, N.; Ansariniya, H.; Zare, F. Possible Role of Leukemia Inhibitory Factor and Inflammatory Cytokines in The Recurrent Spontaneous Abortion: A Case-Control Study. *Int. J. Fertil. Steril.* **2023**, *17*, 140–144. [PubMed]

55. Hambartsoumian, E. Endometrial leukemia inhibitory factor (LIF) as a possible cause of unexplained infertility and multiple failures of implantation. *Am. J. Reprod. Immunol.* **1998**, *39*, 137–143. [CrossRef] [PubMed]

56. Wu, M.; Yin, Y.; Zhao, M.; Hu, L.; Chen, Q. The low expression of leukemia inhibitory factor in endometrium: Possible relevant to unexplained infertility with multiple implantation failures. *Cytokine* **2013**, *62*, 334–339. [CrossRef]

57. Huang, C.; Sun, H.; Wang, Z.; Liu, Y.; Cheng, X.; Liu, J.; Jiang, R.; Zhang, X.; Zhen, X.; Zhou, J.; et al. Increased Kruppel-like factor 12 impairs embryo attachment via downregulation of leukemia inhibitory factor in women with recurrent implantation failure. *Cell Death Discov.* **2018**, *4*, 23. [CrossRef]

58. Pantos, K.; Grigoriadis, S.; Maziotis, E.; Pistola, K.; Xystra, P.; Pantou, A.; Kokkali, G.; Pappas, A.; Lambropoulou, M.; Sfakianoudis, K.; et al. The Role of Interleukins in Recurrent Implantation Failure: A Comprehensive Review of the Literature. *Int. J. Mol. Sci.* **2022**, *23*, 2198. [CrossRef]

59. Dimitriadis, E.; White, C.A.; Jones, R.L.; Salamonsen, L.A. Cytokines, chemokines and growth factors in endometrium related to implantation. *Hum. Reprod. Update* **2005**, *11*, 613–630. [CrossRef]

60. Cork, B.A.; Tuckerman, E.M.; Li, T.C.; Laird, S.M. Expression of interleukin (IL)-11 receptor by the human endometrium in vivo and effects of IL-11, IL-6 and LIF on the production of MMP and cytokines by human endometrial cells in vitro. *Mol. Hum. Reprod.* **2002**, *8*, 841–848. [CrossRef]

61. van Mourik, M.S.; Macklon, N.S.; Heijnen, C.J. Embryonic implantation: Cytokines, adhesion molecules, and immune cells in establishing an implantation environment. *J. Leukoc. Biol.* **2009**, *85*, 4–19. [CrossRef] [PubMed]

62. von Rango, U.; Alfer, J.; Kertschanska, S.; Kemp, B.; Muller-Newen, G.; Heinrich, P.C.; Beier, H.M.; Classen-Linke, I. Interleukin-11 expression: Its significance in eutopic and ectopic human implantation. *Mol. Hum. Reprod.* **2004**, *10*, 783–792. [CrossRef] [PubMed]

63. Bao, L.; Devi, Y.S.; Bowen-Shauver, J.; Ferguson-Gottschall, S.; Robb, L.; Gibori, G. The role of interleukin-11 in pregnancy involves up-regulation of alpha2-macroglobulin gene through janus kinase 2-signal transducer and activator of transcription 3 pathway in the decidua. *Mol. Endocrinol.* **2006**, *20*, 3240–3250. [CrossRef] [PubMed]

64. Linjawi, S.; Li, T.C.; Tuckerman, E.M.; Blakemore, A.I.; Laird, S.M. Expression of interleukin-11 receptor alpha and interleukin-11 protein in the endometrium of normal fertile women and women with recurrent miscarriage. *J. Reprod. Immunol.* **2004**, *64*, 145–155. [CrossRef] [PubMed]

65. Karpovich, N.; Klemmt, P.; Hwang, J.H.; McVeigh, J.E.; Heath, J.K.; Barlow, D.H.; Mardon, H.J. The production of interleukin-11 and decidualization are compromised in endometrial stromal cells derived from patients with infertility. *J. Clin. Endocrinol. Metab.* **2005**, *90*, 1607–1612. [CrossRef]

66. Sabry, D.; Nouh, O.; Marzouk, S.; Hassouna, A. Pilot study on molecular quantitation and sequencing of endometrial cytokines gene expression and their effect on the outcome of in vitro fertilization (IVF) cycle. *J. Adv. Res.* **2014**, *5*, 595–600. [CrossRef]

67. Equils, O.; Kellogg, C.; McGregor, J.; Gravett, M.; Neal-Perry, G.; Gabay, C. The role of the IL-1 system in pregnancy and the use of IL-1 system markers to identify women at risk for pregnancy complicationsdagger. *Biol. Reprod.* **2020**, *103*, 684–694. [CrossRef]

68. Hannum, C.H.; Wilcox, C.J.; Arend, W.P.; Joslin, F.G.; Dripps, D.J.; Heimdal, P.L.; Armes, L.G.; Sommer, A.; Eisenberg, S.P.; Thompson, R.C. Interleukin-1 receptor antagonist activity of a human interleukin-1 inhibitor. *Nature* **1990**, *343*, 336–340. [CrossRef]

69. Gabay, C.; Lamacchia, C.; Palmer, G. IL-1 pathways in inflammation and human diseases. *Nat. Rev. Rheumatol.* **2010**, *6*, 232–241. [CrossRef]

70. Young, A.; Thomson, A.J.; Ledingham, M.; Jordan, F.; Greer, I.A.; Norman, J.E. Immunolocalization of proinflammatory cytokines in myometrium, cervix, and fetal membranes during human parturition at term. *Biol. Reprod.* **2002**, *66*, 445–449. [CrossRef]

71. Romero, R.; Gomez, R.; Galasso, M.; Mazor, M.; Berry, S.M.; Quintero, R.A.; Cotton, D.B. The natural interleukin-1 receptor antagonist in the fetal, maternal, and amniotic fluid compartments: The effect of gestational age, fetal gender, and intrauterine infection. *Am. J. Obstet. Gynecol.* **1994**, *171*, 912–921. [CrossRef] [PubMed]

72. Yang, X.; Tian, Y.; Zheng, L.; Luu, T.; Kwak-Kim, J. The Update Immune-Regulatory Role of Pro- and Anti-Inflammatory Cytokines in Recurrent Pregnancy Losses. *Int. J. Mol. Sci.* **2022**, *24*, 132. [CrossRef] [PubMed]

73. Raghupathy, R.; Makhseed, M.; Azizieh, F.; Hassan, N.; Al-Azemi, M.; Al-Shamali, E. Maternal Th1- and Th2-type reactivity to placental antigens in normal human pregnancy and unexplained recurrent spontaneous abortions. *Cell. Immunol.* **1999**, *196*, 122–130. [CrossRef] [PubMed]

74. Huber, M.; Rutherford, A.; Meister, W.; Weiss, A.; Rollinghoff, M.; Lohoff, M. TCR- and IL-1-mediated co-stimulation reveals an IL-4-independent way of Th2 cell proliferation. *Int. Immunol.* **1996**, *8*, 1257–1263. [CrossRef]

75. von Wolff, M.; Thaler, C.J.; Strowitzki, T.; Broome, J.; Stolz, W.; Tabibzadeh, S. Regulated expression of cytokines in human endometrium throughout the menstrual cycle: Dysregulation in habitual abortion. *Mol. Hum. Reprod.* **2000**, *6*, 627–634. [CrossRef]

76. Lob, S.; Amann, N.; Kuhn, C.; Schmoeckel, E.; Wockel, A.; Zati Zehni, A.; Kaltofen, T.; Keckstein, S.; Mumm, J.N.; Meister, S.; et al. Interleukin-1 beta is significantly upregulated in the decidua of spontaneous and recurrent miscarriage placentas. *J. Reprod. Immunol.* **2021**, *144*, 103283. [CrossRef]

77. Petri, M. Antiphospholipid syndrome. *Transl. Res. J. Lab. Clin. Med.* **2020**, *225*, 70–81. [CrossRef]

78. Mulla, M.J.; Salmon, J.E.; Chamley, L.W.; Brosens, J.J.; Boeras, C.M.; Kavathas, P.B.; Abrahams, V.M. A role for uric acid and the Nalp3 inflammasome in antiphospholipid antibody-induced IL-1beta production by human first trimester trophoblast. *PLoS ONE* **2013**, *8*, e65237. [CrossRef]

79. Kreines, F.M.; Nasioudis, D.; Minis, E.; Irani, M.; Witkin, S.S.; Spandorfer, S. IL-1beta predicts IVF outcome: A prospective study. *J. Assist. Reprod. Genet.* **2018**, *35*, 2031–2035. [CrossRef]

80. Boomsma, C.M.; Kavelaars, A.; Eijkemans, M.J.; Lentjes, E.G.; Fauser, B.C.; Heijnen, C.J.; Macklon, N.S. Endometrial secretion analysis identifies a cytokine profile predictive of pregnancy in IVF. *Hum. Reprod.* **2009**, *24*, 1427–1435. [CrossRef]

81. Inagaki, N.; Stern, C.; McBain, J.; Lopata, A.; Kornman, L.; Wilkinson, D. Analysis of intra-uterine cytokine concentration and matrix-metalloproteinase activity in women with recurrent failed embryo transfer. *Hum. Reprod.* **2003**, *18*, 608–615. [CrossRef] [PubMed]

82. Quenby, S.; Vince, G.; Farquharson, R.; Aplin, J. Recurrent miscarriage: A defect in nature's quality control? *Hum. Reprod.* **2002**, *17*, 1959–1963. [CrossRef] [PubMed]

83. Uchida, H.; Maruyama, T.; Nishikawa-Uchida, S.; Miyazaki, K.; Masuda, H.; Yoshimura, Y. Glycodelin in reproduction. *Reprod. Med. Biol.* **2013**, *12*, 79–84. [CrossRef] [PubMed]

84. Dixit, A.; Karande, A.A. Glycodelin regulates the numbers and function of peripheral natural killer cells. *J. Reprod. Immunol.* **2020**, *137*, 102625. [CrossRef] [PubMed]

85. Toth, B.; Roth, K.; Kunert-Keil, C.; Scholz, C.; Schulze, S.; Mylonas, I.; Friese, K.; Jeschke, U. Glycodelin protein and mRNA is downregulated in human first trimester abortion and partially upregulated in mole pregnancy. *J. Histochem. Cytochem. Off. J. Histochem. Soc.* **2008**, *56*, 477–485. [CrossRef] [PubMed]

86. Lob, S.; Vattai, A.; Kuhn, C.; Schmoeckel, E.; Mahner, S.; Wockel, A.; Kolben, T.; Keil, C.; Jeschke, U.; Vilsmaier, T. Pregnancy Zone Protein (PZP) is significantly upregulated in the decidua of recurrent and spontaneous miscarriage and negatively correlated to Glycodelin A (GdA). *J. Reprod. Immunol.* **2021**, *143*, 103267. [CrossRef] [PubMed]
87. Bastu, E.; Mutlu, M.F.; Yasa, C.; Dural, O.; Nehir Aytan, A.; Celik, C.; Buyru, F.; Yeh, J. Role of Mucin 1 and Glycodelin A in recurrent implantation failure. *Fertil. Steril.* **2015**, *103*, 1059–1064. [CrossRef]
88. Skrzypczak, J.; Wirstlein, P.; Mikolajczyk, M. Is glycodelin an important marker of endometrial receptivity? *Ginekol. Pol.* **2005**, *76*, 770–781.
89. Tu, Z.; Ran, H.; Zhang, S.; Xia, G.; Wang, B.; Wang, H. Molecular determinants of uterine receptivity. *Int. J. Dev. Biol.* **2014**, *58*, 147–154. [CrossRef]
90. Gotestam Skorpen, C.; Hoeltzenbein, M.; Tincani, A.; Fischer-Betz, R.; Elefant, E.; Chambers, C.; da Silva, J.; Nelson-Piercy, C.; Cetin, I.; Costedoat-Chalumeau, N.; et al. The EULAR points to consider for use of antirheumatic drugs before pregnancy, and during pregnancy and lactation. *Ann. Rheum. Dis.* **2016**, *75*, 795–810. [CrossRef]
91. Gunther, V.; Otte, S.V.; Freytag, D.; Maass, N.; Alkatout, I. Recurrent implantation failure—An overview of current research. *Gynecol. Endocrinol. Off. J. Int. Soc. Gynecol. Endocrinol.* **2021**, *37*, 584–590. [CrossRef] [PubMed]

International Journal of
Molecular Sciences

The Landscape of Point Mutations in Human Protein Coding Genes Leading to Pregnancy Loss

Evgeniia M. Maksiutenko, Yury A. Barbitoff *, Yulia A. Nasykhova, Olga V. Pachuliia, Tatyana E. Lazareva, Olesya N. Bespalova and Andrey S. Glotov *

Department of Genomic Medicine, D.O. Ott Research Institute of Obstetrics, Gynaecology and Reproductology, Mendeleevskaya Line 3, 199034 St. Petersburg, Russia; jmrose@yandex.ru (E.M.M.); yulnasa@gmail.com (Y.A.N.); for.olga.kosyakova@gmail.com (O.V.P.); lazata1997@gmail.com (T.E.L.); shiggerra@mail.ru (O.N.B.)
* Correspondence: barbitoff@bk.ru (Y.A.B.); anglotov@mail.ru (A.S.G.)

Abstract: Pregnancy loss is the most frequent complication of a pregnancy which is devastating for affected families and poses a significant challenge for the health care system. Genetic factors are known to play an important role in the etiology of pregnancy loss; however, despite advances in diagnostics, the causes remain unexplained in more than 30% of cases. In this review, we aggregated the results of the decade-long studies into the genetic risk factors of pregnancy loss (including miscarriage, termination for fetal abnormality, and recurrent pregnancy loss) in euploid pregnancies, focusing on the spectrum of point mutations associated with these conditions. We reviewed the evolution of molecular genetics methods used for the genetic research into causes of pregnancy loss, and collected information about 270 individual genetic variants in 196 unique genes reported as genetic cause of pregnancy loss. Among these, variants in 18 genes have been reported by multiple studies, and two or more variants were reported as causing pregnancy loss for 57 genes. Further analysis of the properties of all known pregnancy loss genes showed that they correspond to broadly expressed, highly evolutionary conserved genes involved in crucial cell differentiation and developmental processes and related signaling pathways. Given the features of known genes, we made an effort to construct a list of candidate genes, variants in which may be expected to contribute to pregnancy loss. We believe that our results may be useful for prediction of pregnancy loss risk in couples, as well as for further investigation and revealing genetic etiology of pregnancy loss.

Keywords: miscarriage; pregnancy loss; spontaneous abortion; RPL; genetic variant; mutation

check for
updates

Citation: Maksiutenko, E.M.; Barbitoff, Y.A.; Nasykhova, Y.A.; Pachuliia, O.V.; Lazareva, T.E.; Bespalova, O.N.; Glotov, A.S. The Landscape of Point Mutations in Human Protein Coding Genes Leading to Pregnancy Loss. *Int. J. Mol. Sci.* **2023**, *24*, 17572. https://doi.org/10.3390/ijms242417572

Academic Editor: Giovanni Tossetta

Received: 23 October 2023
Revised: 1 December 2023
Accepted: 6 December 2023
Published: 17 December 2023

1. Introduction: Miscarriage and Genetics

Human infertility is a frequently occurring problem, affecting more than 50 to 80 million couples worldwide. Despite advances in diagnosis and treatment, the disease etiology remains unexplained in more than 30% of cases, strongly suggesting the involvement of genetic, epigenetic, and environmental factors. Besides inability to conceive, many couples face reproductive problems during pregnancy and childbirth, including pregnancy loss.

According to the European Society of Human Reproduction and Embryology (ESHRE), pregnancy loss (PL) is the spontaneous demise of a pregnancy before the fetus reaches viability [1]. Many genetic studies, however, jointly study cases of unintentional pregnancy loss and pregnancy termination for fetal abnormality (hence, we will use the term 'pregnancy loss' to refer to both of these cases). Unintentional pregnancy loss includes spontaneous abortion (SA or miscarriage) defined as fetal death prior to 20 weeks of gestation (though the exact threshold may vary in different sources and countries), and stillbirth defined as fetal death at a later period of gestation [2]. Besides spontaneous pregnancy loss, recurrent pregnancy loss (RPL) is an important reproductive health problem and affects 2–5% of couples. The incidence of RPL may vary between reports because of the differences in the definitions and criteria used, as well as the populations' characteristics [3]. Back in 1976, the World Health Organization (WHO) defined the RPL as three and more consecutive miscarriages before the 22nd week of gestation, or the loss of a fetus weighing <500 g [4]. Later, in 2011, in line with the WHO definition, the Royal College of Obstetricians and Gynecologists (RCOG) guideline defined recurrent miscarriage as the loss of three or more consecutive pregnancies before 24 weeks of gestation, without imposing any limits on the fetal weight [5].

Miscarriages are common among both parous and nulliparous women, with about 43–49% of women reporting having experienced one or more spontaneous miscarriages. One in every seventeen parous women have three or more miscarriages [6,7]. Multiple factors are involved in miscarriage, such as genetic variation, structural abnormalities and infection of the reproductive tract, endocrine factors, immunological factors, drug, poison, maternal systemic diseases, and so on [8,9]. In addition to sporadic miscarriages, termination of pregnancy for fetal anomaly happens at a rate of 4.6 per 1000 births (according to European statistics). In the UK, over 70% of congenital anomalies are detected during pregnancy, and of those around 37% result in termination [10].

Similar to spontaneous miscarriage, multiple factors contribute to RPL, including maternal age (9–75% of RPL cases), endocrine diseases (17–20%), uterine morphological pathologies (10–15%), chromosomal abnormalities (2–8%), thrombophilia, infectious agents (0.5–5%), and autoimmune disorders (20%). Nevertheless, in approximately 50–75% of RPL cases, the exact cause is not clearly identified and therefore remains unexplained (idiopathic) [5].

Determination of the underlying genetic factors involved in spontaneous or recurrent pregnancy loss currently includes cell culture and karyotype analysis, as well as chromosome microarray analysis (CMA) of chorionic villus tissues. In clinical practice, SNP array analysis is sometimes applied for detecting causal copy number variation (CNV) [11,12]. Previous studies showed that 70–80% of sporadic spontaneous abortions were caused by an abnormal embryonic karyotype, with embryonic aneuploidies being one of the most frequent causes of miscarriage before 10 weeks of gestation [13]. Nevertheless, the underlying cause of the condition is not determined by current standard-of-care practices for a substantial number of cases. In RPL, genetic factors such as DNA methylation, sperm DNA fragmentation, chromosome heteromorphisms, and single nucleotide genetic variation, have been shown to play a role in pathogenesis of the condition. However, none was proven to be a stand-alone risk factor for RPL [5].

In this work, we tried to systematically review recent progress in the investigation of the genetic factors of pregnancy loss, including miscarriage, stillbirth or termination of pregnancy for morphological anomalies. In particular, we analyzed the results of studies that have reported point mutations contributing to loss of euploid pregnancies to understand the common properties of human genes involved in these conditions, as well as to compile a list of candidate genes for future research in this field. Taken together, the knowledge of specific genes that contribute to pregnancy loss could be of importance in designing a diagnostic sequencing panel for affected couples, as well as in understanding the biological pathways that can cause any type of reproductive loss.

2. Genetic Research into Pregnancy Loss: The Evolution of Methods

A correlation between chromosomal abnormalities and spontaneous abortion has been observed since the 1960s [14]. In the 1970s, Boue et al. published one of the earliest large cytogenetic studies where almost 1500 samples of fetal tissue were karyotyped, and a chromosomal abnormality rate of over 60% was found [15,16]. Currently, cytogenetic evaluation of the products of conception (POCs) remains one of the most important approaches to determine the genetic cause of pregnancy loss. Cytogenetic studies have revealed that fetal chromosome abnormalities account for at least half of the cases of SA before 12 weeks and nearly a third in the second trimester miscarriages. Most of these abnormalities comprise numerical chromosomal aberrations (86%), whereas a minority of the cases show structural chromosomal abnormalities (6%) and chromosome mosaicism (8%) [17]. The cytogenetic method is helpful to estimate the recurrence risk and give advice for subsequent pregnancies. However, this approach still had many challenges to overcome. Cytogenetic studies of SA or intrauterine fetal deaths rely on conventional tissue culture and karyotyping. This technique has known limitations such as culture failure and selective growth of contaminating maternal cells [18,19].

The technologies for miscarriage analysis have become more and more accurate in the last few decades. Later emerging techniques such as Chromosomal Comparative Genomic Hybridization (CGH), array–Comparative Genomic Hybridization (array-CGH), Fluorescence in situ hybridization (FISH), Multiplex Ligation-dependent Probe Amplification (MLPA), and Quantitative Fluorescent Polymerase Chain reaction (QF-PCR) have overcame some of the disadvantages inherited from conventional cytogenetic techniques, including poor chromosome preparations, culture failure, or maternal cell contamination. These molecular biological techniques offer multiple advantages, including short turnaround time and high resolutions.

CGH uses different fluorescent tags to label control (reference) DNA and test DNA which are then competitively hybridized to metaphase chromosomes. This approach could be used for screening imbalances in the whole genome in a single experiment, and its accuracy and usefulness have been demonstrated in cancer cytogenetics [20,21]. However, CGH is unable to differentiate between diploid, triploid and tetraploid states. Thus, analysis of the ploidy status is sometimes performed by the complementary use of flow cytometry [22] or FISH [23–25]. Submicroscopic chromosomal anomalies also may be diagnosed using FISH, but this is highly limited in being a 'targeted' approach, requiring knowledge that a specific karyotypic rearrangement is associated with a particular embryonic phenotype [26].

aCGH is a modification of CGH which, instead of using metaphase chromosomes, uses slides arrayed with small segments of DNA as the targets for analysis. This method is used both in PL studies and prenatal diagnosis [27]. A recent systematic review reported an increase in prenatal detection rate of 10% (95% confidence interval (CI), 8–13%) from using aCGH compared to conventional karyotyping in fetuses with structural malformations [28]. In the genetic research of pregnancy loss, microarray-based chromosome analysis has been applied to numerous cases. For example, one group of researchers have provided summary of findings from over 3000 miscarriages that underwent chromosome microarray analysis [29–31]. Clinically relevant CNVs were identified in only 1.6% of cases [32], while rare variants of uncertain clinical significance (VUS) were seen in up to 40% of cases [33]. However, despite its common use, aCGH is limited to the detection of CNVs of >10–100 kb in specifically targeted regions [26].

Despite the benefits of the aforementioned genetic testing methods, the etiology of 10 to 60% of spontaneous abortions remains unexplained despite extensive investigation [34,35]. It can be hypothesized that the majority of these cases are caused not by larger chromosomal aberrations or copy-number variants, but by other, smaller-scale genetic events undetectable by these methods. For a long time, there has not been enough evidence supporting the role of such factors as single-gene abnormalities, uniparental disomy, or genomic imprinting in the pathogenesis of miscarriages. Today, however, clinicians are paying more and more attention to the causal single nucleotide substitutions, as well as short insertions and deletions (indels) that may cause pregnancy loss [36]. This change is facilitated by the emergence and rapid spread of Next Generation Sequencing (NGS) technologies that allow for reading the entire sequence of the patient's genome (or its most relevant parts).

NGS has markedly enhanced reproductive medicine, improving prenatal diagnostic yield by identifying pathogenic genetic variants that are below the resolution of aCGH and other methods in clinical use, as well as by localizing breakpoints of cytogenetically balanced chromosomal rearrangements to individual genes [37]. More recently, NGS has become a necessary instrument for studying the genetic basis of spontaneous abortions and RPL [33], discovering new variants which are potentially related to pathogenesis of these conditions [38]. However, incomplete understanding of the molecular pathology of PL may hinders the identification of the genetic causes of PL by NGS.

In practice, NGS can be combined with previously used chromosome microarray analysis (CMA or aCGH), which opens new possibilities for a comprehensive screening of the genetic causes of pregnancy loss. Sanger sequencing is also commonly applied for validation of potential causal genetic variants, identified by NGS methods [38–40].

3. Next Generation Sequencing in the Analysis of Pregnancy Loss Genetics

Several NGS-based strategies are routinely used in medical genetics, differing in their costs and diagnostic yield. Whole-genome sequencing (WGS) is the most comprehensive of such strategies. In the field of pediatric genomics, a diagnostic yield of 42% has been reported for trio-based WGS [41]. However, WGS is expensive and demanding in terms of resources needed for data analysis, storage, and interpretation. Hence, alternative targeted strategies are commonly used, such as whole-exome sequencing (WES) and clinical exome sequencing.

Exome sequencing methods are focused on the coding part of the genome called 'exome', which makes up ≈1% of the human genome, and is estimated to contain more than 80% of the genetic mutations associated with disease [42]. WES was first introduced in 2009 and adopted quickly as a highly effective approach for postnatal and prenatal genetic diagnosis of Mendelian disorders. WES has been shown to provide a valuable diagnostic option in postnatal genetic evaluation because it is not disease- or gene-specific and does not require prior knowledge regarding the potential causal gene(s) for an observed phenotype [43]. Exome sequencing has therefore started to be incorporated into clinical care for pediatric and adult populations. Generally, WES is the preferred method of sequencing because it is cheaper than WGS and has a smaller, more manageable data set while still comprehensively covering the coding regions of DNA [36]. In terms of prenatal diagnosis, if karyotype testing and CMA cannot determine the underlying cause of fetal malformations and structural abnormalities, WES can provide relevant information to aid in current pregnancy management. Currently, WES on the products of miscarriage is helpful to identify lethal genetic variants in key genes, and it expands our knowledge of prenatal phenotypes of many Mendelian disorders [44]. However, multiple studies have been conducted over the years to demonstrate the utility of NGS methods for this purpose.

In 2013, it was hypothesized that exome sequencing was able to detect underlying genetic factors of pregnancy loss, uncovering the association between miscarriage and single or multigenic changes. Shamseldin et al. published the first report of WES on a family with recurrent pregnancy loss due to nonimmune hydrops fetalis, through which they identified a homozygous rare variant in a highly conserved region of the *CHRNA1* gene as a Mendelian cause [45]. After this proof-of-concept report for a single case, WES-based studies that included several (or even dozens of) fetuses from different families began to appear. In 2014, Carss et al. performed exome sequencing on 30 non-aneuploid fetuses and neonates with diverse structural abnormalities detected by prenatal ultrasound. They identified candidate pathogenic variants for some of the cases, concluding that exome sequencing may substantially increase the detection rate of underlying etiologies of prenatal abnormalities. In 3 out of 30 fetuses, they found highly likely causal variants in *FGFR3* and *COL2A1* genes, as well as a de novo 16.8 kb deletion that included most of *OFD1* [26]. In a further proof-of-principle study, a strategy was developed to diagnose rare autosomal recessive lethal disorders by exome sequencing in non-consanguineous couples with a history of multiple affected fetuses. The aim of the study was to obtain a molecular genetic diagnosis and enable prenatal testing in current/future pregnancies. In this work, heterozygous *DYNC2H1* variants were successfully identified as causing short-rib polydactyly (leading to pregnancy termination). Another two families presenting with a current at-risk pregnancy were then studied prospectively, and a molecular genetic diagnosis was obtained in both families through the identification of *GLE1* and *RYR1* variants causing a severe form of fetal akinesia syndrome with arthrogryposis [46].

Further application of WES detected relevant alterations in four out of seven cases of fetal demises in a cohort of American patients [47] and compound heterozygous mutations in *DYNC2H1* and *ALOX15* in two out of four miscarriages among families with recurrent pregnancy loss [33].

In the latest studies, many genetic variants related to pregnancy loss have been found by WES, further proving the contribution of point mutations to the etiology of pregnancy loss. A 2017 WES-based study detected definitely and likely causal variants in 20% and 45% of the 84 fetal death cases with ultrasound anomalies, respectively. The most frequently reported ultrasound anomalies were central nervous system abnormalities (mutations in *PIK3CA*, *FLNA*, *AMER1*, *PIK3R2*, and *L1CAM* genes) and hydrops/edema (affected genes: *PIEZO1*, *HRAS*, *RIPK4*, *FOXP3*, *MRPS22*, *CYP11A1*, *RIT1*). Abnormalities in the cardiovascular system were also observed frequently, though pathogenic or likely pathogenic variants were detected only in a single case in the *FANCB* gene [48]. In 2018, Mengu Fu et al. performed exome sequencing on 19 products of miscarriage of unrelated couples and reported 36 rare variants in 32 genes associated with miscarriage. Gene Ontology analysis of these genes revealed the enrichment of biological processes involved in early embryonic development, including 'chordate embryonic development', 'cell proliferation' and 'forebrain development' [49].

In another large-scale study, targeted sequencing on a panel of 70 genes associated with cardiac channelophies and cardiomyopathies detected pathogenic variants in 12% of 290 cases of stillbirth in Stockholm County [50]. A greater diagnostic yield was reported by Najafi et al. in an exome sequencing study of Iranian RPL cases (25–40%) in whom oligoarray CGH was normal and the maternal causes of miscarriage were ruled out. In most cases, there were no phenotypes other than lethality, and a possible variant was found in 65% of cases. In 45% of cases, the variant was classified as pathogenic/likely pathogenic according to the American College of Medical Genetics and Genomics (ACMG) guidelines [51]. Despite the differences in case selection criteria and inconsistency in variant classification from these studies, accumulated research data indicated that exome sequencing is instrumental in identifying monogenic causes of a significant portion of pregnancy loss cases and should be integrated into the diagnostic practice [52].

The largest NGS-based study conducted to date was published in 2023 by Byrne et al. [34]. In this work, the 'genomic autopsy' using exome and genome sequencing was performed as an adjunct to standard autopsy for miscarriages in 200 families. For 105 of these families who had experienced fetal or newborn death, a definitive or candidate genetic diagnosis was provided, and novel phenotypes and disease genes were described. More importantly, the study showed evidence of severe atypical in utero presentations of known genetic disorders. Pathogenic (P) and likely pathogenic (LP) variants in disease-associated genes were identified in 42 of 200 families (21%). For an additional 10 of 200 families, ACMG-classified VUS were detected, for which proof of pathogenicity was obtained using experimental methods. The majority (57.7%, 30 of 52) of variants leading to definitive diagnoses occurred de novo in the proband. In addition to the 52 families, candidate variants with no additional support of pathogenicity were discovered for further 53 families: (1) novel variants (17 of 53) or phenotypes (10 of 53) not previously described in well-established OMIM disease genes, and (2) predicted deleterious variants (26 of 53) in genes with none to limited evidence of gene–disease relationships.

4. A Systematic Review of Pregnancy Loss Genes

As demonstrated in the previous section, the focus of genetic research in pregnancy loss has shifted over the last decade from single cases to large-scale NGS-based studies. Since then, much information about causal genes and genetic variants in these genes has accumulated, but has not yet been systematically aggregated. To address this issue, we analyzed the results obtained in 31 different studies that were found in PubMed using a set of keywords presented in Figure 1. The analysis was restricted to studies which employed NGS methods for causal variant identification.

First, we compiled a list combining all 270 genetic variants found in these works, with these variants scattered across 196 unique genes (the complete list is presented as Supplementary Table S1). The dataset included genetic variants observed in euploid fetuses at all gestation ages (the exact gestation age is reported in 57% of cases, with approximately one third of these (36%) corresponding to first-trimester miscarriages). This set of variants included variants of three classes defined according to the ACMG criteria [53]—pathogenic (P), likely pathogenic (LP), and VUS. Different strategies for exome sequencing were used to identify these variants, including fetus-only, trio-, and quad-based analysis.

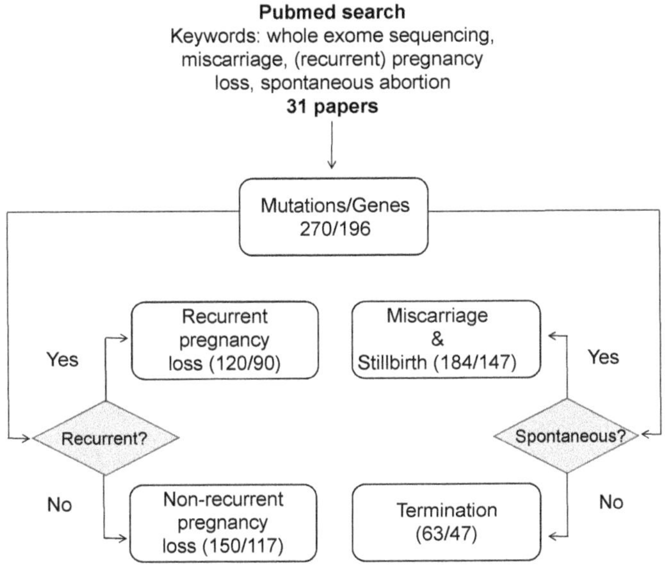

Figure 1. A schematic representation of the workflow used to review known pregnancy loss genes.

We next divided the genes and variants into categories according to spontaneity and recurrence of pregnancy loss reported in each case. Most of the variants were associated with miscarriage and stillbirth (147 unique genes/184 variants), and ≈25% of cases corresponded to willful termination of pregnancies due to developmental defects observed on ultrasound (47 genes/63 variants). For 23 variants, 'fetal demise or termination' was stated as a cause of pregnancy loss, but the exact reason was not stated for each case. Among all cases, 120 distinct variants in 90 genes were reported in recurrent pregnancy loss, while the remaining 150 mutations in 117 genes were causal for a single pregnancy loss, or information about recurrence was not provided in the original study (Figure 1).

During the analysis of these variants, we discovered 57 genes in which mutations were identified two or more times; among them, 18 genes were mentioned in independent studies as causal for pregnancy loss. The genetic variants identified in these genes in various studies are summarized in Table 1. Further analysis of the functions of these genes, as well as the clinical features of the associated pregnancy loss cases, suggested that these genes could be broadly divided into several categories: (i) early embryonic development genes with mutations leading to early lethality; (ii) genes involved in structural development of the embryo, mutations in which cause serious abnormalities that could be detected by ultrasound and can lead to both spontaneous abortion and termination; (iii) genes with important tissue-specific functions, mutations in which could occasionally lead to pregnancy loss and (iv) genes that regulate the structure and function of the uterus or placenta. Below, we briefly summarize the functions of these genes and their role in pregnancy loss.

In total, 2 of the 18 genes exemplify the first group, *PADI6* and *STIL*. *PADI6* variants have no reported postnatal phenotype; nevertheless, homozygous nonsense mutations (e.g., c.1141C>T (p.Gln381*)) were shown to be responsible for early embryonic arrest. Compound heterozygous variants in *PADI6* were also identified in embryos that displayed developmental arrest after an in vitro fertilization [54]. Notably, all cases of pregnancy loss in individuals with *PADI6* variants corresponded to first-trimester miscarriage rather than willful pregnancy termination [49,55]. This observation is consistent with the early embryonic arrest seen in the aforementioned studies and in mouse models [56,57]. *PADI6* is a part of the subcortical maternal complex (SCMC) that has an important role in genomic imprinting (reviewed in [58]), and *PADI6* mutations cause imprinted disorders [59]. Studies suggest that *PADI6* variants act as either recessive or dominant-negative maternal-effect mutations [60]. Notably, *PADI6* variants cause the same spectrum of pregnancy outcomes as variants in other SCMC component-encoding genes, including chromosomal aberrations and disturbed imprinting [61].

STIL1 is a centrosomal protein involved in the maintenance of centrosome integrity, mitotic spindle organization, and positioning—essential for centriole duplication during the cell cycle [62]. It was shown that *STIL1* is ubiquitously expressed in proliferating cells and during mouse embryonic development [63,64]. This information is consistent with data in [65], where compound heterozygous variants in *STIL1* resulted in five spontaneous miscarriages occurring between the 7th and 11th gestational week. Similarly to the case of *PADI6*, all of the reported cases of pregnancy loss due to *STIL1* mutations involved spontaneous abortion rather than termination, in good concordance with the role of the gene in early embryogenesis.

The second subgroup includes 9 genes (*DYNC2H1, FGFR2, FGFR3, FRAS1, GREB1L, LZTR1, PIEZO1, PIK3R2, PTPN11*) associated with severe developmental abnormalities that result in miscarriage or termination based on fetal disorders revealed by ultrasound. For example, the *DYNC2H1* gene encodes a protein involved in ciliary intraflagellar transport, an evolutionarily conserved process that is essential for ciliogenesis and plays a role in cell signaling events. Homozygous and compound heterozygous mutations in the *DYNC2H1* gene have been identified in patients with Jeune asphyxiating thoracic dystrophy and with short rib-polydactyly [66,67]. In the reviewed set of studies, variants in this gene were reported in cases with termination of pregnancy owing to pathologies observed on ultrasound [46] as well as mutations which were prenatally lethal [33].

Another gene in this group, the *FRAS1* gene, encodes an extracellular matrix protein which regulates epidermal-basement membrane adhesion and organogenesis during development [68]. Compound heterozygous missense mutations in this gene may lead to multiple abnormalities including a multicystic dysplastic kidney, distorted ribs and spine, brain defects and bilateral talipes equinovarus in fetus [26]. Knockouts and homozygous mutations in *FRAS1* have been shown to affect kidney and skeletal development in mice and zebrafish [69,70].

PIK3R2 in humans encodes phosphatidylinositol 3-kinase regulatory subunit beta. A notable feature of this gene is the presence of a recurrent mutation, c.1117G>A (p.Gly373Arg), which was shown to cause a spectrum of related megalencephaly syndromes in a clinical analysis of 42 children [71]. The same recurrent mutation was reported as the cause of a fetal demise in one of the recent studies [52], and the c.1690A>G (p.Lys564Glu) variant has also been implicated in miscarriage [48].

Table 1. Pregnancy loss-associated genetic variants in 18 genes reported in several studies.

Gene (Locus)	Associated Diseases	Variant	Variant Origin †	Pregnancy Outcome	Reference
PADI6 chr1p36.13	OMIM:617234	NM_207421.4:c.1793A>G (p.Asn598Ser) NM_207421.4:c.2045G>A (p.Arg682Gln)	Inherited	Miscarriage	[55]
		NM_207421:c.122C>T (p.Ala41Val)	n.a.	Miscarriage	[49]
STIL chr1p33	OMIM:612703	NM_001048166.1:c.1231C>G (p.His411Asp) NM_001048166.1:c.3370A>G (p.Met1124Val)	uncertain	Miscarriage	[65]
		NM_001048166:c. 1012C>T (p.His338Tyr)	Inherited	Miscarriage	[51]
DYNC2H1 chr11q22.3	OMIM:613091	NM_001080463.1:c.2819-14A4G	Inherited	Termination	[46]
		NM_001080463.1:c.7577T4G (p. Ile2526Ser)	Inherited	Termination	[46]
		NM_001377.3:c.6047A>G (p.Tyr2016Cys) NM_001377.3:c.6551A>T (p.Asp2184Val)	Inherited	Miscarriage	[33]
FGFR2 chr10q26.13	14 conditions	NM_000141:c.940-1G>A	n.a.	Fetal demise	[48]
		NM_022970.3:c.764G>A (p.Arg255Gln)	Inherited	Neonatal death	[34]
		NM_022970.3:c.758C>G (p.Pro253Arg)	de novo	Termination	[34]
FGFR3 chr4p16.3	14 conditions	NM_000142:c.1537G>T (p.Asp513Tyr)	n.a.	Miscarriage	[52]
		NM_000142:c.742C>T (p.Arg248Cys)	n.a.	Miscarriage	[52]
		NM_000142:c.1118A>G (p.Tyr373Cys)	n.a.	Termination	[26]
FRAS1 chr4q21.21	OMIM:219000	NM_025074:c.8537C>A (p.Ala2846Asp)	Inherited	Miscarriage	[51]
		NM_025074.7:c.1918C>T (p.Arg640Cys) NM_025074.7:c.5205C>A (p.His1735Gln)	n.a.	Termination	[26]
GREB1L chr18q11.1-q11.2	OMIM:619274 OMIM:617805	NM_001142966.2:c.5614dupA (p.Thr1872Asnfs*)	de novo	Termination	[34]
		NM_001142966:c.1305dupA (p.Asp436Argfs*32)	n.a.	Miscarriage	[52]
LZTR1 chr22q11.21	OMIM:616564 OMIM:605275	ENST00000215739.8:c.902G>T (p.Gly301Val)	de novo	Termination	[34]
		NM_006767:c.2317G>A (p.Val773Met)	n.a.	Miscarriage	[72]
PIEZO1 chr16q24.3	OMIM:194380 OMIM:616843	NM_001142864:c.1264C>T (p.Gln422Ter)	uncertain	Miscarriage	[72]
		NM_001142864:c.2035G>T (p.Glu679X)	uncertain	Termination	[48]
		NM_001142864.3:c.3206G>A (p.Trp1069Ter) NM_001142864.3:c.6208A>C (p.Lys2070Gln)	Inherited	Termination	[73]
		NM_001142864:c.30_31delAC (p.Leu10fs)	uncertain	Miscarriage	[51]
PIK3R2 chr19p13.11	OMIM:603387	NM_005027:c.1117G>A (p.Gly373Arg)	n.a.	Fetal demise	[52]
		NM_005027:c.1690A>G (p.Lys564Glu)	n.a.	Miscarriage	[48]
PTPN11 chr12q24.13	4 conditions	NM_002834:c.174C>A (p.Asn58Lys)	n.a.	Fetal demise	[48]
		NM_002834.4:c.218C>T (p.Thr73Ile)	de novo	Neonatal death	[34]
COL2A1 chr12q13.11	15 conditions	NM_001844.5:c.3864_3865delCT (p.Cys1289Pfs*)	Inherited	Termination	[34]
		NM_001844.5:c.3490G>T (p.Gly1164Cys)	de novo	Miscarriage	[26]
FOXP3 chrXp11.23	OMIM:304790	NM_014009.3:c.1009C>T (p.Arg337Ter)	Inherited	RPL	[74]
		NM_014009.3:c.906delT (p.Asp303fs*87)	Inherited	Fetal death	[75]
		NM_014009.3:c.1009C>T (p.Arg337X)	Inherited	Miscarriage	[76]
		NM_014009.3:c.1033C>T (p.Leu345Phe)	Inherited	Miscarriage	[77]
		NM_014009.3:c.1189CNT (p.Arg397Trp)	Inherited	Miscarriage	[78]
		NM_014009.3:c.319_320delTC	Inherited	Miscarriage	[78]
NEB chr2q23.3	OMIM:619334 OMIM:256030	NM_001164507:c.20974delA (p.Val6993Serfs*8)	uncertain	Miscarriage	[72]
		NM_001271208:c.24094C>T (p.Arg8032Ter) NM_001271208:c.20098C>A (p.Leu6700Ile)	uncertain	Miscarriage	[52]
RYR1 chr19q13.2	4 conditions	NM_000540.2:c.14130-2A>G NM_000540.2:c.9221C>T (p.Ser3074Phe)	Inherited	Termination	[46]
		NM_000540.2:c.6721C>T (p.Arg2241Ter)	Inherited	Miscarriage	[79]
		NM_000540.2:c.2097_2123del (p.Glu699_Gly707del)	Inherited	Termination	[79]
		NM_000540.2:c.7043delGA (p.Glu2347del)	Inherited	Termination	[79]

Table 1. *Cont.*

Gene (Locus)	Associated Diseases	Variant	Variant Origin †	Pregnancy Outcome	Reference
RYR2 chr1q43	OMIM:115000 OMIM:115000	NM_001035.2:c.409C>T (p.Arg137Trp) NM_001035.2:c.4652A>G (p.Asn1551Ser)	Inherited	RPL	[39]
		NM_001035.2:c.12526G>A (p.Val4176Met)	Inherited	Stillbirth	[34]
SCN5A chr3p22.2	9 conditions	NM_001160161:c.3749C>T (p.Thr1250Met)	Inherited	Miscarriage	[51]
		NM_198056:c.5393G>A (p.Trp1798Ter)	n.a.	Fetal demise	[52]
		NM_198056.3:c.1663G>T (p.Glu555Ter)	n.a.	Stillbirth	[50]
		NM_198056.3:c.1858C>T (p.Arg620Cys)	n.a.	Stillbirth	[50]
		NM_198056.3:c.5350G>A (p.Glu1784Lys)	n.a.	Stillbirth	[50]
GBE1 chr3p12.2	OMIM: 232500 OMIM:263570	NM_000158:c.467G>A (p.Arg156His) NM_000158:c.-35_-54del	uncertain	Miscarriage	[51]
		NM_000158:c.1064G>A (p.Arg355His)	Inherited	Fetal demise	[47]
		NM_000158:c.1543C>T (p.Arg515Cys)	Inherited	Fetal demise	[47]

†—the following notations are used: uncertain—parental genotypes were not evaluated, but the variant was homozygous or compound in the fetus; n.a.—variant origin unknown.

Variants in *GREB1L*, a gene which plays a major role in early metanephros and genital development, were associated with perinatal lethality and bilateral renal agenesis [80]. In the reviewed set of studies, fetuses with *GREB1L* mutations were aborted or stillborn due to the severity of malformations [34,52,81].

The *PIEZO1* gene encodes a large mechanosensitive ion channel that acts as a nonselective cation channel [82]. *PIEZO1* is involved in a number of crucial processes in the lungs, bladder, colon, and kidney, as well as in sensing of blood flow in the vasculature system [82]. In good concordance with its broad set of functions, *PIEZO1* mutations have been observed in cases with fetal malformations leading to early miscarriage or termination.

Mutations in the leucine zipper-like transcriptional regulator (*LZTR1*) gene were associated with Noonan syndrome (NS)—one of several RASopathies [83]. NS is a genetic disorder characterized by developmental delays, typical facial gestalt and cardiovascular defects. Mutations in multiple genes have been proven to be involved in the disturbance of the transduction signal through the RAS-MAP Kinase pathway and the manifestation of various subtypes of Noonan syndrome. It was shown that *SOS1, RAF1, KRAS, BRAF, NRAS, MAP2K1,* and *RIT1*, and recently described *SOS2, LZTR1,* and *A2ML1* are involved in the molecular etiology of this disorder. The first gene described as a causal for NS was *PTPN11* that encodes protein-tyrosine phosphatase, nonreceptor-type 11, which is involved in several developmental processes such as limb development, semilunar valvulogenesis, hemopoietic cell differentiation and mesodermal patterning [84]. Notably, *PTPN11* was also among the pregnancy loss genes reported by multiple studies.

Among genes that serve tissue-specific functions not directly linked to development, one example is *NEB* which encodes giant actin filament-associated protein. Mutations in *NEB* were associated with autosomal recessive nemaline myopathy, a disease characterized by generalized skeletal muscle weakness and the presence of electron dense protein accumulations on patient muscle biopsy examination [85]. Previously, it was shown that severe homozygous truncating *NEB* mutations may result in embryonic lethality [86]. This explains multiple observations of pathogenic *NEB* variants in pregnancy loss.

In the same group of genes, several of them lead to pregnancy loss solely in male fetuses. One of these genes, *FOXP3*, is a transcription factor which controls the activity of genes that are involved in regulating the immune system [87]. Mutations in *FOXP3* cause immune dysregulation, polyendocrinopathy, enteropathy, and X-linked (IPEX) syndrome which is a rare X-linked recessive disorder usually diagnosed in males during infancy and often fatal within the first year of life. At the same time, heterozygous female carriers of *FOXP3* mutations are unaffected and otherwise healthy. Clinical manifestations may be highly heterogeneous in patients with the same mutation, different even within the same family. IPEX syndrome may lead to type 1 diabetes mellitus, hypothyroidism, autoimmune hemolytic anemia, thrombocytopenia, lymphadenopathy, hepatitis, and nephritis [75,88]. Among the reviewed studies, there were cases in which mutations in *FOXP3* led to both miscarriage and intrauterine death or pregnancy termination based on ultrasound results (Table 1). Notably, all of the six cases of pregnancy loss with *FOXP3* mutations corresponded to male fetuses, making *FOXP3* a unique male-specific risk factor of miscarriage. A similar male-specific genetic cause of miscarriage was reported in a Chinese couple suffering from recurrent spontaneous abortion in male fetuses. In this case, WES revealed a novel c.790-6C>T mutation in the *NSDHL* gene. This syndrome is an X-linked dominant condition which leads to the lethality of the male fetuses. Females with the *NSDHL* mutation show phenotypic heterogeneity ranging from a normal to a typical CHILD syndrome phenotype. Studies have indicated that the absence of clinical symptoms may be related to X-chromosome inactivation [89,90].

SCN5A, *COL2A1*, *RYR1*, and *RYR2* genes are associated with more than two conditions that affect different body systems, such as cardiovascular, skeletal, muscle and connective tissue. Most likely, the multifunctionality of these genes explains that mutations in them lead to pregnancy loss. It is worth noting that for *COL2A1*, experimental studies showed isoform-specific epigenetic modifications consistent with imprinting [91]. *RYR1* was also reported to be imprinted in some patients with multi-minicore disease which affects muscles used for movement [92]. *RYR2* encodes another ryanodine receptor isoform, a major calcium handling channel located within cardiomyocytes [93]. Mutations in the *RYR2* gene cause the inherited arrhythmia and catecholaminergic polymorphic ventricular tachycardia [93], and, similarly to *SCN5A*, may result in stillbirth.

Mutations in *GBE1*, which encodes the glycogen branching enzyme 1, cause glycogen storage disease type IV. This is a rare metabolic disorder with an autosomal recessive inheritance that involves the liver, neuromuscular, and cardiac systems [94]. It was demonstrated that compound heterozygous mutation in *GBE1* can lead to trophoblastic damage and early fetal loss [95]. Thus, *GBE1* appears to be one of the few cases in which pathogenic variants may have a direct negative impact on the maternal–fetal interface.

It is important to note that the vast majority of the 18 genes described above (with the exception of early embryonic development ones) are associated with inherited diseases that manifest in the postnatal period. This observation indicates that mutations in many of the known PL genes have a heterogeneous phenotypic manifestation, and miscarriage is only one of the possible consequences of these variants in the fetal genotype. On the one hand, this observation is expected given that the identification of causal variants in pregnancy loss is almost inevitably based on known gene–disease relationships. On the other hand, detection of deleterious variants in the same gene both in miscarriage and in patients with monogenic disorders suggests that additional factors determine the actual phenotypic manifestation of these mutations. It may be hypothesized that epistatic genetic interactions or, more likely, the interplay between the maternal and fetal genotypes determines the outcome of pregnancy in these cases.

5. Common Properties of Known Pregnancy Loss Genes

The next goal of our review was the analysis of the complete list of 196 known pregnancy loss genes. We tried to systematically characterize the properties of these genes and define the main groups of genes involved in the pathology of pregnancy loss. First, we analyzed the biological processes they regulate via a gene set enrichment analysis method. Gene Ontology (GO) terms as well as Molecular Signatures Database (MSigDB) [96] canonical pathways were used as reference gene sets for this analysis. The analysis performed using the GO biological process terms showed enrichment for genes involved in epithelial tube morphogenesis (27 genes), control of appendage development (17 genes), as well as the urogenital/renal system and heart development (27 and 21 genes) (Figure 2A). In total, only slightly more than a half of PL genes (105) were annotated with development-related GO terms (all GO terms containing the words "development", "morphogenesis", and "formation" were considered to be related to developmental processes); at the same time, as many as 91 genes from our list did not correspond to any development-related pathway.

This observation confirms our earlier assumption that several distinct mechanisms may lead to pregnancy loss.

In addition to biological process terms, we also investigated the involvement of specific cellular components and proteins with particular molecular functions. Analysis of cellular component enrichment revealed four genes (*ARID1A, ARID1B, SMARCB1, SMARCC2*) among PL genes which are the main components of ATP-dependent chromatin remodeling complexes SWI/SNF (Figure 2B). SWI/SNF (BAF in mammals) are a group of proteins that interact to change DNA packaging and in transcription regulatory complexes. Extensive evidence suggests that SWI/SNF-mediated chromatin remodeling is critical for mammalian embryo development. Down-regulation of *SMARCC2* and *SMARCB1* results in cell cycle arrest and disturbances in gene expression [97]. The expressions of *ARID1A, ARID1B, and SMARCA2* are up-regulated in rhesus monkey blastocyst-stage embryos, implying that these subunits function during cell lineage commitment [98,99]. In good concordance with these findings, enrichment analysis for molecular functions also revealed an over-representation of genes involved in transcription (12 out of 188 genes), as well as growth factor (FGF) binding and function (*COL1A1, COL1A2, COL2A1, FGFR2, FGFR3, LIFR, FLT1, NLRP7, SCN5A*) (Figure 2C). The enrichment for FGF-related proteins is expected given the known role of FGF signaling pathways in development; during embryogenesis, FGF signaling plays an important role in the induction/maintenance of the mesoderm and the neuroectoderm, the control of morphogenetic movements, anteroposterior (AP) patterning, somitogenesis and the development of various organs [100–102].

Analysis of our gene set against MSigDB curated gene sets (C2) also revealed multiple enriched gene sets (Figure 2D), with one of the the most significant enrichments corresponding to the Ras signaling pathway (14 out of 190 genes). The Ras pathway is one of the best characterized signal transduction pathways in cell biology. The function of this pathway is to transduce signals from the extracellular milieu to the cell nucleus where specific genes are activated for cell growth, division and differentiation. The Ras/Raf/MAPK pathway is also involved in cell cycle regulation, wound healing and tissue repair, and cell migration. Finally, the Ras/Raf/MAPK pathway is able to stimulate angiogenesis through changes in expression of genes directly involved in the formation of new blood vessels [103]. All of the roles played by the Ras/Raf/MAPK pathway in development make its involvement in PL pathogenesis perfectly reasonable. It is worth noting that in Luo et al. studied where screening for novel biomarkers in the endometrium associated with RPL was conducted; differentially expressed genes of RPL patients, compared to control, were also enriched by genes involved in the Ras signaling pathway [104].

Having performed the functional analysis of pregnancy loss genes, we next turned our attention to other properties of the genes that might be used to identify most likely candidate genes for future research. Prior to this analysis, however, we split the whole set of 196 genes into subcategories depending on the specific features of the cases in which respective causal genetic variants were described (Figure 1). For all the aforementioned gene groups, we assessed several important gene-level metrics that characterize the gene in terms of its functional significance and role in the cell. These included the degree of evolutionary conservation of genes in these groups, as well as the properties of the gene expression pattern (its breadth across tissues and absolute expression levels).

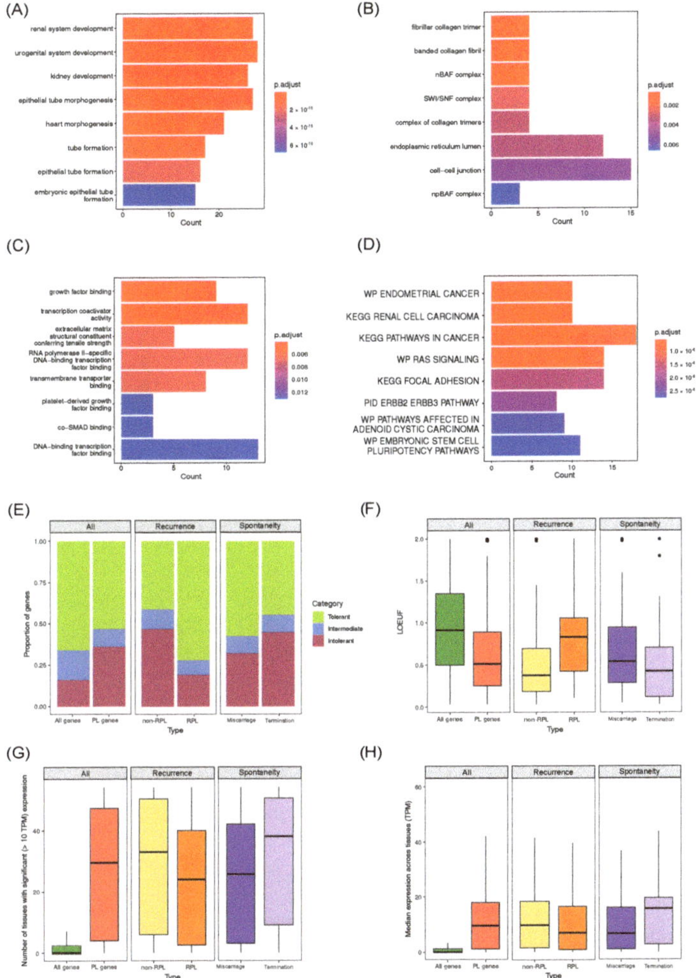

Figure 2. Analysis of the common properties of pregnancy loss genes. (**A–D**). Barplots showing gene set enrichment analysis results for the whole set of PL genes using the Gene Ontology terms and MSigDB pathways. (**A**) GO biological processes, (**B**) GO cellular components, (**C**) GO molecular functions, and (**D**) curated gene sets and canonical pathways from MSigDB. The color gradient represents the adjusted significance level. (**E,F**) Evolutionary constraint analysis of different subsets of PL genes. In (**E**), genes were grouped into three classes based on pLI value (tolerant, pLI < 0.1; intermediate, 0.1 < pLI < 0.9, and intolerant, pLI > 0.9). In (**F**), boxplots show the gene-level LOEUF values obtained from gnomAD. (**G,H**) Boxplots showing the results of gene expression level analysis for different subgroups of PL genes. In (**G**), the number of tissues with expression at the level of at least 10 transcripts per million (TPM) is shown. In (**H**), median expression across tissues is shown. In (**E–H**), All genes—all genes in the genome, PL genes—the full set of 196 genes linked to pregnancy loss.

The degree of evolutionary constraint for human genes is commonly assessed using the population frequencies of a specific class of variants that may cause functional knock-outs or knock-downs of a gene (putative loss-of-function (pLoF) variants). Such variants may substantially impact disease risk in their carriers [105]; hence, individuals carrying a pLoF allele likely have lower ability to survive and reproduce in their environment, leaving fewer descendants. Consequently, the frequency of a LoF allele is expected to be lower in the population. Therefore, observing a reduced quantity of pLoF variants in a gene compared to putatively neutral variants is indicative of their deleteriousness [106]. Measures of 'mutation intolerance' effectively rank genes by the deficit of pLoF variants in large samples [107], and several widely used measures are based on this principle, including pLI [108] and LOEUF [109]. The first metric, pLI, estimates the probability that a pLoF allele present in one copy of the gene causes a haploinsufficient phenotype, and estimates the likelihood that a gene falls into the category of LoF-haploinsufficient genes. pLI separates genes into those intolerant (pLI $\geq$ 0.9) or tolerant (pLI $\leq$ 0.1) to LoF [110]. The second metric, 'the loss-of-function observed/expected upper bound fraction' (LOEUF), represents a conservative estimate of the ratio of observed to expected pLoF variants. Both pLI and LOEUF were initially proposed by the Genome Aggregation Database (gnomAD), and the reference values for human genes are taken from this source.

We performed the evolutionary constraint analysis of the genes involved in PL using the aforementioned pLI and LOEUF metrics. A comparison with all human genes showed that, on average, the genes involved in the pathogenesis of pregnancy loss are more conserved: for these genes, the median LOEUF value was 0.52 (Figure 2F). Curiously, we observed substantial differences for a group of genes found only in cases of non-recurrent miscarriages (non-RPL) compared to other (RPL) genes. For RPL, the proportion of tolerant and intolerant genes was 0.73/0.18, while this ratio was markedly different for non-RPL—0.42/0.46, (Figure 2E). At the same time, the median LOEUF (Figure 2F) value for RPL-associated genes was 0.795, and it was much lower for non-RPL genes (0.376). The value for non-RLP genes was close to the margin of haploinsufficiency, indicating that the genes involved in a single miscarriage tend to be much more evolutionary constrained. A similar trend is seen for genes found in studies where pregnancy resulted in forced termination, with these genes being more conserved than those connected to spontaneous abortion. Comparisons of conservation metrics for SA across trimesters were performed; however, the results are not representative due to the lack of data on gestational age across studies.

To evaluate the gene expression pattern of subsets of genes involved in pregnancy loss, we used the median by-tissue expression data from the Genotype-Tissue Expression (GTEx) consortium (https://www.gtexportal.org/home/, accessed on 1 September 2023) [111], the largest available source of information on gene expression levels in various human tissues. We discovered that pregnancy loss genes are broadly expressed across tissues, with a median value of 29 tissue per gene (Figure 2G). This observation is consistent with a greater degree of evolutionary conservation of these genes. Genes associated with RPL, however, tend to be expressed in fewer tissues (24), though the number is still much higher than for an average gene (which is not expressed in any tissue at the level of 10 TPM). This observation is in perfect consistency with the results of evolutionary constraint analysis, which also showed a lower level of functional significance of these genes. It is also worth noting that causal miscarriage genes are less broadly expressed in tissues (Figure 2G) than pregnancy termination-associated genes, and their median expression level is almost two times lower (Figure 2G,H). For all investigated groups, median expression levels (Figure 2H) across all tissues were from 9 to 16 times higher than for all human genes, highlighting the ubiquitous expression of these genes.

Taken together, our analysis revealed that genes which are involved in the pathogenesis of PL are broadly expressed, highly evolutionarily conserved and involved in crucial cell differentiation and developmental processes and related signaling pathways. At the same time, relatively higher degree of evolutionary conservation and broader expression profile of genes involved in non-recurrent PL may indicate that non-genetic factors play a relatively greater role in RPL. At the same time, comparison between different groups of PL genes may be confounded by biases of individual studies.

6. Construction of a List of Candidate Pregnancy Loss Genes

Given the common properties of genes implicated in miscarriage and recurrent pregnancy loss, we next set off to use the acquired knowledge to identify a set of possible candidate genes with a similar set of properties that may be used to search for genetic causes of pregnancy loss in future studies. To achieved this, we used a multi-step strategy which is summarized in Figure 3.

At the first stage, among all human genes, we selected the ones with a degree of evolutionary constraint similar to that of the known pregnancy loss genes (the LOEUF value was below the median value for the studied list of genes). After this, broadly expressed genes were selected among the remaining 5140 ones. For this, we filtered out genes with a median expression across all tissues below the median level of known PL genes (11.2 TPM). This step narrowed the set of candidate genes down to 2777 genes. Next, we analyzed the number of the remaining genes that physically interact with known PL genes. Filtering using known protein–protein interaction data from BioGRID [112] removed firther 475 genes, leaving a broad set of 2302 genes.

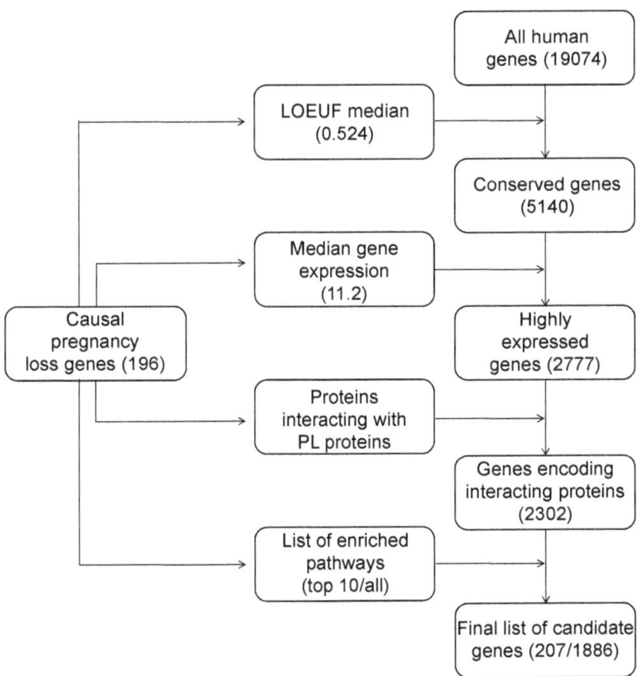

Figure 3. A schematic representation of the workflow used for prediction of candidate genes which may be involved in pregnancy loss. Numbers in barckets represent gene counts.

To further narrow down this gene list, we went on to select genes involved in pathways we identify during the gene set enrichment analysis (either all pathways or 10 most significant pathways were used in this step). This step resulted in a final set of 207 (for top 10 enriched pathways) or 1886 genes.In the narrow set of 207 genes, 21 overlapped with the known ones, and the remaining 186 could be viewed as confident candidates for investigation in future studies related to pregnancy loss (Supplementary Table S2). Out of this set of candidate genes, 103 have previously been reported as causal genes for Mendelian diseases. As many as 83 genes, however, have not yet been reported as causal for any disorder, making them the most important targets for future research.

7. Conclusions

A whole range of studies have evaluated the genetic variants which can cause pregnancy loss, including miscarriage, fetal death, stillbirth, or termination of pregnancy for severe fetal malformations. Identification of genetic variants leading to these conditions is extremely important both to deepen the understanding of their pathogenesis and to enable accurate prediction of the risk for each individual couple. Furthermore, accumulation of data supporting the causal role of short genetic variants in pregnancy loss emphasizes the need to change standard practices of genetic testing in pregnancy loss, both spontaneous and especially recurrent. While karyotyping and/or analysis of chromosomal abnormalities are commonly performed, NGS on POCs is still mostly used in research rather than in a clinical setting. Introduction of NGS methods into clinical practice for couples who suffered a loss of pregnancy may greatly enhance family planning, inform the usage of assisted reproductive technologies and consequently decrease the risk of further pregnancy losses for such couples. This is especially important given the deep psychological effects of pregnancy loss on women (demonstrated in multiple studies, e.g. [113]).

Our systematic analysis of NGS-based studies in pregnancy loss revealed a total of 196 genes with point mutations associated with PL, and 18 of these genes were reported as causal in several studies, while others were discovered only once. In-depth evaluation of this gene set suggested that the genes could be classified into several groups depending on the gestational age at pregnancy loss, as well as the molecular functions of the gene product. Further enrichment analysis of the complete list of 196 genes confirmed that these genes are involved in a broad range of pathways. While developmental genes are the dominant group, almost a half of our gene set did not correspond to any developmental biological process. This finding highlights the fact that, while disturbance of essential embryological processes may explain the majority of early miscarriages, there seems to be a plethora of other biological events that may prevent a successful pregnancy outcome. Nevertheless, further research efforts are indeed required both for the analysis of the genetic causes of early pregnancy loss and for creation of an exhaustive list of point mutations leading to all types of pregnancy losses.

In light of the latter goal, in this study, we made an effort to expand the list of candidate genes for further investigation in sequencing-based studies of pregnancy loss, with a total of 83 predicted candidate genes found to have no known disease associations. We believe that such efforts would increase the effectiveness of early detection of risk alleles and prevention of pregnancy loss.

Supplementary Materials: The supporting information can be downloaded at: https://www.mdpi.com/article/10.3390/ijms242417572/s1.

Author Contributions: E.M.M., Y.A.B., Y.A.N. and A.S.G. conceptualized the project; E.M.M., Y.A.B. and T.E.L. analyzed the data; Y.A.B., O.V.P., O.N.B., Y.A.N. and A.S.G. curated the data; E.M.M. wrote the review draft and prepared the figures. All authors participated in manuscript review and editing. All authors have read and agreed to the published version of the manuscript.

Funding: This work was supported by the Ministry of Science and Higher Education of Russian Federation (project "Multicenter research bioresource collection "Human Reproductive Health"" contract No. 075-15-2021-1058 from 28 September 2021).

Institutional Review Board Statement: Not applicable.

Informed Consent Statement: Not applicable.

Data Availability Statement: Data and code pertinent to the analyses presented in this review are available in the repository at https://github.com/mrbarbitoff/pregnancy_loss_genes.

Conflicts of Interest: The authors declare no conflict of interest.

Abbreviations

The following abbreviations are used in this manuscript:

gnomAD	genome aggregation database
pLI	probability of loss-of-function intolerance
pLoF	putative loss-of-function variants
ACMG	American College of Medical Genetics and Genomics
GO	gene ontology
GTEx	genotype tissue expression
LOEUF	loss-of-function observed-to-expected upper fraction
MSigDB	molecular signatures database
PL	pregnancy loss
POC	product of conception
RPL	recurrent pregnancy loss
SA	spontaneous abortion
TPM	transcripts per million
VUS	variant of uncertain significance

References

1. The ESHRE Guideline Group on RPL; Bender Atik, R.; Christiansen, O.B.; Elson, J.; Kolte, A.M.; Lewis, S.; Middeldorp, S.; Nelen, W.; Peramo, B.; Quenby, S.; et al. ESHRE guideline: Recurrent pregnancy loss. *Hum. Reprod. Open* **2018**, *2018*, hoy004. [CrossRef] [PubMed]
2. Tetruashvili, N.; Domar, A.; Bashiri, A. Prevention of Pregnancy Loss: Combining Progestogen Treatment and Psychological Support. *J. Clin. Med.* **2023**, *12*, 1827. [CrossRef] [PubMed]
3. El Hachem, H.; Crepaux, V.; May-Panloup, P.; Descamps, P.; Legendre, G.; Bouet, P.E. Recurrent pregnancy loss: Current perspectives. *Int. J. Women Health* **2017**, *9*, 331–345. [CrossRef] [PubMed]
4. WHO. Recommended definitions, terminology and format for statistical tables related to the perinatal period and use of a new certificate for cause of perinatal deaths. Modifications recommended by FIGO as amended October 14, 1976. *Acta Obstet. Gynecol. Scand.* **1977**, *56*, 247–253.
5. Turesheva, A.; Aimagambetova, G.; Ukybassova, T.; Marat, A.; Kanabekova, P.; Kaldygulova, L.; Amanzholkyzy, A.; Ryzhkova, S.; Nogay, A.; Khamidullina, Z.; et al. Recurrent Pregnancy Loss Etiology, Risk Factors, Diagnosis, and Management. Fresh Look into a Full Box. *J. Clin. Med.* **2023**, *12*, 4074. [CrossRef]
6. Cohain, J.S.; Buxbaum, R.E.; Mankuta, D. Spontaneous first trimester miscarriage rates per woman among parous women with 1 or more pregnancies of 24 weeks or more. *BMC Pregnancy Childbirth* **2017**, *17*, 437. [CrossRef]
7. Linnakaari, R.; Helle, N.; Mentula, M.; Bloigu, A.; Gissler, M.; Heikinheimo, O.; Niinimäki, M. Trends in the incidence, rate and treatment of miscarriage-nationwide register-study in Finland, 1998–2016. *Hum. Reprod.* **2019**, *34*, 2120–2128. [CrossRef]
8. Larsen, E.C.; Christiansen, O.B.; Kolte, A.M.; Macklon, N. New insights into mechanisms behind miscarriage. *BMC Med.* **2013**, *11*, 154. [CrossRef]
9. Qu, S.; Wang, L.; Cai, A.; Cui, S.; Bai, N.; Liu, N.; Kong, X. Exploring the cause of early miscarriage with SNP-array analysis and karyotyping. *J. Matern. Fetal Neonatal Med.* **2019**, *32*, 1–10. [CrossRef]
10. Heaney, S.; Tomlinson, M.; Aventin, Á. Termination of pregnancy for fetal anomaly: A systematic review of the healthcare experiences and needs of parents. *BMC Pregnancy Childbirth* **2022**, *22*, 441. [CrossRef]
11. Cai, M.; Lin, N.; Xu, L.; Huang, H. Comparative clinical genetic testing in spontaneous miscarriage: Insights from a study in Southern Chinese women. *J. Cell. Mol. Med.* **2021**, *25*, 5721–5728. [CrossRef]
12. Manning, M.; Hudgins, L.; Professional Practice and Guidelines Committee. Array-based technology and recommendations for utilization in medical genetics practice for detection of chromosomal abnormalities. *Genet. Med.* **2010**, *12*, 742–745. [CrossRef]
13. Suzumori, N.; Sugiura-Ogasawara, M. Genetic factors as a cause of miscarriage. *Curr. Med. Chem.* **2010**, *17*, 3431–3437. [CrossRef]
14. Bowen, P.; Lee, C.S. Spontaneous abortion. Chromosome studies on 41 cases and an analysis of maternal age and duration of pregnancy in relation to karyotype. *Am. J. Obstet. Gynecol.* **1969**, *104*, 973–983. [CrossRef] [PubMed]
15. Boué, J.; Boué, A.; Lazar, P. Retrospective and prospective epidemiological studies of 1500 karyotyped spontaneous human abortions. *Teratology* **1975**, *12*, 11–26. [CrossRef] [PubMed]
16. Menasha, J.; Levy, B.; Hirschhorn, K.; Kardon, N.B. Incidence and spectrum of chromosome abnormalities in spontaneous abortions: New insights from a 12-year study. *Genet. Med.* **2005**, *7*, 251–263. [CrossRef] [PubMed]
17. Goddijn, M.; Leschot, N. Genetic aspects of miscarriage. *Best Pract. Res. Clin. Obstet. Gynaecol.* **2000**, *14*, 855–865. [CrossRef]
18. Griffin, D.K.; Millie, E.A.; Redline, R.W.; Hassold, T.J.; Zaragoza, M.V. Cytogenetic analysis of spontaneous abortions: Comparison of techniques and assessment of the incidence of confined placental mosaicism. *Am. J. Med Genet.* **1997**, *72*, 297–301. [CrossRef]
19. Lebedev, I.N.; Ostroverkhova, N.V.; Nikitina, T.V.; Sukhanova, N.N.; Nazarenko, S.A. Features of chromosomal abnormalities in spontaneous abortion cell culture failures detected by interphase FISH analysis. *Eur. J. Hum. Genet. EJHG* **2004**, *12*, 513–520. [CrossRef]

20. Kallioniemi, A.; Kallioniemi, O.P.; Sudar, D.; Rutovitz, D.; Gray, J.W.; Waldman, F.; Pinkel, D. Comparative genomic hybridization for molecular cytogenetic analysis of solid tumors. *Science* **1992**, *258*, 818–821. [CrossRef]

21. Kallioniemi, O.P.; Kallioniemi, A.; Piper, J.; Isola, J.; Waldman, F.M.; Gray, J.W.; Pinkel, D. Optimizing comparative genomic hybridization for analysis of DNA sequence copy number changes in solid tumors. *Genes Chromosom. Cancer* **1994**, *10*, 231–243. [CrossRef] [PubMed]

22. Lomax, B.; Tang, S.; Separovic, E.; Phillips, D.; Hillard, E.; Thomson, T.; Kalousek, D.K. Comparative genomic hybridization in combination with flow cytometry improves results of cytogenetic analysis of spontaneous abortions. *Am. J. Hum. Genet.* **2000**, *66*, 1516–1521. [CrossRef] [PubMed]

23. Lestou, V.S.; Desilets, V.; Lomax, B.L.; Barrett, I.J.; Wilson, R.D.; Langlois, S.; Kalousek, D.K. Comparative genomic hybridization: A new approach to screening for intrauterine complete or mosaic aneuploidy. *Am. J. Med Genet.* **2000**, *92*, 281–284. [CrossRef]

24. Dória, S.; Carvalho, F.; Ramalho, C.; Lima, V.; Francisco, T.; Machado, A.P.; Brandão, O.; Sousa, M.; Matias, A.; Barros, A. An efficient protocol for the detection of chromosomal abnormalities in spontaneous miscarriages or foetal deaths. *Eur. J. Obstet. Gynecol. Reprod. Biol.* **2009**, *147*, 144–150. [CrossRef] [PubMed]

25. Xu, J.; Chen, M.; Liu, Q.Y.; Hu, S.Q.; Li, L.R.; Li, J.; Ma, R.M. Detecting trisomy in products of conception from first-trimester spontaneous miscarriages by next-generation sequencing (NGS). *Medicine* **2020**, *99*, e18731. [CrossRef] [PubMed]

26. Carss, K.J.; Hillman, S.C.; Parthiban, V.; McMullan, D.J.; Maher, E.R.; Kilby, M.D.; Hurles, M.E. Exome sequencing improves genetic diagnosis of structural fetal abnormalities revealed by ultrasound. *Hum. Mol. Genet.* **2014**, *23*, 3269–3277. [CrossRef]

27. Van den Veyver, I.B.; Patel, A.; Shaw, C.A.; Pursley, A.N.; Kang, S.L.; Simovich, M.J.; Ward, P.A.; Darilek, S.; Johnson, A.; Neill, S.E.; et al. Clinical use of array comparative genomic hybridization (aCGH) for prenatal diagnosis in 300 cases. *Prenat. Diagn.* **2009**, *29*, 29–39. [CrossRef]

28. Hillman, S.C.; McMullan, D.J.; Hall, G.; Togneri, F.S.; James, N.; Maher, E.J.; Meller, C.H.; Williams, D.; Wapner, R.J.; Maher, E.R.; et al. Use of prenatal chromosomal microarray: Prospective cohort study and systematic review and meta-analysis. *Ultrasound Obstet. Gynecol.* **2013**, *41*, 610–620. [CrossRef]

29. Rajcan-Separovic, E.; Qiao, Y.; Tyson, C.; Harvard, C.; Fawcett, C.; Kalousek, D.; Stephenson, M.; Philipp, T. Genomic changes detected by array CGH in human embryos with developmental defects. *Mol. Hum. Reprod.* **2010**, *16*, 125–134. [CrossRef]

30. Rajcan-Separovic, E.; Diego-Alvarez, D.; Robinson, W.P.; Tyson, C.; Qiao, Y.; Harvard, C.; Fawcett, C.; Kalousek, D.; Philipp, T.; Somerville, M.J.; et al. Identification of copy number variants in miscarriages from couples with idiopathic recurrent pregnancy loss. *Hum. Reprod.* **2010**, *25*, 2913–2922. [CrossRef]

31. Bagheri, H.; Mercier, E.; Qiao, Y.; Stephenson, M.D.; Rajcan-Separovic, E. Genomic characteristics of miscarriage copy number variants. *Mol. Hum. Reprod.* **2015**, *21*, 655–661. [CrossRef] [PubMed]

32. Levy, B.; Sigurjonsson, S.; Pettersen, B.; Maisenbacher, M.K.; Hall, M.P.; Demko, Z.; Lathi, R.B.; Tao, R.; Aggarwal, V.; Rabinowitz, M. Genomic imbalance in products of conception: Single-nucleotide polymorphism chromosomal microarray analysis. *Obstet. Gynecol.* **2014**, *124*, 202–209. [CrossRef] [PubMed]

33. Qiao, Y.; Wen, J.; Tang, F.; Martell, S.; Shomer, N.; Leung, P.C.K.; Stephenson, M.D.; Rajcan-Separovic, E. Whole exome sequencing in recurrent early pregnancy loss. *Mol. Hum. Reprod.* **2016**, *22*, 364–372. [CrossRef] [PubMed]

34. Byrne, A.B.; Arts, P.; Ha, T.T.; Kassahn, K.S.; Pais, L.S.; O'Donnell-Luria, A.; Broad Institute Center for Mendelian Genomics.; Babic, M.; Frank, M.S.B.; Feng, J.; et al. Genomic autopsy to identify underlying causes of pregnancy loss and perinatal death. *Nat. Med.* **2023**, *29*, 180–189. [CrossRef] [PubMed]

35. Normand, E.A.; Braxton, A.; Nassef, S.; Ward, P.A.; Vetrini, F.; He, W.; Patel, V.; Qu, C.; Westerfield, L.E.; Stover, S.; et al. Clinical exome sequencing for fetuses with ultrasound abnormalities and a suspected Mendelian disorder. *Genome Med.* **2018**, *10*, 74. [CrossRef] [PubMed]

36. Colley, E.; Hamilton, S.; Smith, P.; Morgan, N.V.; Coomarasamy, A.; Allen, S. Potential genetic causes of miscarriage in euploid pregnancies: A systematic review. *Hum. Reprod. Update* **2019**, *25*, 452–472. [CrossRef] [PubMed]

37. Talkowski, M.E.; Ordulu, Z.; Pillalamarri, V.; Benson, C.B.; Blumenthal, I.; Connolly, S.; Hanscom, C.; Hussain, N.; Pereira, S.; Picker, J.; et al. Clinical diagnosis by whole-genome sequencing of a prenatal sample. *N. Engl. J. Med.* **2012**, *367*, 2226–2232. [CrossRef]

38. Quintero-Ronderos, P.; Laissue, P. Genetic Variants Contributing to Early Recurrent Pregnancy Loss Etiology Identified by Sequencing Approaches. *Reprod. Sci.* **2020**, *27*, 1541–1552. [CrossRef]

39. Wang, X.; Shi, W.; Zhao, S.; Gong, D.; Li, S.; Hu, C.; Chen, Z.J.; Li, Y.; Yan, J. Whole exome sequencing in unexplained recurrent miscarriage families identified novel pathogenic genetic causes of euploid miscarriage. *Hum. Reprod.* **2023**, *38*, 1003–1018. [CrossRef]

40. Xiang, H.; Wang, C.; Pan, H.; Hu, Q.; Wang, R.; Xu, Z.; Li, T.; Su, Y.; Ma, X.; Cao, Y.; et al. Exome-Sequencing Identifies Novel Genes Associated with Recurrent Pregnancy Loss in a Chinese Cohort. *Front. Genet.* **2021**, *12*, 746082. [CrossRef]

41. Wright, C.F.; FitzPatrick, D.R.; Firth, H.V. Paediatric genomics: Diagnosing rare disease in children. *Nat. Rev. Genet.* **2018**, *19*, 253–268. [CrossRef] [PubMed]

42. Barbitoff, Y.A.; Polev, D.E.; Glotov, A.S.; Serebryakova, E.A.; Shcherbakova, I.V.; Kiselev, A.M.; Kostareva, A.A.; Glotov, O.S.; Predeus, A.V. Systematic dissection of biases in whole-exome and whole-genome sequencing reveals major determinants of coding sequence coverage. *Sci. Rep.* **2020**, *10*, 2057. [CrossRef] [PubMed]

43. Goh, G.; Choi, M. Application of whole exome sequencing to identify disease-causing variants in inherited human diseases. *Genom. Inform.* **2012**, *10*, 214–219. [CrossRef] [PubMed]
44. Xiao, Q.; Li, Z.; Lu, J. Advances of Genetic Testing Technology in Etiology Diagnosis of Recurrent Spontaneous Abortion. *Yangtze Med.* **2023**, *7*, 76–86. [CrossRef]
45. Shamseldin, H.E.; Swaid, A.; Alkuraya, F.S. Lifting the lid on unborn lethal Mendelian phenotypes through exome sequencing. *Genet. Med.* **2013**, *15*, 307–309. [CrossRef] [PubMed]
46. Ellard, S.; Kivuva, E.; Turnpenny, P.; Stals, K.; Johnson, M.; Xie, W.; Caswell, R.; Lango Allen, H. An exome sequencing strategy to diagnose lethal autosomal recessive disorders. *Eur. J. Hum. Genet. EJHG* **2015**, *23*, 401–404. [CrossRef]
47. Alamillo, C.L.; Powis, Z.; Farwell, K.; Shahmirzadi, L.; Weltmer, E.C.; Turocy, J.; Lowe, T.; Kobelka, C.; Chen, E.; Basel, D.; et al. Exome sequencing positively identified relevant alterations in more than half of cases with an indication of prenatal ultrasound anomalies. *Prenat. Diagn.* **2015**, *35*, 1073–1078. [CrossRef]
48. Yates, C.L.; Monaghan, K.G.; Copenheaver, D.; Retterer, K.; Scuffins, J.; Kucera, C.R.; Friedman, B.; Richard, G.; Juusola, J. Whole-exome sequencing on deceased fetuses with ultrasound anomalies: Expanding our knowledge of genetic disease during fetal development. *Genet. Med.* **2017**, *19*, 1171–1178. [CrossRef]
49. Fu, M.; Mu, S.; Wen, C.; Jiang, S.; Li, L.; Meng, Y.; Peng, H. Whole-exome sequencing analysis of products of conception identifies novel mutations associated with missed abortion. *Mol. Med. Rep.* **2018**, *18*, 2027–2032. [CrossRef]
50. Sahlin, E.; Gréen, A.; Gustavsson, P.; Liedén, A.; Nordenskjöld, M.; Papadogiannakis, N.; Pettersson, K.; Nilsson, D.; Jonasson, J.; Iwarsson, E. Identification of putative pathogenic single nucleotide variants (SNVs) in genes associated with heart disease in 290 cases of stillbirth. *PLoS ONE* **2019**, *14*, e0210017. [CrossRef]
51. Najafi, K.; Mehrjoo, Z.; Ardalani, F.; Ghaderi-Sohi, S.; Kariminejad, A.; Kariminejad, R.; Najmabadi, H. Identifying the causes of recurrent pregnancy loss in consanguineous couples using whole exome sequencing on the products of miscarriage with no chromosomal abnormalities. *Sci. Rep.* **2021**, *11*, 6952. [CrossRef] [PubMed]
52. Zhao, C.; Chai, H.; Zhou, Q.; Wen, J.; Reddy, U.M.; Kastury, R.; Jiang, Y.; Mak, W.; Bale, A.E.; Zhang, H.; et al. Exome sequencing analysis on products of conception: A cohort study to evaluate clinical utility and genetic etiology for pregnancy loss. *Genet. Med.* **2021**, *23*, 435–442. [CrossRef] [PubMed]
53. Richards, S.; Aziz, N.; Bale, S.; Bick, D.; Das, S.; Gastier-Foster, J.; Grody, W.W.; Hegde, M.; Lyon, E.; Spector, E.; et al. Standards and guidelines for the interpretation of sequence variants: A joint consensus recommendation of the American College of Medical Genetics and Genomics and the Association for Molecular Pathology. *Genet. Med.* **2015**, *17*, 405–424. [CrossRef] [PubMed]
54. Xu, Y.; Shi, Y.; Fu, J.; Yu, M.; Feng, R.; Sang, Q.; Liang, B.; Chen, B.; Qu, R.; Li, B.; et al. Mutations in PADI6 Cause Female Infertility Characterized by Early Embryonic Arrest. *Am. J. Hum. Genet.* **2016**, *99*, 744–752. [CrossRef] [PubMed]
55. Qian, J.; Nguyen, N.M.P.; Rezaei, M.; Huang, B.; Tao, Y.; Zhang, X.; Cheng, Q.; Yang, H.; Asangla, A.; Majewski, J.; et al. Biallelic PADI6 variants linking infertility, miscarriages, and hydatidiform moles. *Eur. J. Hum. Genet. EJHG* **2018**, *26*, 1007–1013. [CrossRef] [PubMed]
56. Liu, X.; Morency, E.; Li, T.; Qin, H.; Zhang, X.; Zhang, X.; Coonrod, S. Role for PADI6 in securing the mRNA-MSY2 complex to the oocyte cytoplasmic lattices. *Cell Cycle* **2017**, *16*, 360–366. [CrossRef] [PubMed]
57. Yurttas, P.; Vitale, A.M.; Fitzhenry, R.J.; Cohen-Gould, L.; Wu, W.; Gossen, J.A.; Coonrod, S.A. Role for PADI6 and the cytoplasmic lattices in ribosomal storage in oocytes and translational control in the early mouse embryo. *Development* **2008**, *135*, 2627–2636. [CrossRef]
58. Eggermann, T. Maternal Effect Mutations: A Novel Cause for Human Reproductive Failure. *Geburtshilfe Frauenheilkd.* **2021**, *81*, 780–788. [CrossRef]
59. Begemann, M.; Rezwan, F.I.; Beygo, J.; Docherty, L.E.; Kolarova, J.; Schroeder, C.; Buiting, K.; Chokkalingam, K.; Degenhardt, F.; Wakeling, E.L.; et al. Maternal variants in NLRP and other maternal effect proteins are associated with multilocus imprinting disturbance in offspring. *J. Med Genet.* **2018**, *55*, 497–504. [CrossRef]
60. Cubellis, M.V.; Pignata, L.; Verma, A.; Sparago, A.; Del Prete, R.; Monticelli, M.; Calzari, L.; Antona, V.; Melis, D.; Tenconi, R.; et al. Loss-of-function maternal-effect mutations of PADI6 are associated with familial and sporadic Beckwith-Wiedemann syndrome with multi-locus imprinting disturbance. *Clin. Epigenetics* **2020**, *12*, 139. [CrossRef]
61. Eggermann, T.; Kadgien, G.; Begemann, M.; Elbracht, M. Biallelic PADI6 variants cause multilocus imprinting disturbances and miscarriages in the same family. *Eur. J. Hum. Genet. EJHG* **2021**, *29*, 575–580. [CrossRef] [PubMed]
62. Vulprecht, J.; David, A.; Tibelius, A.; Castiel, A.; Konotop, G.; Liu, F.; Bestvater, F.; Raab, M.S.; Zentgraf, H.; Izraeli, S.; et al. STIL is required for centriole duplication in human cells. *J. Cell Sci.* **2012**, *125*, 1353–1362. [CrossRef] [PubMed]
63. Izraeli, S.; Colaizzo-Anas, T.; Bertness, V.L.; Mani, K.; Aplan, P.D.; Kirsch, I.R. Expression of the SIL gene is correlated with growth induction and cellular proliferation. *Cell Growth Differ.* **1997**, *8*, 1171–1179. [PubMed]
64. Izraeli, S.; Lowe, L.A.; Bertness, V.L.; Good, D.J.; Dorward, D.W.; Kirsch, I.R.; Kuehn, M.R. The SIL gene is required for mouse embryonic axial development and left-right specification. *Nature* **1999**, *399*, 691–694. [CrossRef] [PubMed]
65. Cristofoli, F.; De Keersmaecker, B.; De Catte, L.; Vermeesch, J.R.; Van Esch, H. Novel STIL Compound Heterozygous Mutations Cause Severe Fetal Microcephaly and Centriolar Lengthening. *Mol. Syndromol.* **2017**, *8*, 282–293. [CrossRef] [PubMed]
66. Schmidts, M.; Arts, H.H.; Bongers, E.M.H.F.; Yap, Z.; Oud, M.M.; Antony, D.; Duijkers, L.; Emes, R.D.; Stalker, J.; Yntema, J.B.L.; et al. Exome sequencing identifies DYNC2H1 mutations as a common cause of asphyxiating thoracic dystrophy (Jeune syndrome) without major polydactyly, renal or retinal involvement. *J. Med. Genet.* **2013**, *50*, 309–323. [CrossRef]

67. Dagoneau, N.; Goulet, M.; Geneviève, D.; Sznajer, Y.; Martinovic, J.; Smithson, S.; Huber, C.; Baujat, G.; Flori, E.; Tecco, L.; et al. DYNC2H1 mutations cause asphyxiating thoracic dystrophy and short rib-polydactyly syndrome, type III. *Am. J. Hum. Genet.* **2009**, *84*, 706–711. [CrossRef]

68. Pitera, J.E.; Scambler, P.J.; Woolf, A.S. Fras1, a basement membrane-associated protein mutated in Fraser syndrome, mediates both the initiation of the mammalian kidney and the integrity of renal glomeruli. *Hum. Mol. Genet.* **2008**, *17*, 3953–3964. [CrossRef]

69. Vrontou, S.; Petrou, P.; Meyer, B.I.; Galanopoulos, V.K.; Imai, K.; Yanagi, M.; Chowdhury, K.; Scambler, P.J.; Chalepakis, G. Fras1 deficiency results in cryptophthalmos, renal agenesis and blebbed phenotype in mice. *Nat. Genet.* **2003**, *34*, 209–214. [CrossRef]

70. Talbot, J.C.; Walker, M.B.; Carney, T.J.; Huycke, T.R.; Yan, Y.L.; BreMiller, R.A.; Gai, L.; Delaurier, A.; Postlethwait, J.H.; Hammerschmidt, M.; et al. fras1 shapes endodermal pouch 1 and stabilizes zebrafish pharyngeal skeletal development. *Development* **2012**, *139*, 2804–2813. [CrossRef]

71. Rivière, J.B.; Mirzaa, G.M.; O'Roak, B.J.; Beddaoui, M.; Alcantara, D.; Conway, R.L.; St-Onge, J.; Schwartzentruber, J.A.; Gripp, K.W.; Nikkel, S.M.; et al. De novo germline and postzygotic mutations in AKT3, PIK3R2 and PIK3CA cause a spectrum of related megalencephaly syndromes. *Nat. Genet.* **2012**, *44*, 934–940. [CrossRef] [PubMed]

72. Shamseldin, H.E.; Kurdi, W.; Almusafri, F.; Alnemer, M.; Alkaff, A.; Babay, Z.; Alhashem, A.; Tulbah, M.; Alsahan, N.; Khan, R.; et al. Molecular autopsy in maternal-fetal medicine. *Genet. Med.* **2018**, *20*, 420–427. [CrossRef] [PubMed]

73. Datkhaeva, I.; Arboleda, V.A.; Senaratne, T.N.; Nikpour, G.; Meyerson, C.; Geng, Y.; Afshar, Y.; Scibetta, E.; Goldstein, J.; Quintero-Rivera, F.; et al. Identification of novel PIEZO1 variants using prenatal exome sequencing and correlation to ultrasound and autopsy findings of recurrent hydrops fetalis. *Am. J. Med Genet. Part A* **2018**, *176*, 2829–2834. [CrossRef] [PubMed]

74. Rae, W.; Gao, Y.; Bunyan, D.; Holden, S.; Gilmour, K.; Patel, S.; Wellesley, D.; Williams, A. A novel FOXP3 mutation causing fetal akinesia and recurrent male miscarriages. *Clin. Immunol.* **2015**, *161*, 284–285. [CrossRef] [PubMed]

75. Shehab, O.; Tester, D.J.; Ackerman, N.C.; Cowchock, F.S.; Ackerman, M.J. Whole genome sequencing identifies etiology of recurrent male intrauterine fetal death. *Prenat. Diagn.* **2017**, *37*, 1040–1045. [CrossRef] [PubMed]

76. Reichert, S.L.; McKay, E.M.; Moldenhauer, J.S. Identification of a novel nonsense mutation in the FOXP3 gene in a fetus with hydrops–Expanding the phenotype of IPEX syndrome. *Am. J. Med Genet. Part A* **2016**, *170A*, 226–232. [CrossRef]

77. Vasiljevic, A.; Poreau, B.; Bouvier, R.; Lachaux, A.; Arnoult, C.; Fauré, J.; Cordier, M.P.; Ray, P.F. Immune dysregulation, polyendocrinopathy, enteropathy, X-linked syndrome and recurrent intrauterine fetal death. *Lancet* **2015**, *385*, 2120. [CrossRef]

78. Xavier-da Silva, M.M.; Moreira-Filho, C.A.; Suzuki, E.; Patricio, F.; Coutinho, A.; Carneiro-Sampaio, M. Fetal-onset IPEX: Report of two families and review of literature. *Clin. Immunol.* **2015**, *156*, 131–140. [CrossRef]

79. McKie, A.B.; Alsaedi, A.; Vogt, J.; Stuurman, K.E.; Weiss, M.M.; Shakeel, H.; Tee, L.; Morgan, N.V.; Nikkels, P.G.J.; van Haaften, G.; et al. Germline mutations in RYR1 are associated with foetal akinesia deformation sequence/lethal multiple pterygium syndrome. *Acta Neuropathol. Commun.* **2014**, *2*, 148. [CrossRef]

80. De Tomasi, L.; David, P.; Humbert, C.; Silbermann, F.; Arrondel, C.; Tores, F.; Fouquet, S.; Desgrange, A.; Niel, O.; Bole-Feysot, C.; et al. Mutations in GREB1L Cause Bilateral Kidney Agenesis in Humans and Mice. *Am. J. Hum. Genet.* **2017**, *101*, 803–814. [CrossRef]

81. Schrauwen, I.; Liaqat, K.; Schatteman, I.; Bharadwaj, T.; Nasir, A.; Acharya, A.; Ahmad, W.; Van Camp, G.; Leal, S.M. Autosomal Dominantly Inherited GREB1L Variants in Individuals with Profound Sensorineural Hearing Impairment. *Genes* **2020**, *11*, 687. [CrossRef] [PubMed]

82. Karamatic Crew, V.; Tilley, L.A.; Satchwell, T.J.; AlSubhi, S.A.; Jones, B.; Spring, F.A.; Walser, P.J.; Martins Freire, C.; Murciano, N.; Rotordam, M.G.; et al. Missense mutations in PIEZO1, which encodes the Piezo1 mechanosensor protein, define Er red blood cell antigens. *Blood* **2023**, *141*, 135–146. [CrossRef] [PubMed]

83. Sewduth, R.N.; Pandolfi, S.; Steklov, M.; Sheryazdanova, A.; Zhao, P.; Criem, N.; Baietti, M.F.; Lechat, B.; Quarck, R.; Impens, F.; et al. The Noonan Syndrome Gene Lztr1 Controls Cardiovascular Function by Regulating Vesicular Trafficking. *Circ. Res.* **2020**, *126*, 1379–1393. [CrossRef] [PubMed]

84. El Bouchikhi, I.; Belhassan, K.; Moufid, F.Z.; Iraqui Houssaini, M.; Bouguenouch, L.; Samri, I.; Atmani, S.; Ouldim, K. Noonan syndrome-causing genes: Molecular update and an assessment of the mutation rate. *Int. J. Pediatr. Adolesc. Med.* **2016**, *3*, 133–142. [CrossRef] [PubMed]

85. Yuen, M.; Ottenheijm, C.A.C. Nebulin: Big protein with big responsibilities. *J. Muscle Res. Cell Motil.* **2020**, *41*, 103–124. [CrossRef] [PubMed]

86. Shamseldin, H.E.; Tulbah, M.; Kurdi, W.; Nemer, M.; Alsahan, N.; Al Mardawi, E.; Khalifa, O.; Hashem, A.; Kurdi, A.; Babay, Z.; et al. Identification of embryonic lethal genes in humans by autozygosity mapping and exome sequencing in consanguineous families. *Genome Biol.* **2015**, *16*, 116. [CrossRef]

87. Hori, S.; Nomura, T.; Sakaguchi, S. Control of regulatory T cell development by the transcription factor Foxp3. *Science* **2003**, *299*, 1057–1061. [CrossRef]

88. Barzaghi, F.; Passerini, L.; Bacchetta, R. Immune dysregulation, polyendocrinopathy, enteropathy, x-linked syndrome: A paradigm of immunodeficiency with autoimmunity. *Front. Immunol.* **2012**, *3*, 211. [CrossRef]

89. Zhuang, J.; Luo, Q.; Xie, M.; Chen, Y.; Jiang, Y.; Zeng, S.; Wang, Y.; Xie, Y.; Chen, C. Etiological identification of recurrent male fatality due to a novel NSDHL gene mutation using trio whole-exome sequencing: A rare case report and literature review. *Mol. Genet. Genom. Med.* **2023**, *11*, e2121. [CrossRef]

90. König, A.; Happle, R.; Bornholdt, D.; Engel, H.; Grzeschik, K.H. Mutations in the NSDHL gene, encoding a 3beta-hydroxysteroid dehydrogenase, cause CHILD syndrome. *Am. J. Med Genet.* **2000**, *90*, 339–346. [CrossRef]

91. Jamieson, S.E.; de Roubaix, L.A.; Cortina-Borja, M.; Tan, H.K.; Mui, E.J.; Cordell, H.J.; Kirisits, M.J.; Miller, E.N.; Peacock, C.S.; Hargrave, A.C.; et al. Genetic and epigenetic factors at COL2A1 and ABCA4 influence clinical outcome in congenital toxoplasmosis. *PloS ONE* **2008**, *3*, e2285. [CrossRef] [PubMed]

92. Zhou, H.; Brockington, M.; Jungbluth, H.; Monk, D.; Stanier, P.; Sewry, C.A.; Moore, G.E.; Muntoni, F. Epigenetic allele silencing unveils recessive RYR1 mutations in core myopathies. *Am. J. Hum. Genet.* **2006**, *79*, 859–868. [CrossRef] [PubMed]

93. Hopton, C.; Tijsen, A.J.; Maizels, L.; Arbel, G.; Gepstein, A.; Bates, N.; Brown, B.; Huber, I.; Kimber, S.J.; Newman, W.G.; et al. Characterization of the mechanism by which a nonsense variant in RYR2 leads to disordered calcium handling. *Physiol. Rep.* **2022**, *10*, e15265. [CrossRef] [PubMed]

94. Li, Y.; Tian, C.; Huang, S.; Zhang, W.; Liutang, Q.; Wang, Y.; Ma, G.; Chen, R. Case report: Familial glycogen storage disease type IV caused by novel compound heterozygous mutations in a glycogen branching enzyme 1 gene. *Front. Genet.* **2022**, *13*, 1033944. [CrossRef]

95. Dainese, L.; Adam, N.; Boudjemaa, S.; Hadid, K.; Rosenblatt, J.; Jouannic, J.M.; Heron, D.; Froissart, R.; Coulomb, A. Glycogen Storage Disease Type IV and Early Implantation Defect: Early Trophoblastic Involvement Associated with a New GBE1 Mutation. *Pediatr. Dev. Pathol.* **2016**, *19*, 512–515. [CrossRef] [PubMed]

96. Liberzon, A.; Birger, C.; Thorvaldsdóttir, H.; Ghandi, M.; Mesirov, J.P.; Tamayo, P. The Molecular Signatures Database (MSigDB) hallmark gene set collection. *Cell Syst.* **2015**, *1*, 417–425. [CrossRef]

97. Phelan, M.L.; Sif, S.; Narlikar, G.J.; Kingston, R.E. Reconstitution of a core chromatin remodeling complex from SWI/SNF subunits. *Mol. Cell* **1999**, *3*, 247–253. [CrossRef]

98. Tseng, Y.C.; Cabot, B.; Cabot, R.A. ARID1A, a component of SWI/SNF chromatin remodeling complexes, is required for porcine embryo development. *Mol. Reprod. Dev.* **2017**, *84*, 1250–1256. [CrossRef]

99. Zheng, P.; Patel, B.; McMenamin, M.; Paprocki, A.M.; Schramm, R.D.; Nagl, N.G.; Wilsker, D.; Wang, X.; Moran, E.; Latham, K.E. Expression of genes encoding chromatin regulatory factors in developing rhesus monkey oocytes and preimplantation stage embryos: Possible roles in genome activation. *Biol. Reprod.* **2004**, *70*, 1419–1427. [CrossRef]

100. Böttcher, R.T.; Niehrs, C. Fibroblast growth factor signaling during early vertebrate development. *Endocr. Rev.* **2005**, *26*, 63–77. [CrossRef]

101. McIntosh, I.; Bellus, G.A.; Jab, E.W. The pleiotropic effects of fibroblast growth factor receptors in mammalian development. *CEll Struct. Funct.* **2000**, *25*, 85–96. [CrossRef] [PubMed]

102. Dorey, K.; Amaya, E. FGF signalling: Diverse roles during early vertebrate embryogenesis. *Development* **2010**, *137*, 3731–3742. [CrossRef] [PubMed]

103. Molina, J.R.; Adjei, A.A. The Ras/Raf/MAPK pathway. *J. Thorac. Oncol.* **2006**, *1*, 7–9. [CrossRef] [PubMed]

104. Luo, Y.; Zhou, Y. Identification of novel biomarkers and immune infiltration features of recurrent pregnancy loss by machine learning. *Sci. Rep.* **2023**, *13*, 10751. [CrossRef] [PubMed]

105. MacArthur, D.G.; Balasubramanian, S.; Frankish, A.; Huang, N.; Morris, J.; Walter, K.; Jostins, L.; Habegger, L.; Pickrell, J.K.; Montgomery, S.B.; et al. A systematic survey of loss-of-function variants in human protein-coding genes. *Science* **2012**, *335*, 823–828. [CrossRef] [PubMed]

106. Agarwal, I.; Fuller, Z.L.; Myers, S.R.; Przeworski, M. Relating pathogenic loss-of-function mutations in humans to their evolutionary fitness costs. *eLife* **2023**, *12*, e83172. [CrossRef] [PubMed]

107. Petrovski, S.; Wang, Q.; Heinzen, E.L.; Allen, A.S.; Goldstein, D.B. Genic intolerance to functional variation and the interpretation of personal genomes. *PLoS Genet.* **2013**, *9*, e1003709. [CrossRef]

108. Lek, M.; Karczewski, K.J.; Minikel, E.V.; Samocha, K.E.; Banks, E.; Fennell, T.; O'Donnell-Luria, A.H.; Ware, J.S.; Hill, A.J.; Cummings, B.B.; et al. Analysis of protein-coding genetic variation in 60,706 humans. *Nature* **2016**, *536*, 285–291. [CrossRef]

109. Karczewski, K.J.; Francioli, L.C.; Tiao, G.; Cummings, B.B.; Alföldi, J.; Wang, Q.; Collins, R.L.; Laricchia, K.M.; Ganna, A.; Birnbaum, D.P.; et al. The mutational constraint spectrum quantified from variation in 141,456 humans. *Nature* **2020**, *581*, 434–443. [CrossRef]

110. Oved, J.H.; Babushok, D.V.; Lambert, M.P.; Wolfset, N.; Kowalska, M.A.; Poncz, M.; Karczewski, K.J.; Olson, T.S. Human mutational constraint as a tool to understand biology of rare and emerging bone marrow failure syndromes. *Blood Adv.* **2020**, *4*, 5232–5245. [CrossRef]

111. GTEx Consortium. The Genotype-Tissue Expression (GTEx) project. *Nat. Genet.* **2013**, *45*, 580–585. [CrossRef] [PubMed]

112. Stark, C.; Breitkreutz, B.J.; Reguly, T.; Boucher, L.; Breitkreutz, A.; Tyers, M. BioGRID: A general repository for interaction datasets. *Nucleic Acids Res.* **2006**, *34*, D535–D539. [CrossRef] [PubMed]

113. Schwerdtfeger, K.L.; Shreffler, K.M. Trauma of pregnancy loss and infertility among mothers and involuntarily childless women in the United States. *J. Loss Trauma* **2009**, *14*, 211–227. [CrossRef] [PubMed]

International Journal of
Molecular Sciences

Article

Features and Comparative Characteristics of Fucosylated Glycans Expression in Endothelial Glycocalyx of Placental Terminal Villi in Patients with Preeclampsia Treated with Different Antihypertensive Regimens

Marina M. Ziganshina [1,*], **Galina V. Kulikova** [2], **Kamilla T. Muminova** [3], **Alexander I. Shchegolev** [2], **Ekaterina L. Yarotskaya** [4], **Zulfiya S. Khodzhaeva** [3] and **Gennady T. Sukhikh** [1,5]

[1] Laboratory of Clinical Immunology, National Medical Research Center for Obstetrics, Gynecology, and Perinatology Named after Academician V.I. Kulakov of the Ministry of Health of the Russian Federation, Oparina Str. 4, 117997 Moscow, Russia; sekretariat@oparina4.ru

[2] Department of Perinatal Pathology, National Medical Research Center for Obstetrics, Gynecology, and Perinatology Named after Academician V.I. Kulakov of the Ministry of Health of the Russian Federation, Oparina Str. 4, 117997 Moscow, Russia; gvkulikova@gmail.com (G.V.K.); ashegolev@oparina4.ru (A.I.S.)

[3] High Risk Pregnancy Department, National Medical Research Center for Obstetrics, Gynecology, and Perinatology Named after Academician V.I. Kulakov of the Ministry of Health of the Russian Federation, Oparina Str. 4, 117997 Moscow, Russia; kamika91@mail.ru (K.T.M.); zkhodjaeva@mail.ru (Z.S.K.)

[4] Department of International Cooperation, National Medical Research Center for Obstetrics, Gynecology, and Perinatology Named after Academician V.I. Kulakov of the Ministry of Health of the Russian Federation, Oparina Str. 4, 117997 Moscow, Russia; inter_otdel@mail.ru

[5] Department of Obstetrics, Gynecology, Perinatology and Reproductology, Faculty for Postgraduate and Advanced Training of Physicians, I.M. Sechenov First Moscow State Medical University of the Ministry of Health of the Russian Federation (Sechenov University), 119991 Moscow, Russia

* Correspondence: mmz@mail.ru or m_ziganshina@oparina4.ru; Tel.: +7-(495)-4381183

Citation: Ziganshina, M.M.; Kulikova, G.V.; Muminova, K.T.; Shchegolev, A.I.; Yarotskaya, E.L.; Khodzhaeva, Z.S.; Sukhikh, G.T. Features and Comparative Characteristics of Fucosylated Glycans Expression in Endothelial Glycocalyx of Placental Terminal Villi in Patients with Preeclampsia Treated with Different Antihypertensive Regimens. *Int. J. Mol. Sci.* **2023**, *24*, 15611. https://doi.org/10.3390/ijms242115611

Academic Editor: Giovanni Tossetta

Received: 14 September 2023
Revised: 18 October 2023
Accepted: 23 October 2023
Published: 26 October 2023

Abstract: Antihypertensive therapy is an essential part of management of patients with preeclampsia (PE). Methyldopa (Dopegyt®) and nifedipine (Cordaflex®) are basic medications of therapy since they stabilize blood pressure without affecting the fetus. Their effect on the endothelium of placental vessels has not yet been studied. In this study, we analyzed the effect of antihypertensive therapy on the expression of fucosylated glycans in fetal capillaries of placental terminal villi in patients with early-onset PE (EOPE) and late-onset PE (LOPE), and determined correlation between their expression and mother's hemodynamic parameters, fetoplacental system, factors reflecting inflammatory response, and destructive processes in the endothelial glycocalyx (eGC). A total of 76 women were enrolled in the study: the comparison group consisted of 15 women with healthy pregnancy, and the main group comprised 61 women with early-onset and late-onset PE, who received one-component or two-component antihypertensive therapy. Hemodynamic status was assessed by daily blood pressure monitoring, dopplerometry of maternal placental and fetoplacental blood flows, and the levels of IL-18, IL-6, TNFα, galectin-3, endocan-1, syndecan-1, and hyaluronan in the blood of the mother. Expression of fucosylated glycans was assessed by staining placental sections with AAL, UEA-I, LTL lectins, and anti-Le^Y MAbs. It was found that (i) expression patterns of fucosylated glycans in eGC capillaries of placental terminal villi in EOPE and LOPE are characterized by predominant expression of structures with a type 2 core and have a similar pattern of quantitative changes, which seems to be due to the impact of one-component and two-component antihypertensive therapy on their expression; (ii) correlation patterns indicate interrelated changes in the molecular composition of eGC fucoglycans and indicators reflecting changes in maternal hemodynamics, fetoplacental hemodynamics, and humoral factors associated with eGC damage. The presented study is the first to demonstrate the features of placental eGC in women with PE treated with antihypertensive therapy. This study also considers placental fucoglycans as a functional part of the eGC, which affects hemodynamics in the mother–placenta–fetus system.

Keywords: preeclampsia; placenta; fucosylated glycans; antihypertensive therapy; dopegyt; cordaflex; endothelial glycocalyx

1. Introduction

Preeclampsia (PE) belongs to a group of severe complications of pregnancy, with the placenta playing a central role in its pathogenesis. Two main clinical phenotypes of PE have been described. Impaired trophoblast invasion causes early-onset PE (EOPE) with manifestation before 34 weeks; the leading pathogenetic factor is placental. The maternal factor, i.e., inherent maternal cardiovascular dysfunction, causes the development of late-onset PE (LOPE), with manifestation starting from 34 weeks [1]. LOPE comprises around 80% to 95% of all PE cases, while EOPE, although less common, is associated with a higher maternal morbidity and fetal growth restriction (FGR) or neonatal mortality rates [2]. Both clinical phenotypes are characterized by ischemic-hypoxic placental lesions of varying severity. Recent studies of PE pathogenesis demonstrated that EOPE is associated with a proinflammatory placental state, while LOPE is associated with systemic inflammation in the mother. Both subtypes are associated with maternal endothelial dysfunction [3–5]. Endothelial dysfunction together with systemic inflammatory response cause maladaptation of maternal and placental hemodynamics, which clinically show as changes in the blood flow in the fetoplacental system and impaired regulation of systemic blood pressure (BP), and affect obstetric and perinatal outcomes, as well as long-term outcomes for mother and child [2].

The range of antihypertensive medications used for treatment of arterial hypertension (AH) in pregnant women is limited because, along with the decrease in the BP in the mother, the drugs must not cause any negative impact on the fetus. One-component antihypertensive therapy with the use of Dopegyt® (methyldopa), and two-component therapy—a combination of Dopegyt® and, most often, Cordaflex® (nifedipine)—are the common regimens of antihypertensive therapy currently used in clinical practice in Russia, since they meet both requirements mentioned above [6,7].

The mechanisms of effects of Dopegyt® (central-acting agent) and Cordaflex® (a long-acting calcium channel blocker) are well known [8–10]. However, their effects on the endothelial glycocalyx (eGC), the key structure that determines vascular "health" and regulates vascular tone by stimulating production of the endogenous vasodilator, nitric oxide (NO), have not yet been studied. Pro-inflammatory stimuli are among the main factors affecting the eGC, causing its shedding, which affects the mechanosensitivity of endothelial cells, changing their biochemical response, including the synthesis of NO [11]. Shedding causes exfoliation of the upper layer of the eGC, exposing hidden glycan structures [12]. Therefore, the production of pro-inflammatory cytokines and the expression of fucoglycans, either implicit or present as terminal groups, are interconnected, since both of these reflect the severity of the inflammatory response and resulting destruction of the eGC. Our previous studies showed that signs of endothelial dysfunction persist in EOPE and LOPE under both antihypertensive therapy regimens. However, a positive effect of therapy on eGC stabilization and reduction of its "desquamation", confirmed by a decrease in the content of structural components of the eGC in the mother's blood, was found only in the late PE [13]. We have also demonstrated high production of proinflammatory cytokines in the background of both antihypertensive regimens, especially in LOPE [14]. In this study, we evaluated the effect of antihypertensive therapy on the expression of fucosylated glycans of the eGC in the fetal capillaries of placental terminal villi in patients with EOPE and LOPE. Fucosylated glycans were chosen as target glycans because of their important role in the placenta [15–17], in particular, their involvement in angiogenesis and intercellular communication [18]. The study was focused on fucoglycans detected by UEA-I, LTL, and AAL lectins. According to previous publications, these lectins bind specifically to terminal clustered fucose residues linked to $\alpha 1,6$ and $\alpha 1,3$ to N-acetylglucosamine, or to fucose-containing glycans with core type 1 or core type 2 [15]. The study determined correlation patterns for each PE phenotype and antihypertensive therapy regimen; this made it possible to determine a correlation between the expression of fucosylated glycans, hemodynamic parameters of the mother and the fetoplacental system, and factors reflecting the inflammatory response and destructive processes in the eGC.

2. Results

2.1. Clinical Characteristics of the Patients Included in the Study

The patients of the study groups, according to the rank analysis of variances using the Kruskal–Wallis H test (Table 1) for "age" and "BMI" parameters, were comparable and did not significantly differ in the parameters known as the cofounders of the study. Mean values of MAP (mean arterial pressure) and DAP (diastolic arterial pressure), along with the characteristics of newborns, showed a pattern of significantly higher blood pressure and lower neonatal body weight and Apgar score in patients with pregnancy complicated by PE, regardless of gestational age and type of antihypertensive therapy, compared to control. Pairwise comparison with the Mann–Whitney U test of the groups of PE patients under two antihypertensive therapy regimens showed that patients with early- and late-onset PE who received one- and two-component therapy did not differ significantly in gestational age at delivery. This made it possible to compare clinical and laboratory data of the groups of PE patients receiving different therapy regimens at similar gestational ages.

Table 1. Clinical characteristics of the patients included in the study.

Parameter	Group 0 (NP) ($n = 15$)	Group 1 ($n = 13$)	Group 2 ($n = 16$)	Group 3 ($n = 16$)	Group 4 ($n = 16$)	p-Level
Age, years	34.0 (28–43)	32.0 (23–44)	34.0 (23–41)	31.5 (26–42)	30.5 (23–42)	$p^1 = 0.6330$
SAD med, mm Hg	119 (108–126)	145 (131–154)	143 (136–157)	141 (131–154)	137 (132–148)	$p^1 < 0.0001$
DAD med, mm Hg	75 (70–82)	94 (84–105)	100 (92–109)	101 (91–105)	91 (91–102)	$p^1 < 0.0001$
BMI, kg/m^2	27.0 (23.0–31.0)	24.0 (19.0–34.3)	25.0 (20.0–51.0)	27.0 (20.0–42.0)	27.0 (17.0–51.0)	$p^1 = 0.3184$
Gestational age, weeks	37.0 (34.0–39.0)	30.1 (26.1–33.3)	30.35 (25.3–33.4)	37.55 (36.0–39.3)	37.0 (34.0–40.3)	$p^1 < 0.0001$ $p^2 = 0.5982$ $p^3 = 0.4644$
Mean dose of medication used, mg	-	1500	2000 + 120	1000	2000 + 80	-
Newborn weight, gramms	3310 (2485–3948)	1220 (670–1840)	1335 (440–2300)	2981 (2130–3777)	2726 (1770–3920)	$p^1 < 0.0001$
Apgar 1, score	8 (7–8)	7 (4–7)	7 (2–8)	8 (8–8)	8 (6–8)	$p^1 < 0.0001$
Apgar 5, score	9 (8–9)	8 (6–8)	8 (5–9)	9 (8–9)	8 (7–9)	$p^1 = 0.0001$

p^1—significance level calculated by the Kruskal–Wallis H test. p^2—significance level calculated by the Mann–Whitney U test when comparing patients receiving antihypertensive therapy for EOPE. p^3—significance level calculated by the Mann–Whitney U test when comparing patients receiving antihypertensive therapy for LOPE. Data are presented as median (minimum, maximum value).

2.2. Description of the Specificity Profile of Lectins Used for Placental Tissue Staining

The carbohydrate structures that bind to AAL and LTL lectins were characterized in detail using a glycochip and are presented in Figure 1. The characterization of UEA-I lectin specificity was carried out previously on a chip of similar format [19]. A more detailed review of the carbohydrate specificity of these three lectins was reported by Shilova et al., 2023 [20]. The results of lectin binding to microarray glycans showed LTL interaction with fucosylated oligosaccharides that have an N-acetyllactosamine Galβ1-4GlcNAc core (type 2 carbohydrate structures), including Le^Y as well as H type 6 (Figure 1A). The specificity of LTL appeared to be similar to that of UEA-I described earlier [16]. UEA-I is also known to bind to type 2 carbohydrate structures, but the spectrum of structures is broader and includes, in addition to Le^Y and H type 6 glycans, the H type 2 disaccharide and other oligosaccharides. Despite sharing common structures in their specificity profiles, these two lectins have individual differences in the glycans that bind to them. Lectin AAL interacts with a broader spectrum of fucosylated oligosaccharides than LTL and UEA-I: both N-acetyllactosamine core and isolactosamine core (Galβ1-3GlcNAc, type 1 structures), as well as terminal fragments of blood group H, A, and B antigens (Figure 1B). Characterization of the specificity of commercial anti-Le^Y MAbs had been carried out by Ziganshina et al., 2021 [21]. Notably, anti-Le^Y MAbs was found to bind only to the difucosylated Le^Y oligosaccharide (Fucα1-2Galβ1-4(Fucα1-3)GlcNAcβ-R) [21].

2.3. Expression of Fucosylated Glycans in the Endothelium of Placental Terminal Villi

In patients with early-onset PE, significant differences in the expression of the endothelial glycocalyx of the capillaries of placental villi of all studied fucosylated glycans were revealed, depending on the applied antihypertensive therapy. A low level of expression of AAL-stained fucoglycans was detected in the endothelium of capillaries of placental villi of the patients with early-onset PE (Figure 2A). However, a comparison of groups 1 and 2 revealed significantly higher levels of fucoglycans in patients with two-component therapy. High expression of fucoglycans stained with UEA-I was found in the capillary endothelium of patients receiving single-component antihypertensive therapy. In contrast, the fucoglycan expression was low in two-component therapy (Figure 2B). Le^Y glycan, which is present in the specificity profiles of the UEA-I and LTL lectins, was more expressed in the endothelial glycocalyx of patients with two-component therapy (Figure 2C). Fucoglycans stained by the LTL (Figure 2D) showed the same pattern. Therefore, a significantly higher level of expression of Le^Y and LTL- and AAL-stained fucoglycans, but a lower level of fucoglycans stained with UEA-I, were found in the eGC of placental terminal villi of patients with early-onset PE treated with two-component therapy, than in patients taking Dopegyt alone.

The expression levels of fucoglycans in the eGC of the placental capillary villi at later terms were compared between the groups with normal and complicated pregnancy. The expression of fucoglycans stained with LTL was found to be comparable to a normal level in placenta tissue of patients with single-component therapy (Figure 2D); the findings were similar for fucoglycans stained with anti-Le^Y antibodies in the placenta tissue of patients with PE under both regimens of antihypertensive therapy (Figure 2C). Fucoglycans stained with UEA-I showed significantly increased concentrations in the eGC of placental villi in patients with one-component therapy, and decreased levels in patients with two-component therapy (Figure 2B). Relatively high expression of AAL- and LTL-anchored glycans was detected in the endothelium of patients with two-component therapy, with minimal detection of AAL-anchored fucoglycans in single-component therapy (Figure 2A,D). Thus, the expression patterns of fucoglycans in the eGC of placental terminal villi in patients with late-onset PE taking different antihypertensive therapy regimens showed multidirectional changes compared to the normal pattern. In single-component therapy, the pattern was characterized by low expression of AAL-stained fucoglycans and high expression of UEA-I-stained fucoglycans. In two-component therapy, high expression of AAL- and LTL-anchored fucoglycans was combined with low expression of UEA-I-anchored fucoglycans.

Photomicrographs demonstrating glycoconjugate staining in eGC of fetal capillaries of placental terminal villi are presented in Appendix A (Figures A1–A4).

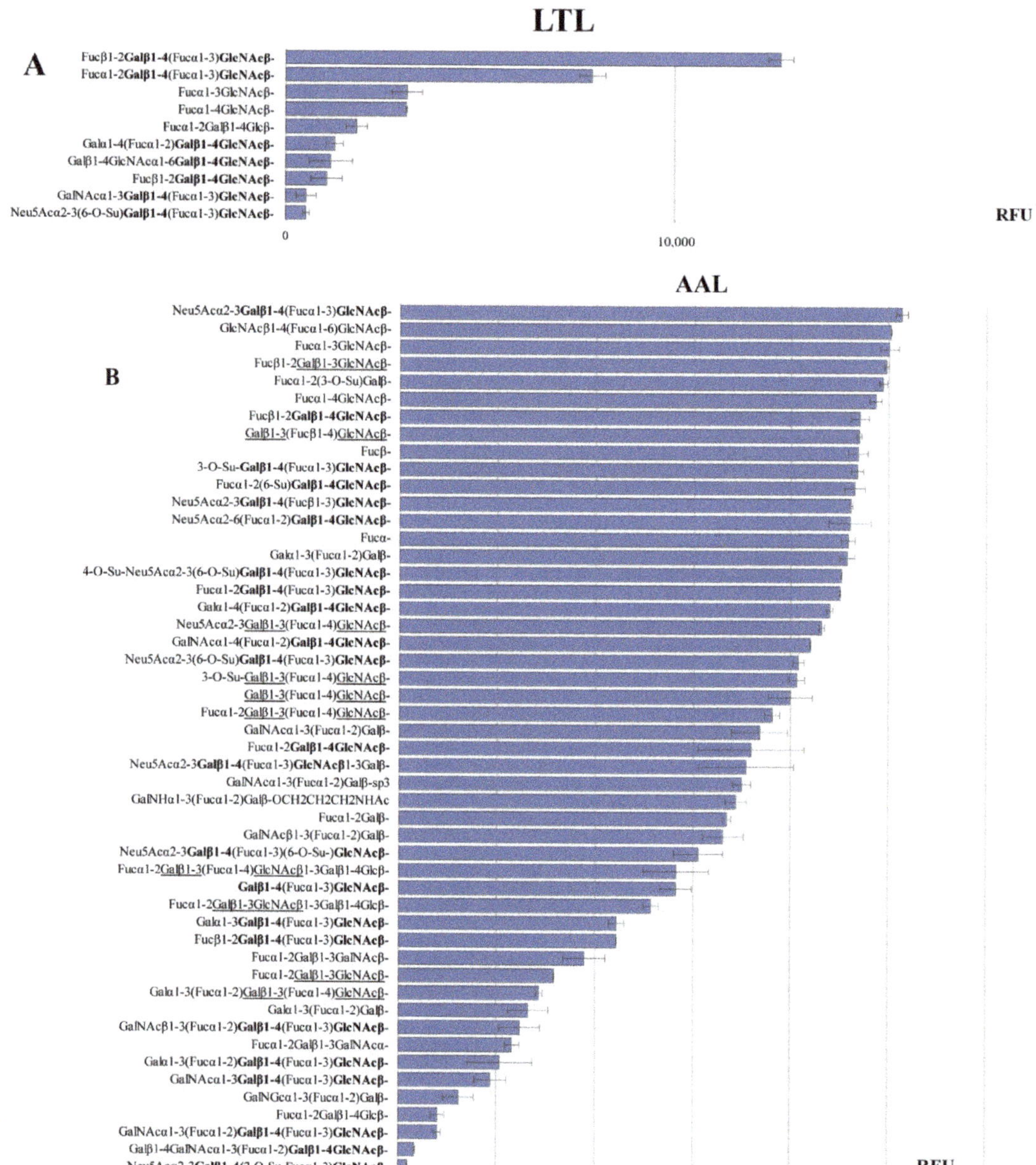

Figure 1. Carbohydrate specificity of plant lectins LTL (**A**) and AAL (**B**) determined on a glycochip. Core type 1 is underlined, core type 2 is in bold. X-axis indicates median of relative fluorescence units (RFUs) of six replicates of glycan. The median deviation was measured as an interquartile range. A signal with fluorescence intensity. RFU scale range is 0–65,535 RFU.

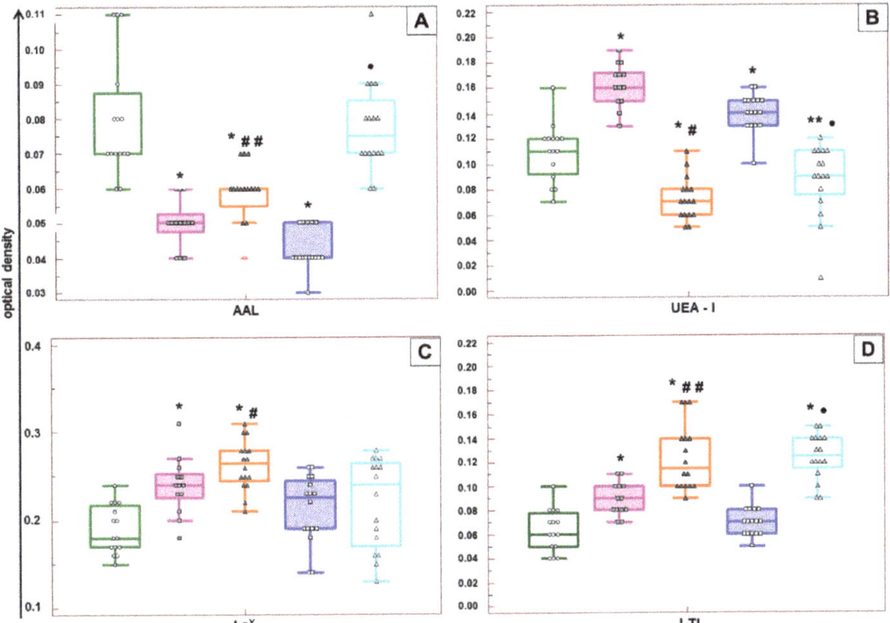

Figure 2. Contents of fucoglycans stained with AAL (**A**), UEA-I (**B**), LTL (**C**), and anti-Le^Y antibodies (**D**) in the endothelium of placental terminal villi in patients with normal pregnancy (green boxplots, $n = 15$), in patients with early-onset PE receiving single- (pink boxplots, $n = 13$) or dual-component (orange boxplots, $n = 16$) therapy; and in patients with late-onset PE receiving single- (blue boxplots, $n = 16$) or dual-component (light blue boxplots, $n = 16$) therapy. * Comparison with the level of fucoglycans in the placental samples from patients with normal pregnancy, $p \leq 0.0010$; ** comparison with the level of fucoglycans in the placental samples from patients with normal pregnancy, $p = 0.0200$; # comparison with the level of fucoglycans in the placental samples from patients with pregnancy complicated by early-onset PE and receiving single-component therapy, $p < 0.0500$; ## comparison with the level of fucoglycans in the placental samples from patients with pregnancy complicated by early-onset PE and receiving single-component therapy, $p \leq 0.0010$; • comparison with the level of fucoglycans in the placental samples from patients with pregnancy complicated by late-onset PE and receiving single-component therapy, $p < 0.0001$.

2.4. Hemodynamic Status of Early- and Late-Onset PE Patients Receiving Single- or Dual-Component Therapy

The analysis of the patients' hemodynamics showed significant changes in the daily trends of blood pressure, i.e., an increase in the maximum and average values of SAD and DAD in patients with PE with both antihypertensive therapy regimens. The main indicators of arterial stiffness—PWVao and RWTT—as well as the AIx augmentation index, did not show significant changes between the groups (Table 2). At the same time, there was a significant difference in the following indicators between the groups: (dP/dt)$_{max}$, which indirectly reflects myocardial contractility, total stiffness of the main arteries, and dynamic loading; and ED, which reflects the duration of the left ventricular ejection (Table 2). Both indices were elevated compared to Group 0 and also differed in patients at similar gestational term receiving different antihypertensive therapy regimens.

Table 2. Hemodynamic parameters of the study patients.

Parameter	Group 0 (NP) (*n* = 15)	Group 1 (*n* = 13)	Group 2 (*n* = 16)	Group 3 (*n* = 16)	Group 4 (*n* = 16)	*p*-Level
AASI	0.516 (0.213–0.810)	0.415 (0.313–0.651)	0.436 (0.179–0.729)	0.585 (0.162–0.712)	0.526 (0.170–0.836)	0.7172
AIx, %	−56 (−72–(−9))	−52 (−71–(−19))	−43 (−71–(−13))	−46 (−75–(−12))	−40.5 (−71–(−3))	0.3593
(dP/dt)$_{max}$, mm Hg/s	352 (173–598)	486 (352–681)	561 (173–861)	590 (352–973)	469 (246–811)	**0.0017**
ED, ms	271 (235–304)	343 (292–409)	304 (136–381)	336 (273–398)	322.5 (125–382)	**0.0008**
PPA, %	117 (100–189)	137 (113–189)	129 (107–151)	133 (114–146)	132 (30–144)	0.5558
PWVao, m/s	7.1 (5.8–9.8)	6.8 (5.6–8.1)	7.0 (5.8–7.8)	7.2 (5.0–9.4)	7.3 (5.8–8.2)	0.7586
RWTT, m/s	127 (105–165)	134 (105–136)	130 (105–155)	128 (103–189)	134 (99–180)	0.9402
SERV, %	101 (31–294)	102.5 (31–196)	129 (51–310)	119 (101–152)	116 (98–144)	0.1516
DAD max, mm Hg	81 (74–89)	108 (86–120)	108.5 (102–130)	105 (87–141)	100.5 (92–112)	**<0.0001**
SAD max, mm Hg	129 (124–139)	159 (134–179)	169 (142–194)	162 (146–179)	149.5 (137–188)	**<0.0001**
DAD min, mm Hg	69 (59–73)	74 (57–80)	76 (49–89)	76 (46–83)	63.5 (48–85)	0.1091
SAD min, mm Hg	105 (97–110)	112 (87–124)	113.5 (89–129)	110 (90–127)	105.5 (88–129)	0.0939

p—significance level calculated by the Kruskal–Wallis H test. AASI, arterial stiffness index; AIx, augmentation index; (dP/dt)$_{max}$, maximal blood pressure increase velocity; ED, ejection duration; max DAD, maximal aortal diastolic blood pressure; min DAD, minimal aortal diastolic blood pressure; min SAD, minimal aortal systolic blood pressure; med SAD, mean aortal systolic blood pressure; med DAD, mean aortal diastolic blood pressure; max SAD, maximal aortal systolic blood pressure; PPA, pulse pressure amplification; RWTT, reflected wave transit time; PWVao, aortic pulse wave velocity; SEVR, subendocardial viability ratio.

2.5. Evaluation of Hemodynamics in the Mother-Placenta-Fetus System

Doppler velocimetry findings (Table 3) showed a significant increase in the mean PI values for uterine arteries, umbilical arteries (but not for fetal cerebral artery), and CPR in all study groups. However, a pairwise comparison revealed that there were no significant differences in CPR and PI of umbilical arteries in patients with either early- or late-onset PE receiving one- and two-component therapy. There were also no differences found in pairwise comparisons for uterine artery PI in women receiving single- or two-component therapy in late-onset PE, or for CPR in early-onset PE (Table 3).

2.6. Determination of Cytokines, Soluble Structural Components of Endothelial Glycocalyx, and Associated Proteins in Maternal Blood

Study of the factors reflecting inflammatory reaction in the peripheral blood of the patients (Table 4) demonstrated that the IL-6 level was significantly higher in patients with PE. In addition, the highest cytokine levels were detected in women with PE using two-component therapy. TNFα levels were also elevated in patients with PE, with maximum levels found in patients with late-onset PE.

Table 3. Mean velocity indices of blood flow in uterine, umbilical cord, and fetal cerebral arteries.

Parameter	Group 0 (NP) (n = 15)	Group 1 (n = 13)	Group 2 (n = 16)	Group 3 (n = 16)	Group 4 (n = 16)	p-Level
UtA-PI	0.76 (0.55–0.91)	1.40 (1.01–1.9)	1.18 (0.48–1.98)	1.00 (0.50–2.55)	1.12 (0.04–2.13)	p^1 = 0.0010 p^2 = 0.0369 p^3 = 0.6945
UmA-PI	0.86 (0.63–1.19)	1.31 (0.81–2.00)	1.15 (0.91–3.10)	0.98 (0.71–1.09)	0.86 (0.60–1.36)	p^1 < 0.0001 p^2 = 0.2442 p^3 = 0.2216
MCA-PI	1.51 (1.20–1.96)	1.58 (0.93–2.30)	1.67 (1.13–2.63)	1.67 (1.31–2.18)	1.40 (1.25–1.55)	p^1 = 0.0990 p^2 = 0.2345 **p^3 = 0.0064**
CPR	2.06 (1.71–2.55)	0.90 (0.59–1.95)	1.63 (0.39–2.74)	1.67 (1.10–2.41)	1.58 (0.96–2.38)	**p^1 = 0.0005** p^2 = 0.1077 p^3 = 0.4491

UtA-PI—uterine artery mean pulsatility index; UmA-PI—umbilical artery pulsatility index; MCA-PI—fetal middle cerebral artery pulsatility index; CPR—cerebro–placental ratio. p^1—significance level calculated by the Kruskal–Wallis H test; p^2—significance level calculated by the Mann–Whitney U test when comparing patients receiving antihypertensive therapy for EOPE; p^3—significance level calculated by the Mann–Whitney U test when comparing patients receiving antihypertensive therapy for LOPE. Data are presented as median (minimum, maximum value).

Table 4. Concentration of cytokines, glycans, and associated proteins in maternal peripheral blood.

Parameter	Group 0 (NP) (n = 15)	Group 1 (n = 13)	Group 2 (n = 16)	Group 3 (n = 16)	Group 4 (n = 16)	p-Level
Galectin-3, ng/mL	15.17 (7.83–26.77)	13.76 (6.51–40.73)	12.97 (8.37–26.03)	13.09 (9.27–24.43)	14.82 (9.39–83.24)	p = 0.8444
IL-18, pg/mL	46.26 (14.52–242.10)	80.82 (22.41–161.60)	77.64 (12.50–195.05)	75.49 (15.40–176.12)	81.86 (24.39–207.05)	p = 0.1930
IL-6, pg/mL	0.09 (0.05–3.20)	2.58 (0.09–58.55)	3.17 (0.09–207.30)	2.81 (0.09–45.74)	3.95 (0.10–569.37)	**p = 0.0021**
TNFα, pg/mL	0.33 (0.27–0.37)	0.34 (0.33–0.41)	0.34 (0.31–0.39)	0.36 (0.29–2.80)	0.37 (0.33–2.40)	**p = 0.0111**
ESM-1, ng/mL	0.05 (0.04–0.15)	0.06 (0.04–1.01)	0.07 (90.04–0.13)	0.05 (0.04–0.14)	0.07 (0.04–0.13)	p = 0.0662
HA, ng/mL	175.89 (57.62–381.60)	124.69 (51.67–389.97)	99.10 (6.74–376.68)	185.23 (41.03–513.07)	115.23 (4.03–343.38)	p = 0.6241
SDC-1, pg/mL	1.29 (0.70–20.16)	3.29 (1.03–26.68)	1.72 (0.79–11.80)	5.72 (1.25–15.22)	2.57 (0.96–8.36)	p = 0.2555

p—significance level calculated by the Kruskal–Wallis H test. Data are presented as median (minimum, maximum value).

2.7. Correlation Analysis between the Levels of Fucosylated Glycans in the Endothelium of Placental Terminal Villi and Indicators of the Maternal Hemodynamic Profile, Fetoplacental System, and Humoral Factors of Maternal Peripheral Blood

To identify pathogenetic regularities and trends reflecting the effect of antihypertensive therapy on the expression of fucoglycans in the placenta and the relationship of these expression changes with the disorders of uteroplacental blood flow, maternal hemodynamics, and signs of sterile inflammation in blood, we conducted a study of correlations between the relative levels of glycoconjugates stained with lectins and antibodies, and a complex of indicators reflecting (1) hemodynamic status of the mother (Figure 3A); (2) functioning of the fetoplacental system (Figure 3B); and (3) the signs of sterile inflammation in maternal blood (Figure 3C). The results of correlation analysis were compared between the patients with early-onset PE treated with single-component (Group 1) and two-component (Group 2) antihypertensive regimens; and between patients at later gestational age, with physiological pregnancy (Group 0), with late-onset PE receiving single- (Group 3) and two-component (Group 4) antihypertensive therapy. The results of correlation analysis are presented in Figure 3.

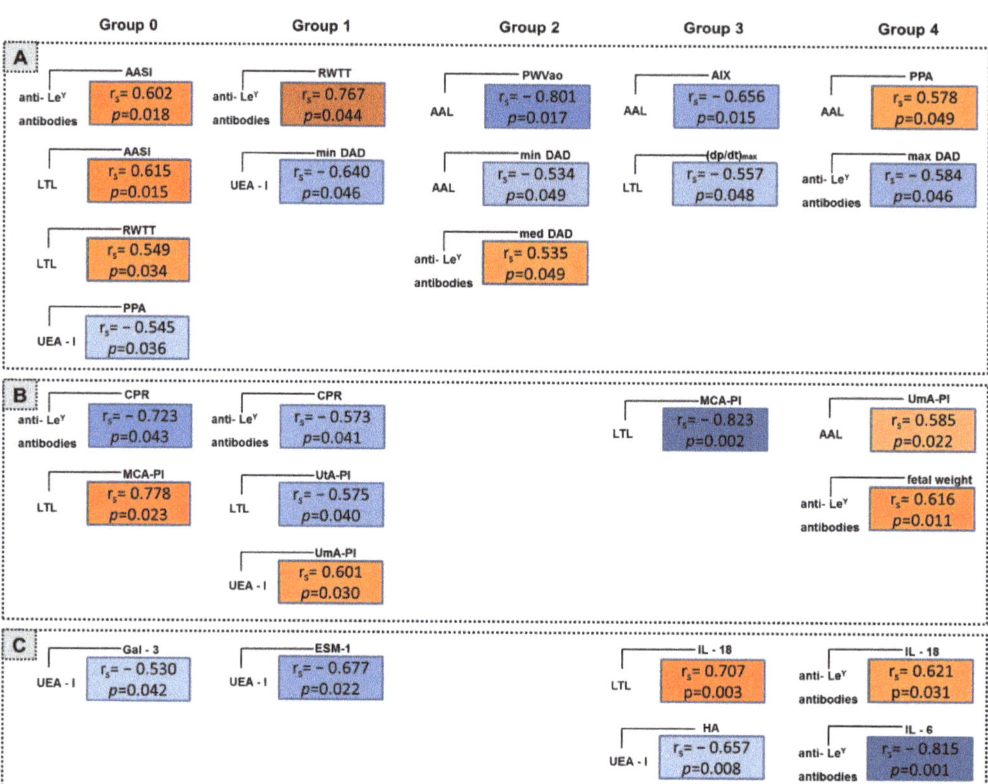

Figure 3. Correlations between the relative levels of fucoglycans stained with lectins and anti-Le^Y antibodies in the endothelium of placental terminal villi: and maternal hemodynamic parameters (**A**); fetal Dopplerometry (**B**); and the contents of factors reflecting the inflammatory reaction in peripheral blood (**C**). Group 0—patients with normal pregnancy (n = 15); Group 1—patients with early-onset PE receiving single-component therapy (n = 13); Group 2—patients with early-onset PE receiving dual-component therapy (n = 16); Group 3—patients with late-onset PE receiving single-component therapy (n = 16); Group 4—patients with late-onset PE receiving dual-component therapy (n = 16).

Different patterns of correlations were observed in patients with early-onset PE receiving single-component (Group 1) and two-component (Group 2) therapy (Figure 3). In Group 1, the associations of medium intensity (both direct and reciprocal) between the expression levels of fucoglycans stained with UEA-I, LTL, and anti-Le^Y antibodies, and various parameters from the three groups of factors presented in Figure 3, were noted. This suggests related changes in the molecular composition of the eGC of fetal placental vessels and changes in (1) maternal hemodynamics; (2) factors that are the signs of destruction of the eGC, and factors that cause this destruction; and (3) indicators of fetoplacental blood flow (Figure 3A–C). In patients with early-onset PE receiving dual-component therapy, only the associations between the level of glycoconjugates stained with AAL and anti-Le^Y antibodies and maternal hemodynamic parameters were found (Figure 3A). Correlation patterns in patients with early-onset PE included strong connection of changes in the expression of fucoglycans, especially fucoglycans with N-acetyllactosamine core (Galβ1-4GlcNAc, type 2 carbohydrate structures), with the changes in blood pressure, arterial stiffness, and destruction of maternal vascular glycocalyx, and with changes in uteroplacental and fetal-placental blood flow in patients with single-component therapy (Figure 3A–C). With two-component therapy, the pattern of correlations includes interrelated changes in the expression of fucoglycans having both core types—isolactosamine Galβ1-3GlcNAc (type 1 structures) and N-acetyllactosamine Galβ1-4GlcNAc (type 2 carbohydrate structures)—with maternal hemodynamic changes, including blood pressure and arterial stiffness parameters (Figure 3A).

The correlations in later terms of pregnancy (Groups 0, 3, and 4) had different patterns (Figure 3A–C). Patients with one-component therapy predominantly showed reciprocal connections, while patients with two-component therapy, on the contrary, showed direct connections. The patterns of correlations in Group 3 mostly involved the fucoglycans stained with LTL. In particular, reciprocal relationships were found with the pulsation index in the fetal middle cerebral artery and with the hemodynamic index $(dP/dt)_{max}$, which collectively characterizes myocardial function, dynamic load, and arterial stiffness, and is significant for intergroup comparisons. In a normal physiological state, the relationship between the mentioned indicators was also noted, but its character was opposite. Both connections were strong and, apparently, may be pathigenetically significant. Reciprocal relationships of medium strength were found between UEA-I-anchored fucoglycans and hyaluronic acid content in blood, and between AAL-anchored fucoglycans and AIx augmentation index. The only direct relationship in this pattern was found between LTL-anchored glycans and IL-18.

In contrast, in patients with two-component therapy, the pattern of correlation predominantly showed direct connections. Moreover, fucoglycans stained with anti-Le^Y antibodies exhibited both reciprocal connections with the maximum value of DAP, as well as with the IL-6 blood level (strong connection), and direct connections with the IL-18 level and the neonate's birthweight. AAL-anchored fucoglycans positively correlated with pulse pressure amplification (PPA) and the pulsatile index in the umbilical artery.

3. Discussion

Fucosylated glycans are key molecules in intercellular interactions, and are involved in signal transduction in the cell since they are a part of the cell's receptor apparatus [15,22]. The carbohydrate part of the endothelial cell receptor is a component of the eGC, a structure which determines the main functions of endothelial cells: regulation of vascular tone and vascular permeability, and adhesive interactions with cells and blood proteins. Damage, shedding, and disruption of the eGC structure critically affect the properties of endothelial cells, which change their biochemical response under the shear stress; the outcomes are hemodynamic disorders, capillary leak syndrome, and risk of the formation of blood clots. A number of studies have shown a correlation between eGC destruction and clinical symptoms of PE [23–26]. In previous studies, we have found association between eGC damage and impaired maternal hemodynamics, and characterized the molecular-functional patterns of vessels in EOPE and LOPE, which confirmed the correlation of circulating components of eGC with indicators of central hemodynamics, arterial stiffness, and myocardial changes [27]. However, eGC has been studied only in part: the majority of eGC glycans are glycosaminoglycans and carbohydrate chains of proteoglycans, but the input of other carbohydrate structures has not been investigated.

Fucosylated glycans, as a functional component of the eGC, have not been studied so far, despite the fact that their expression has been found in endothelial cells, particularly in the endothelium of the fetoplacental system [28–33]. Our previous studies suggest that the expression of fucoglycans is different in the endothelium of stem and terminal villi of normal and pathologic placentas [19,34]; this is evidence that their expression is dependent on pathophysiologic factors complicating pregnancy. Hypoxia may be one of these factors, since it (i) results from placental pathology, which is caused by morphologic or occlusive-stenotic malformations of the vessels of fetoplacental complex, and is a characteristic sign of early-onset and late-onset PE; (ii) is both a factor that stimulates the development of sterile systemic inflammation and endothelial dysfunction in response to placental damage [35,36], and a factor destabilizing the eGC [37–39], thus leading to hemodynamic disorders; (iii) is one of the key stimulating factors of angiogenesis [40,41], which is impaired in patients with placenta-associated complications of pregnancy, in particular, PE and FGR [42,43]; and (iv) is a regulator of glycan expression in human placental structures [44] (in particular, hypoxia stimulates the expression of fucosylated glycans on the surface of endothelial cells) [45,46]. It should also be noted that, in addition to hypoxia, changes in the expression of glycans, including fucoglycans, activate pro-inflammatory factors [47,48]. Based on the above facts, we suggested that the expression of fucosylated glycans in the eGC of fetal capillaries of the placental terminal villi is connected with maternal and fetoplacental hemodynamic parameters, as well as with the factors reflecting the degree of development of systemic inflammatory response and endothelial dysfunction. Since antihypertensive therapy is a necessary part of the management of patients with PE, we divided the patients according to the applied therapy, which additionally made it possible to assess the differences in the fucoglycan expression under different therapy regimens.

Histochemical and immunohistochemical studies of the placental tissue revealed similar patterns of changes in the expression of fucosylated glycans in EOPE and LOPE during one-component and two-component antihypertensive therapy; this assumes a similar impact of treatment on the expression of fucoglycans in the eGC of fetal capillaries of placental terminal villi. In particular, at both earlier and later pregnancy terms, the median expression of glycoconjugates stained with AAL, LTL lectins, and anti-Le^Y antibodies in patients treated with one-component therapy was lower than that in patients treated with two-component therapy; the opposite effect was observed in UEA-I-stained glycans. In the patients receiving two-component therapy, which was prescribed for more severe and persistent forms of hypertension, the fucoglycan expression in the endothelium of the terminal villi differed slightly only from that in the normal placenta. These results are of significant interest since the evaluation of glycan expression of the fetoplacental system in the eGC during antihypertensive therapy including Dopegyt® and Cordaflex® has not been performed before.

The specificity of lectins defined with PGA allowed us to determine that the fucoglycans with N-acetyllactosamine core Galβ1-4GlcNAc were predominantly expressed in the glycocalyx of the endothelium of placental terminal villi. The expression (in relative units of optical density) of Le^Y stained with anti-Le^Y antibodies (specific only to Le^Y glycan) was the highest in all groups, while in patients with LOPE it did not differ from normal under both treatment regimens. Despite binding to a similar type of glycan (based on type 2 structures), and the presence of common glycans in the specificity profiles, the binding activity patterns of LTL and UEA-I lectins were different. While UEA-I has a high binding activity, mainly to glycans H type 2, Le^Y, and a number of sulfated and sialylated N-acetyllactosamine derivatives, LTL has a top specificity to glycans that differ from those detected in the UEA-I profile. Moreover, LTL weakly binds to Le^Y and H disaccharide (Fucα1-4GlcNAcβ-). Despite the presence of Le^Y in the UEA-I profile, the expression patterns of UEA-I-stained glycoconjugates differ from those detected by staining with anti-Le^Y antibodies. Notably, while the expression of UEA-I- stained glycans in the endothelium of fetal capillaries of terminal villi was higher with one-component therapy than with two-component therapy, for the LTL-stained glycans the correlation was opposite. This contrast may be due to the revealed differences in the specificity profiles of the mentioned lectins, and indicates that the variety of glycans with an N-acetyllactosamine-based carbohydrate core, expressed in fetal capillary eGC, is vast and not limited only to H type 2 glycans and the difucosylated oligosaccharide Le^Y, which expression in PE has been described previously [30,33]. Also of note is a weak staining of placenta tissue with AAL, despite the broad profile of glycans to which it binds on PGA, and the large number of top glycans (i.e., showing high binding activity) in its profile. Moreover, both structures with N-acetyllactosamine core and isolactosamine core (Galβ1-3GlcNAc, type 1 structures) are present in the AAL specificity profile. However, despite the fact that H type 2 and Le^Y glycans are present in the AAL binding profile and show high binding activity to this lectin on PGA, the lectin's target in the tissues of the pathological placenta seems to have a specific composition and structure, which makes the detection of the glycans difficult. Weak binding of AAL may also be due to the lack of expression of fucoglycans with isolactosamine-based core in the eGC of capillaries of placental terminal villi. We found a similar picture in the placental tissues from women whose pregnancies were complicated by FGR: there was no staining of the fetal capillary endothelium with the bacterial lectin BL2LC-Nt, which specifically binds to fucoglycans with type 1 core [19].

To interpret the functional significance of the revealed peculiarities of fucoglycan expression, the study of correlations with certain indicators of patient's hemodynamics, hemodynamics in the fetoplacental system, and the severity of pro-inflammatory response was carried out. Intergroup analysis revealed the signs of activation of systemic inflammatory response (high level of IL-6 and TNFα) in patients with PE treated with antihypertensive drugs. Moreover, in previous studies, we found that pro-inflammatory response was more expressed in patients with LOPE [14]. This is consistent with the studies that suggest a more pronounced activation of the systemic inflammatory response, detected by maternal inflammatory markers, in LOPE than in EOPE [3].

The correlation analysis in each group revealed unique patterns of correlation between the studied factors. In particular, in patients with healthy pregnancy (Group 0), the pattern of correlations suggests that the expression of Le^Y and LTL-stained placental fucoglycans regulates maternal and fetoplacental hemodynamics. This regulation may be processed through the maintenance of the optimal expression of fucoglycans, as confirmed by the detected direct correlations with certain maternal hemodynamic parameters, in particular, arterial stiffness (AASI, RWTT) and fetal arterial hemodynamics (MCA-PI). Since an increase in blood flow, assessed by an elevation in MCA-PI, is a compensatory adaptive mechanism in hypoxia [49], and an increase in AASI and RWTT indirectly reflects an augmentation of arterial stiffness, a direct correlation between the fucoglycan expression and these parameters indicates the existence of a mechanism limiting their expression beyond a certain level. This suggestion is indirectly supported by the published data on the increased expression of Le^Y (a factor with angiogenic activity in certain pathologic conditions) [50,51], which can be explained as a compensatory effect or a mechanism of placental adaptation to hypoxia [42,52,53].

The study also revealed the reciprocal relationships in Group 0 between UEA-I-stained glycans and (i) the pulse pressure amplification (PPA), which reflects changes in central aortic pressure, and (ii) the blood levels of galectin-3, a carbohydrate-binding protein, which binds to lactosamine and its derivatives (our unpublished data), is associated with glycocalyx [54,55], and is a marker of cardiac remodeling and heart failure [56,57]. These relationships also confirm an important role of maintaining a certain level of fucoglycan expression in the eGC of placental terminal villi.

Correlation analysis for Groups 1 and 2 demonstrated certain differences in the correlation patterns. No correlation patterns were found for healthy pregnancy, but there were many correlations between the fucoglycan expression and maternal hemodynamic parameters, hemodynamics of the fetoplacental system, and the factors associated with the development of inflammatory response in Group 1, while in Group 3 the fucoglycan expression correlated only with maternal hemodynamic parameters. These data, together with a significantly different level of fucoglycan expression in placental tissues of patients in Groups 1 and 2, which was found in intergroup comparisons, indicate various types of regulatory interactions between the studied factors, possibly determined by the type of therapy.

The correlation analysis in Group 3 and Group 4 showed correlation patterns that were different from those in Group 0. Interestingly, in Group 3 a reciprocal strong correlation between the expression level of LTL-stained fucoglycans and MCA-PI was found. Moreover, a similar direct strong correlation was found in Group 0 (discussed above). Taking into account that the expression of LTL-stained fucoglycans in Group 3 was comparable to normal; the reciprocal correlation between the fucoglycan expression and the parameters that depend on arterial stiffness, $(dP/dt)_{max}$; and the strong direct correlation between the fucoglycan expression and the level of IL-18, the pathogenetic significance of these relationships was suggested. It should be noted that in LOPE, correlation patterns in both therapies included markers of inflammation and destabilization of the eGC; these findings confirm a substantial pro-inflammatory response in patients with LOPE, which does not change with antihypertensive therapy.

4. Materials and Methods

4.1. Selection of Patients for the Study

The interventional longitudinal pilot study included 76 patients. The study was conducted at the National Medical Research Center for Obstetrics, Gynecology, and Perinatology named after academician V.I. Kulakov of the Ministry of Health of the Russian Federation (hereinafter referred to as the Center) in accordance with the principles of the World Medical Association Declaration of Helsinki. The study design was approved by the local ethical committee of the Center (protocol No. 5 of 27 May 2021). All patients signed an informed consent for participation in the study. The main group included 61 patients who were divided by gestational age and type of antihypertensive therapy. A total of 29 patients with early-onset PE (up to 34 weeks) were included: 13 patients receiving one-component treatment with Dopegyt® (Group 1) and 16 patients receiving two-component antihypertensive therapy with Dopegyt® + Cordaflex® (Group 2). A total of 32 patients with late-onset PE (after 34 weeks) were enrolled: 16 patients receiving one-component (Group 3); and 16 patients receiving two-component antihypertensive therapy with Dopegyt® + Cordaflex® (Group 4). The comparison group included 15 pregnant women with physiological pregnancy at 36–39 weeks of gestation (Group 0).

According to clinical recommendations of the Ministry of Health of the Russian Federation and international protocols, PE is defined as elevation of blood pressure ≥ 140 mmHg systolic and/or 90 mmHg diastolic arterial pressure, emerging at or after 20 weeks of gestation, usually accompanied by proteinuria ≥ 0.3 g/L and/or maternal acute kidney failure, liver dysfunction, neurological signs, hemolysis or thrombocytopenia, and/or fetal growth restriction. Inclusion criterion for the main group was PE, and for the comparison group was healthy pregnancy. Exclusion criteria were pregnancy resulting from ART, severe somatic pathology, history of organ transplantation, immunotherapy during pregnancy. Withdrawal criteria were HELLP syndrome, and acute and chronic infectious and viral diseases during pregnancy. Patients were included in the study using the pair matching method based on age, BMI, and gestational age.

4.2. Description of the Antihypertensive Therapy Regimens

Antihypertensive therapy included central-acting medication Dopegyt® (average initial daily dose 750 mg, at later terms 2000 mg). Additional use of prolonged calcium channel blocker medication Cordaflex® (average initial daily dose 40 mg, at later terms up to 120 mg) was prescribed if hypertension persisted despite Dopegyt® at maximal dose intakes. The average duration of therapy in both groups was no less than 16 days. For adjustment of antihypertensive therapy, 24 h ambulatory blood pressure monitoring (ABPM) was performed using a BPLab® device (Petr Telegin LLC, Nizhny Novgorod, Russia), which is recommended for pregnant women [58].

4.3. Microarray Chip Analysis

Specificity analysis of Aleuria aurantia lectin (AAL, 2BScientific, cat. no. B-1395-1), Ulex europaeus agglutinin-I (UEA-I, Vector Laboratories, cat. no. B-1065-2), Lotus tetragonolobus lectin (LTL, Vector Laboratories, cat. no. B-1325-2), and primary mouse monoclonal Anti-Blood Group Lewis Y antibody (LWY/1463, cat. no. ab219336, ABCAM) (anti-Le^Y MAbs are monoclonal antibodies that recognize difucosylated blood group related antigen Lewis-Y (Le^Y) [Fucα1-2Galβ1-4(Fucα1-3)GlcNAcβ-R]) was performed on a polymer-coated glass slide with an N-hydroxysuccinimide-derivatized surface, produced by Schott-Nexterion (Jena, Germany), with 651 spacered glycans in 50 μM solutions in 6 replicates, as described in [21]. The main steps of microarray chip analysis included (i) microarray staining with biotinylated lectins; (ii) microarray staining with fluorescent-labeled streptavidin; (iii) obtaining microarray images using a fluorescence reader (intensity of fluorescence reflects interaction between glycan and lectin); and (iv) processing of images using software to obtain quantitative data. Microarray analysis methodology is described in detail by Knirel et al., 2014 [59].

4.4. Lectin Histochemistry and Immunohistochemistry of Placental Tissues

Placenta examination was performed on paraffin sections of samples of the paracentral area. Placentas obtained after cesarean section were subjected to macroscopic and microscopic evaluation according to the recommendations of the Amsterdam Consensus [60]. The samples stained with hematoxylin and eosin were microscopically assessed to exclude from the study tissue fragments with hemorrhages, calcificates, and massive fibrinoid deposits.

Lectin histochemistry was performed according to the manufacturer's protocol "Detection of Glycoproteins Using Lectins in Histochemistry", Vector Laboratories [61]. Biotinylated lectins were used at concentrations of AAL—5 μg/mL, UEA-I—10 μg/mL, and LTL—20 μg/mL in PBS. The tissues of mature placenta after healthy pregnancy served as positive controls; negative control reactions were performed without adding lectins to the incubation medium.

Immunohistochemical reactions were carried out on paraffin tissue sections using an automated immunohistochemical stainer (Ventana BenchMark XT, Ventana Medical Systems S.A., Kaysersberg, France), according to the manufacturer's protocol. Detection was performed using the Ventana ultraVIEW DAB Detection kit (Ventana Medical Systems, Inc.). The 2.5 μm tissue sections were deparaffinized using the EZ Prep solution (Ventana Medical Systems, Inc.; cat. no. 05279771001). Heat-induced antigen retrieval was performed using the Cell Conditioning 1 solution (Ventana Medical Systems, Inc.; cat. no. 05424569001) at 95 °C for 30 min. Endogenous peroxidase activity was blocked by treatment with the ultraVIEW inhibitor (Ventana Medical Systems, Inc.; cat no. 05269806001) in 3% H_2O_2 for 4 min.

The slides were incubated with the anti-Le^Y MAbs (2.5 μg/mL) for 64 min. The ultraVIEW Universal DAB Detection kit incorporates multimer technology, whereby the ultraVIEW streptavidin horseradish peroxidase (HRP) enzyme is directly conjugated to the secondary antibody.

Slides were incubated with a secondary antibody of ultraVIEW HRP Multimer (Ventana Medical Systems, Inc.; cat. no. 05269806001) at 37 °C for 8 min and a diaminobenzidine + H_2O_2 substrate for 8 min, which was followed by counterstaining with hematoxylin II (Ventana Medical Systems, Tucson, AZ, USA, cat. no. 05277965001) and bluing reagent (Ventana Medical Systems, Tucson, AZ, USA, cat. no. 05266769001) for 2 and 3 min, respectively. Slides were washed with Tris buffer (pH 7.6) and mounted using a xylene-based mounting media.

Stained lung carcinoma slices were considered as a positive control, while reactions without anti-Le^Y MAb antibody were considered as a negative control.

Quantitative evaluation of the results of lectin histochemistry and immunohistochemistry was performed by measuring the optical density of the stained products in the endothelial membrane of the capillaries of the placental terminal villi. To obtain representative data, measurements were performed in 10 randomly selected equally distant fields of view. A Nikon eclipse 80i light microscope, Nikon DS-Fi1 digital camera (Nicon Corporation, Tokyo, Japan) and NIS Elements Advanced Research 4.1 image analysis software were used for image analysis.

4.5. Evaluation of Hemodynamics

The oscillometric method of BP measurement was applied. Twenty-four-hour BP monitoring was performed using the BPLab® monitor (Petr Telegin LLC, Nizhny Novgorod, Russia), which meets international standards of accuracy for oscillometric BP recorders and is recommended for pregnant women [58]. Oscillograms were analyzed using Vasotens software, BPLab® V.06.02.00 (Nizhny Novgorod, Russia). Mean central systolic and diastolic BP were measured during the day and night time, within 24 h. Regular intervals between measurements were 30 min at daytime and 60 min at night. The parameters characterizing changes in blood pressure (BP) were determined: max DAD; max SAD; min DAD; max SAD; med DAD; med SAD. Parameters determined by Vasotens software, BPLab® V.06.02.00 were similar to those used in the study by [27] and included the following:

1. Reflected wave transit time (RWTT): the return time of the wave reflected from aorta;
2. Aortic pulse wave velocity (PWVao): the velocity at which blood pressure pulse propagates into the aorta;
3. Augmentation index (AIx): a noninvasive measure of pulse wave reflection;
4. Arterial stiffness index (AASI);
5. Ambulatory rigidity index (AASI = 1 − (inclination BPdiastolic − BPsystolic));
6. Maximal BP increase velocity ($(dP/dt)_{max}$): indirectly represents myocardial contractility, total vascular resistance, and dynamic load of pulse wave on vascular walls;
7. Pulse pressure amplification (PPA): the increase in pulse pressure (PP) amplitude when pressure waves propagate distally in the systemic network, accompanied by morphological alterations of pressure waveforms;
8. Ejection duration (ED): an interval of blood flow from the start of pulsation till the closure of the aortic valve;
9. Subendocardial viability ratio (SEVR), defined as diastolic to systolic pressure–time integral ratio.

4.6. Arterial Doppler

Dopplerometry of maternal–placental and fetoplacental blood flows was performed using transabdominal transducer of the expert ultrasound machines (The Voluson E8 ultrasound system, GE Healthcare Austria GmbH&Co OG, (Tiefenbac, Austria)). Uterine arteries (UtA), umbilical artery (UA), middle cerebral artery (MCA), and ductus venosus pulsatility index (PI), as well as cerebroplacental ratio, were measured. When evaluating umbilical artery PI, positive or absent diastolic flow was assessed. Ductus venosus was analyzed through the cross-section and sagittal plane of the fetal abdomen using Doppler color flow mapping. The curve was defined as normal if the A-wave was positive, or abnormal in case of absent or negative A-wave.

4.7. Determination of Soluble Factors in Maternal Blood

Blood sampling was performed after fasting. Serum samples were collected into vacuum blood-collecting tubes, S-Monovette® Serum, 4.9 mL, cap white, (L×Ø): 90 × 13 mm. Serum preparation for the study was performed according to the standard operating procedure of the Center's Biobank, where the samples were stored at −80 °C. The study of humoral factors was carried out via the ELISA method using commercial test systems for the determination of cytokines: IL-18 (BMS267-2, Bender MedSystem GmbH, Wien, Austria), IL-6 (A-8768, Vector Best, Novosibirsk, Russia), TNFα (BMS223-4, Bender MedSystem GmbH, Wien, Austria), and galectin-3 (BMS279-4, Bender MedSystem GmbH, Wien, Austria). Determination of desquamated structural components of eGC was performed on test systems manufactured by Cloud-Clone Corp., Katy, TX, USA: SEC463Hu (determination of endocan-1); SEB966Hu (determination of syndecan-1); CEA182Ge (determination of hyaluronic acid).

4.8. Statistical Analysis

The software used for statistical analysis was "Statistica 10.0" (StatSoft, Inc., Tulsa, OH, USA) and "MedCalc version 16.4" (MedCalc Software Ltd, Ostend, Belgium). Microarray data are presented using descriptive statistics. The data were processed using ScanArray Express 4.0 (PerkinElmer Life & Analytical Sciences, Shelton, CT, USA) with measurement of median of relative fluorescent units (RFU) of 6 replicates of each glycan on the microarray. The median deviation was measured as an interquartile range. A signal with fluorescence intensity 5 times exceeding the background value was considered significant as described in [62].

The evaluation of clinical, (immuno)histochemical, and biochemical data was conducted using non-parametric methods, since the assessments of skewness and kurtosis in the analyzed distributions significantly deviated from zero. The Kolmogorov–Smirnov test denied the normality of the majority of the distributions. The data are presented as median (minimum, maximum value); the differences in comparisons of three or more groups by a quantitative indicator, whose distribution differed from normal, were calculated using the Kruskal–Wallis analysis of variance with Bonferroni correction ($p < 0.025$). This method was used to compare three groups in later terms of pregnancy. A posteriori comparisons were performed using the Mann–Whitney U test. For pairwise comparisons of data in earlier terms of pregnancy, only the Mann–Whitney U test was used. Differences between parameters were considered significant at $p < 0.05$.

5. Conclusions

The patterns of expression of fucosylated glycans in the eGC of the capillaries of placental terminal villi of patients with EOPE and LOPE have a specific fucoglycan composition, with the predominant expression of fucoglycans with type 2 core, which are detected by lectins with different carbohydrate specificity profiles. Regardless of the fucoglycan composition, their expression patterns in the eGC show similar quantitative changes at earlier and later terms of pregnancy with PE, which, apparently, is due to the similar effects of one-component and two-component antihypertensive therapy. The correlation patterns indicate interrelated changes in the molecular composition of eGC fucoglycans in fetal capillaries of the terminal villi and indicators of the maternal hemodynamics, fetoplacental hemodynamics, and humoral factors associated with eGC damage. The present study is the first to demonstrate the features of placental eGC in women with PE treated with antihypertensive therapy; the study also considers placental fucoglycans as a functional part of the eGC, which has an influence on the hemodynamics in the mother–placenta–fetus system.

Author Contributions: Conceptualization, M.M.Z., Z.S.K. and G.T.S.; methodology, M.M.Z. and Z.S.K.; formal analysis, M.M.Z., G.V.K., K.T.M., A.I.S., E.L.Y., Z.S.K. and G.T.S.; investigation, M.M.Z., G.V.K. and K.T.M.; data curation, M.M.Z., G.V.K., K.T.M. and A.I.S.; writing—original draft preparation, M.M.Z., G.V.K. and K.T.M.; writing—review and editing, M.M.Z., K.T.M., E.L.Y. and G.T.S.; supervision, M.M.Z., Z.S.K. and G.T.S. All authors have read and agreed to the published version of the manuscript.

Funding: This research was funded by of the State Task of the Ministry of Health of the Russian Federation № 121040600435-0 "Justification of personalized approaches to antihypertensive therapy in HDP and PE".

Institutional Review Board Statement: The study was conducted in accordance with the Declaration of Helsinki, and approved by the Institutional Ethics Committee of the National Medical Research Center for Obstetrics, Gynecology, and Perinatology named after academician V.I. Kulakov of the Ministry of Health of the Russian Federation (protocol No. 5 of 27 May 2021).

Informed Consent Statement: Informed consent was obtained from all subjects involved in the study.

Data Availability Statement: The datasets generated during and/or analyzed during the current study are available from the corresponding author on reasonable request.

Acknowledgments: The authors wish to thank the Laboratory for Collection and Storage of Biological Material (Biobank) and the staff of the Department of Perinatal Pathology for providing blood samples and for their assistance in histochemical studies. The team of authors is grateful to Khasbiullina N.R., a researcher of the Laboratory of Clinical Immunology of the National Medical Research Center for Obstetrics, Gynecology, and Perinatology named after Academician V.I. Kulakov, for determining the specificity of lectins and antibodies on PGA.

Conflicts of Interest: All authors declare that they have no conflict or financial interest. The funders had no role in the design of the study; in the collection, analyses, or interpretation of data; in the writing of the manuscript; or in the decision to publish the results.

Appendix A. Figures A1–A4

AAL

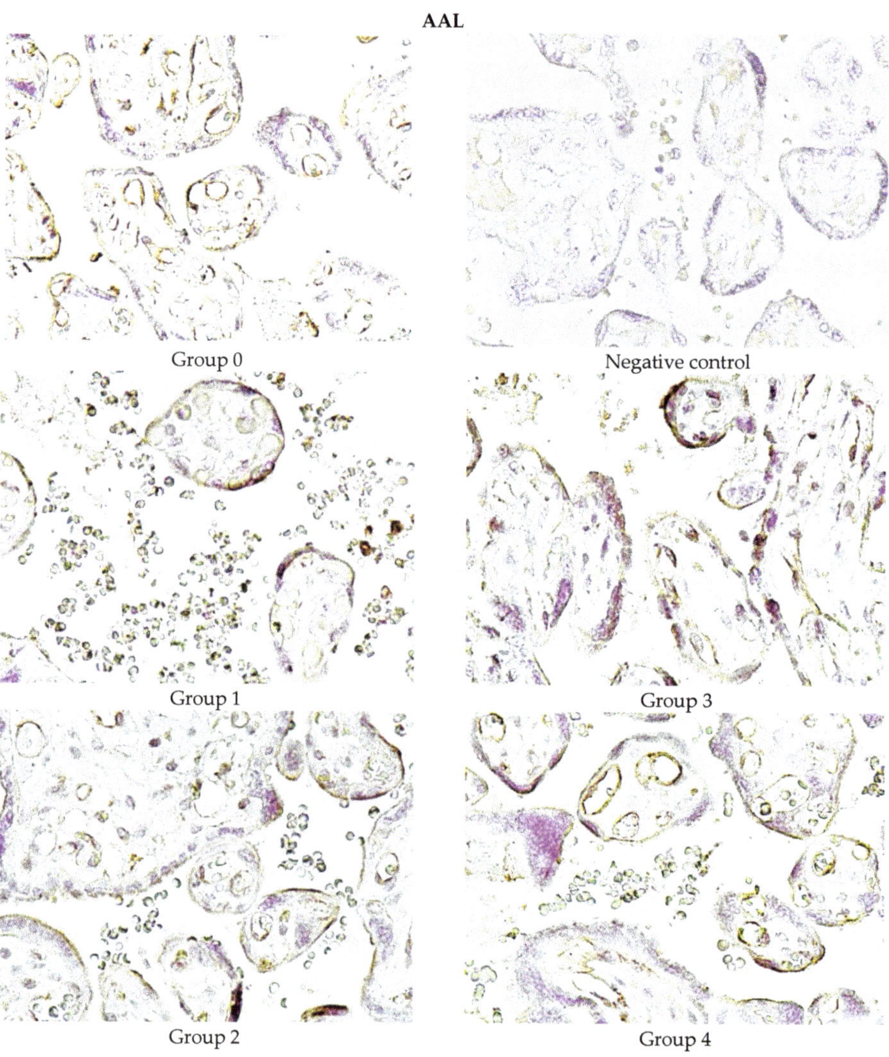

Figure A1. Expression of fucosylated glycans stained with AAL lectin in the endothelial glycocalyx of the capillaries of placental terminal villi. Magnification: ×400.

UEA-I

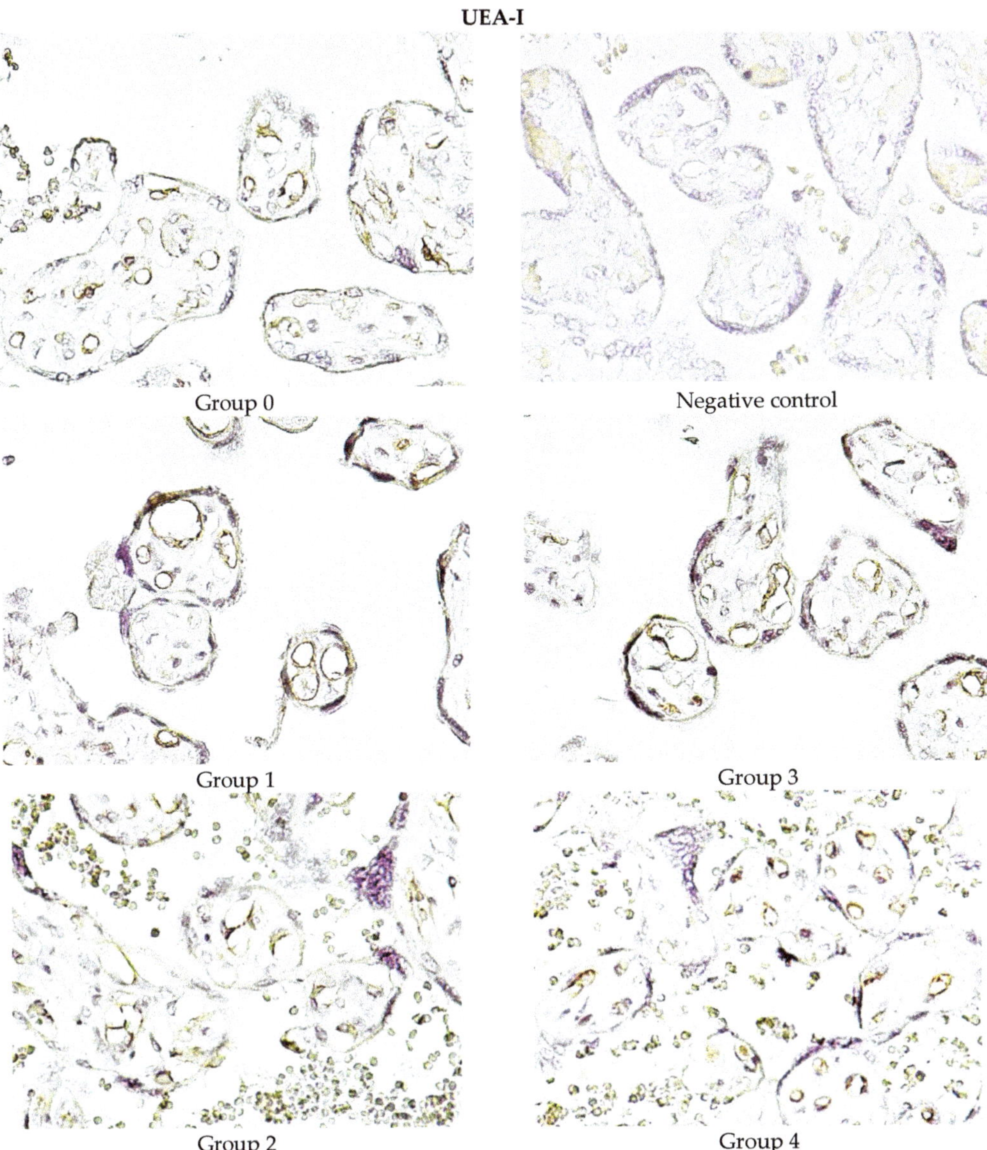

Figure A2. Expression of fucosylated glycans stained with UEA-I lectin in the endothelial glycocalyx of the capillaries of placental terminal villi. Magnification: ×400.

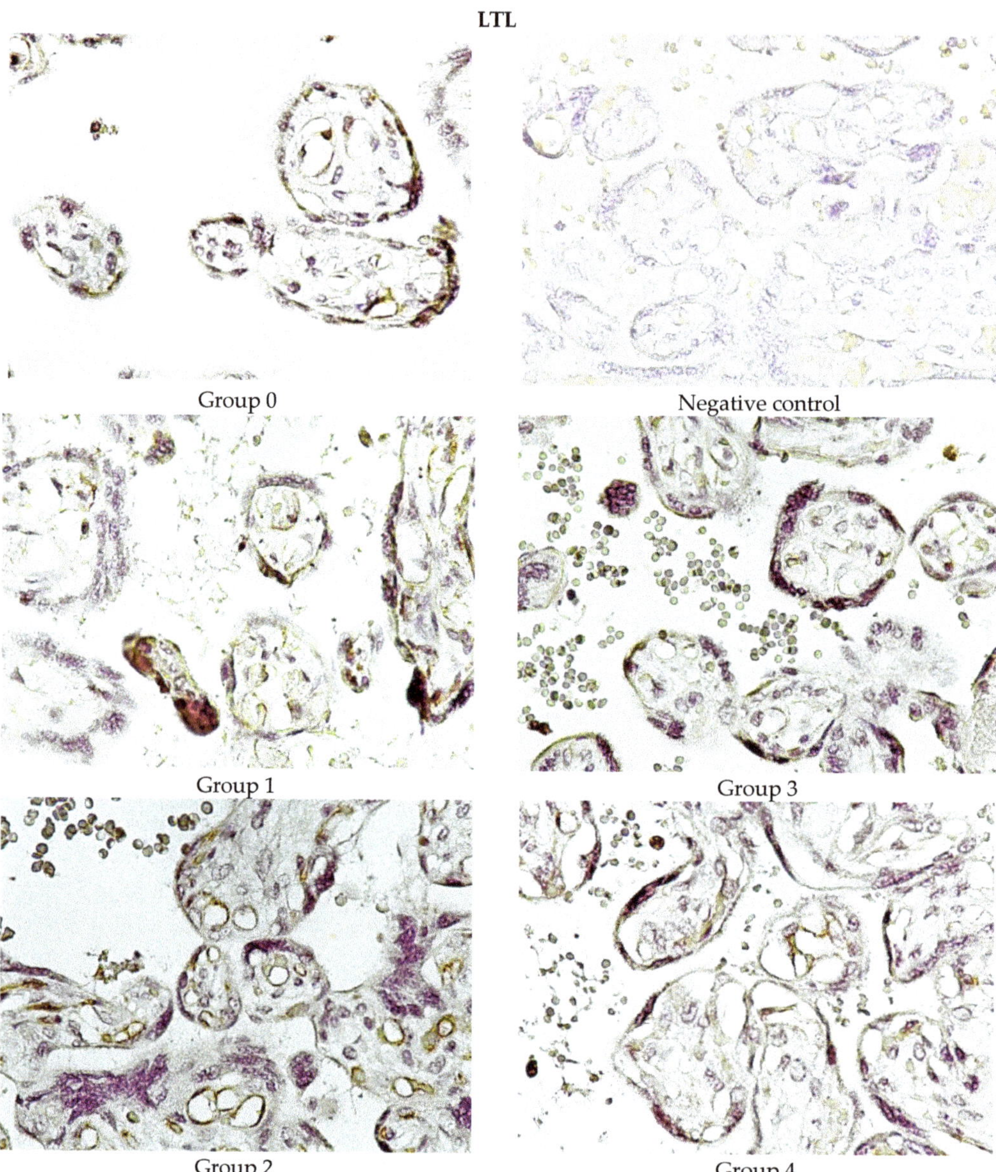

Figure A3. Expression of fucosylated glycans stained with LTL lectin in the endothelial glycocalyx of the capillaries of placental terminal villi. Magnification: ×400.

Lewis Y

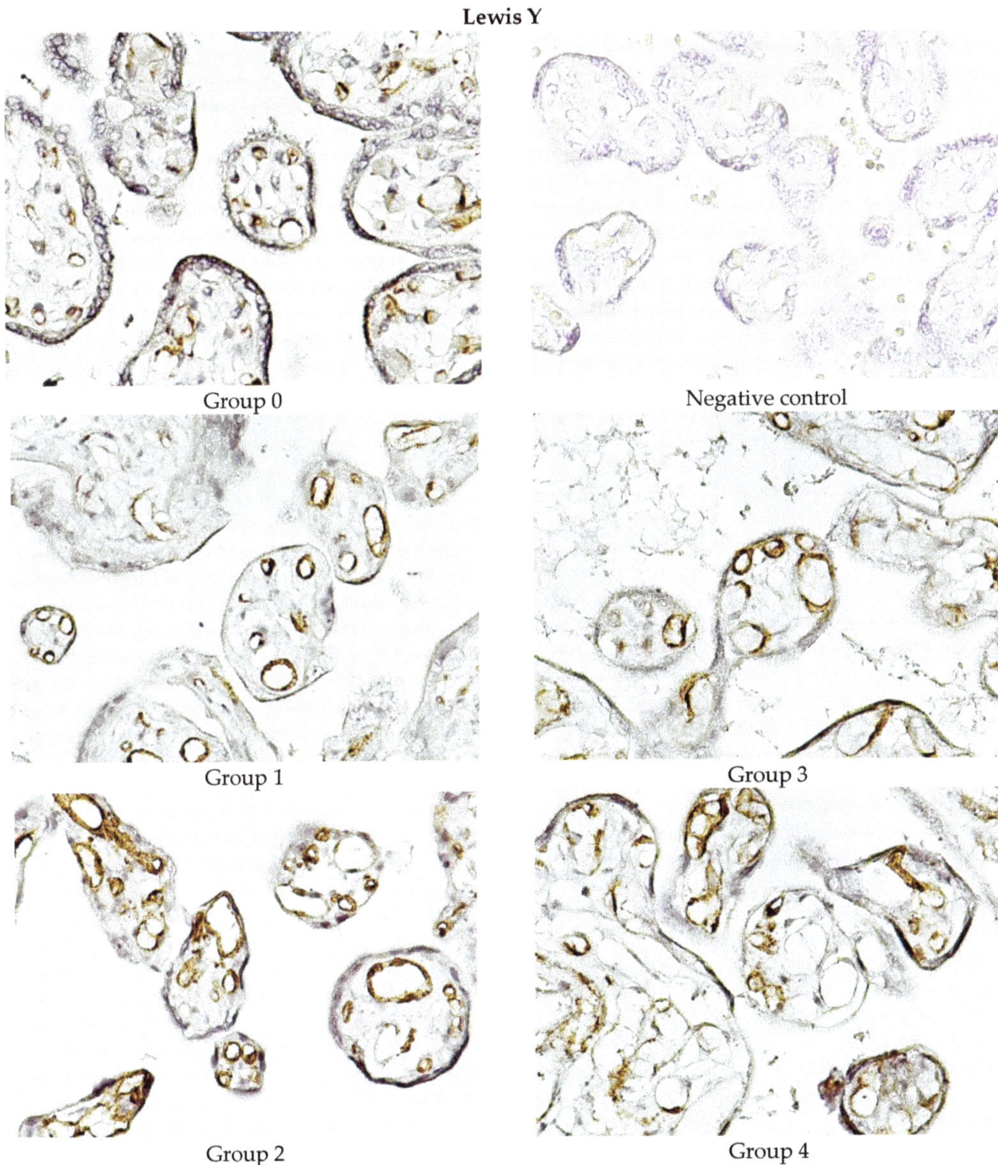

Figure A4. Expression of fucosylated glycans stained with anti-Le^Y MAbs in the endothelial glycocalyx of the capillaries of placental terminal villi. Magnification: ×400.

References

1. Rolink, D.L.; de Carvalho, M.H.B.; Lobo, G.A.R.; Verlohren, S.; Poon, L.; Baschat, A.; Hyett, J.; Thilaganathan, B.; Bujold, E.; da Silva Costa, F. Preeclampsia: Universal Screening or Universal Prevention for Low and Middle-Income Settings? *Rev. Bras. Ginecol. Obstet.* **2021**, *43*, 334–338. [CrossRef] [PubMed]
2. Masini, G.; Foo, L.F.; Tay, J.; Wilkinson, I.B.; Valensise, H.; Gyselaers, W.; Lees, C.C. Preeclampsia has two phenotypes which require different treatment strategies. *Am. J. Obstet. Gynecol.* **2022**, *226*, 1006–1018. [CrossRef] [PubMed]
3. Valencia-Ortega, J.; Zárate, A.; Saucedo, R.; Hernández-Valencia, M.; Cruz, J.G.; Puello, E. Placental Proinflammatory State and Maternal Endothelial Dysfunction in Preeclampsia. *Gynecol. Obstet. Investig.* **2019**, *84*, 12–19. [CrossRef]

4. Vilotić, A.; Nacka-Aleksić, M.; Pirković, A.; Bojić-Trbojević, Ž.; Dekanski, D.; Krivokuća, M.J. IL-6 and IL-8: An Overview of Their Roles in Healthy and Pathological Pregnancies. *Int. J. Mol. Sci.* **2022**, *23*, 14574. [CrossRef] [PubMed]
5. Aneman, I.; Pienaar, D.; Suvakov, S.; Simic, T.P.; Garovic, V.D.; McClements, L. Mechanisms of Key Innate Immune Cells in Early- and Late-Onset Preeclampsia. *Front. Immunol.* **2020**, *18*, 1864. [CrossRef]
6. Awaludin, A.; Rahayu, C.; Daud, N.A.A.; Zakiyah, N. Antihypertensive Medications for Severe Hypertension in Pregnancy: A Systematic Review and Meta-Analysis. *Healthcare* **2022**, *10*, 325. [CrossRef]
7. Cífková, R. Hypertension in Pregnancy: A Diagnostic and Therapeutic Overview, High Blood Press. *Cardiovasc. Prev.* **2023**, *30*, 289–303. [CrossRef]
8. Gupta, M.; Al Khalili, Y. Methyldopa. In *StatPearls*; StatPearls Publishing: Treasure Island, FL, USA, 2023.
9. Easterling, T.; Mundle, S.; Bracken, H.; Parvekar, S.; Mool, S.; Magee, L.A.; von Dadelszen, P.; Shochet, T.; Winikoff, B. Oral antihypertensive regimens (nifedipine retard, labetalol, and methyldopa) for management of severe hypertension in pregnancy: An open-label, randomised controlled trial. *Lancet* **2019**, *394*, 1011–1021. [CrossRef]
10. Odigboegwu, O.; Pan, L.J.; Chatterjee, P. Use of Antihypertensive Drugs During Preeclampsia. *Front. Cardiovasc. Med.* **2018**, *29*, 50. [CrossRef]
11. Pillinger, N.L.; Kam, P. Endothelial glycocalyx: Basic science and clinical implications. *Anaesth. Intensive Care* **2017**, *45*, 295–307. [CrossRef]
12. Willhauck-Fleckenstein, M.; Moehler, T.M.; Merling, A.; Pusunc, S.; Goldschmidt, H.; Schwartz-Albiez, R. Transcriptional regulation of the vascular endothelial glycome byangiogenic and inflammatory signaling. *Angiogenesis* **2010**, *13*, 25–42. [CrossRef]
13. Muminova, K.T.; Ziganshina, M.M.; Khodzhaeva, Z.S. Evaluation of the effect of different strategies of antihypertensive treatment on endothelial glycocalyx in patients with preeclampsia. *Exp. Clin. Pharmacol.* **2022**, *85*, 4–10. [CrossRef]
14. Muminova, K.T.; Khodzhaeva, Z.S.; Yarotskaya, E.L.; Ziganshina, M.M. Analysis of factors associated with sterile inflammation in women with pe receiving different antihypertensive treatment strategies. *Med. Immunol.* **2023**, *25*, 1183–1190. [CrossRef]
15. Aplin, J.D.; Jones, C.J. Fucose, placental evolution and the glycocode. *Glycobiology* **2012**, *22*, 470–478. [CrossRef]
16. Miyoshi, E.; Moriwaki, K.; Nakagawa, T. Biological function of fucosylation in cancer biology. *J. Biochem.* **2008**, *143*, 725–729. [CrossRef]
17. Silva-Filho, A.F.; Sena, W.L.B.; Lima, L.R.A.; Carvalho, L.V.N.; Pereira, M.C.; Santos, L.G.S.; Santos, R.V.C.; Tavares, L.B.; Pitta, M.G.R.; Rêgo, M.J.B.M. Glycobiology Modifications in Intratumoral Hypoxia: The Breathless Side of Glycans Interaction. *Cell. Physiol. Biochem.* **2017**, *41*, 1801–1829. [CrossRef]
18. Croci, D.O.; Cerliani, J.P.; Pinto, N.A.; Morosi, L.G.; Rabinovich, G.A. Regulatory role of glycans in the control of hypoxia-driven angiogenesis and sensitivity to anti-angiogenic treatment. *Glycobiology* **2014**, *24*, 1283–1290. [CrossRef] [PubMed]
19. Ziganshina, M.M.; Kulikova, G.V.; Fayzullina, N.M.; Yarotskaya, E.L.; Shchegolev, A.I.; Le Pendu, J.; Breiman, A.; Shilova, N.V.; Khasbiullina, N.R.; Bovin, N.V.; et al. Expression of fucosylated glycans in endothelial glycocalyces of placental villi at early and late fetal growth restriction. *Placenta* **2020**, *90*, 98–102. [CrossRef] [PubMed]
20. Shilova, N.; Galanina, O.; Polyakova, S.; Nokel, A.; Pazynina, G.; Golovchenko, V.; Patova, O.; Mikshina, P.; Gorshkova, T.; Bovin, M.N. Specificity of widely used lectins as probed with oligosaccharide and plant polysasccharide arrays. *J. Histochem. Cytochem.* 2023; *submitted.*
21. Ziganshina, M.M.; Dolgushina, N.V.; Kulikova, G.V.; Fayzullina, N.M.; Yarotskaya, E.L.; Khasbiullina, N.R.; Abdurakhmanova, N.F.; Asaturova, A.V.; Shchegolev, A.I.; Dovgan, A.A.; et al. Epithelial apical glycosylation changes associated with thin endometrium in women with infertility—A pilot observational study. *Reprod. Biol. Endocrinol.* **2021**, *19*, 73. [CrossRef]
22. Becker, D.J.; Lowe, J.B. Fucose: Biosynthesis and biological function in mammals. *Glycobiology* **2003**, *13*, 41–53. [CrossRef] [PubMed]
23. Kornacki, J.; Wirstlein, P.; Wender-Ozegowska, E. Serum levels of soluble FMS-like tyrosine kinase 1 and endothelial glycocalyx components in early- and late-onset preeclampsia. *J. Matern. Fetal. Neonatal Med.* **2022**, *35*, 7466–7470. [CrossRef] [PubMed]
24. Carlberg, N.; Cluver, C.; Hesse, C.; Thörn, S.E.; Gandley, R.; Damén, T.; Bergman, L. Circulating concentrations of glycocalyx degradation products in preeclampsia. *Front. Physiol.* **2022**, *13*, 1022770. [CrossRef] [PubMed]
25. Weissgerber, T.L.; Garcia-Valencia, O.; Milic, N.M.; Codsi, E.; Cubro, H.; Nath, M.C.; White, W.M.; Nath, K.A.; Garovic, V.D. Early Onset Preeclampsia Is Associated with Glycocalyx Degradation and Reduced Microvascular Perfusion. *J. Am. Heart. Assoc.* **2019**, *8*, e010647. [CrossRef]
26. Magee, L.A.; Rey, E.; Asztalos, E.; Hutton, E.; Singer, J.; Helewa, M.; Lee, T.; Logan, A.G.; Ganzevoort, W.; Welch, R.; et al. Management of non-severe pregnancy hypertension—A summary of the CHIPS Trial (Control of Hypertension in Pregnancy Study) research publications. *Pregnancy Hypertens.* **2019**, *18*, 156–162. [CrossRef]
27. Ziganshina, M.M.; Muminova, K.T.; Khasbiullina, N.R.; Khodzhaeva, Z.S.; Yarotskaya, E.L.; Sukhikh, G.T. Characterization of Vascular Patterns Associated with Endothelial Glycocalyx Damage in Early- and Late-Onset Preeclampsia. *Biomedicines* **2022**, *10*, 2790. [CrossRef]
28. Tatsuzuki, A.; Ezaki, T.; Makino, Y.; Matsuda, Y.; Ohta, H. Characterization of the sugar chain expression of normal term human placental villi using lectin histochemistry combined with immunohistochemistry. *Arch. Histol. Cytol.* **2009**, *72*, 35–49. [CrossRef]
29. Ma, Z.; Yang, H.; Peng, L.; Kuhn, C.; Chelariu-Raicu, A.; Mahner, S.; Jeschke, U.; von Schönfeldt, V. Expression of the Carbohydrate Lewis Antigen, Sialyl Lewis A, Sialyl Lewis X, Lewis X, and Lewis Y in the Placental Villi of Patients with Unexplained Miscarriages. *Front. Immunol.* **2021**, *12*, 679424. [CrossRef]

30. Marini, M.; Bonaccini, L.; Thyrion, G.D.; Vichi, D.; Parretti, E.; Sgambati, E. Distribution of sugar residues in human placentas from pregnancies complicated by hypertensive disorders. *Acta Histochem.* **2011**, *113*, 815–825. [CrossRef]

31. Sgambati, E.; Biagiotti, R.; Marini, M.; Brizzi, E. Lectin histochemistry in the human placenta of pregnancies complicated by intrauterine growth retardation based on absent or reversed diastolic flow. *Placenta* **2002**, *23*, 503–515. [CrossRef]

32. Seidmann, L.; Suhan, T.; Kamyshanskiy, Y.; Nevmerzhitskaya, A.; Gerein, V.; Kirkpatrick, C.J. CD15—A new marker of pathological villous immaturity of the term placenta. *Placenta* **2014**, *35*, 925–931. [CrossRef] [PubMed]

33. Sukhikh, G.T.; Ziganshina, M.M.; Nizyaeva, N.V.; Kulikova, G.V.; Volkova, J.S.; Yarotskaya, E.L.; Kan, N.E.; Shchyogolev, A.I.; Tyutyunnik, V.L. Differences of glycocalyx composition in the structural elements of placenta in preeclampsia. *Placenta* **2016**, *43*, 69–76. [CrossRef]

34. Kulikova, G.V.; Ziganshina, M.M.; Shchegolev, A.I.; Sukhikh, G.T. Comparative Characteristics of the Expression of Fucosylated Glycans and Morphometric Parameters of Terminal Placental Villi Depending on the Severity of Preeclampsia. *Bull. Exp. Biol. Med.* **2021**, *172*, 90–95. [CrossRef]

35. Baker, B.C.; Heazell, A.E.P.; Sibley, C.; Wright, R.; Bischof, H.; Beards, F.; Guevara, T.; Girard, S.R.L. Jones Hypoxia and oxidative stress induce sterile placental inflammation in vitro. *Sci Rep.* **2021**, *11*, 7281. [CrossRef]

36. Zhao, M.; Wang, S.; Zuo, A.; Zhang, J.; Wen, W.; Jiang, W.; Chen, H.; Liang, D.; Sun, J.; Wang, M. HIF-1α/JMJD1A signaling regulates inflammation and oxidative stress following hyperglycemia and hypoxia-induced vascular cell injury. *Cell. Mol. Biol. Lett.* **2021**, *26*, 40. [CrossRef] [PubMed]

37. Jackson-Weaver, O.; Friedman, J.K.; Rodriguez, L.A.; Hoof, M.A.; Drury, R.H.; Packer, J.T.; Smith, A.; Guidry, C.; Duchesne, J.C. Hypoxia/reoxygenation decreases endothelial glycocalyx via reactive oxygen species and calcium signaling in a cellular model for shock. *J. Trauma Acute Care Surg.* **2019**, *87*, 1070–1076. [CrossRef]

38. Johansson, P.I.; Bergström, A.; Aachmann-Andersen, N.J.; Meyer, M.A.; Ostrowski, S.R.; Nordsborg, N.B.; Olsen, N.V. Effect of acute hypobaric hypoxia on the endothelial glycocalyx and digital reactive hyperemia in humans. *Front. Physiol.* **2014**, *24*, 459. [CrossRef]

39. Becker, B.F.; Fischer, J.; Hartmann, H.; Chen, C.C.; Sommerhoff, C.P.; Tschoep, J.; Conzen, P.C.; Annecke, T. Inosine; not adenosine, initiates endothelial glycocalyx degradation in cardiac ischemia and hypoxia. *Nucleosides Nucleotides Nucleic Acids* **2011**, *30*, 1161–1167. [CrossRef] [PubMed]

40. Moeller, B.J.; Cao, Y.; Vujaskovic, Z.; Li, C.Y.; Haroon, Z.A.; Dewhirst, M.W. The relationship between hypoxia and angiogenesis, *Semin. Radiat. Oncol.* **2004**, *14*, 215–221. [CrossRef]

41. Krock, B.L.; Skuli, N.; Simon, M.C. Hypoxia-induced angiogenesis: Good and evil. *Genes Cancer* **2011**, *2*, 1117–1133. [CrossRef]

42. Barut, F.; Barut, A.; Gun, B.D.; Kandemir, N.O.; Harma, M.I.; Harma, M.; Aktunc, E.; Ozdamar, S.O. Intrauterine growth restriction and placental angiogenesis. *Diagn. Pathol.* **2010**, *5*, 24. [CrossRef] [PubMed]

43. Junaid, T.O.; Brownbill, P.; Chalmers, N.; Johnstone, E.D.; Aplin, J.D. Fetoplacental vascular alterations associated with fetal growth restriction. *Placenta* **2014**, *35*, 808–815. [CrossRef] [PubMed]

44. Ermini, L.; Bhattacharjee, J.; Spagnoletti, A.; Bechi, N.; Aldi, S.; Ferretti, C.; Bianchi, L.; Bini, L.; Rosati, F.; Paulesu, L.; et al. Oxygen governs Galβ1-3GalNAc epitope in human placenta. *Am. J. Physiol. Cell Physiol.* **2013**, *305*, 931–940. [CrossRef] [PubMed]

45. Maller, S.M.; Cagnoni, A.J.; Bannoud, N.; Sigaut, L.; Sáez, J.M.P.; Pietrasanta, L.I.; Yang, R.Y.; Liu, F.T.; Croci, D.O.; Di Lella, S.; et al. An adipose tissue galectin controls endothelial cell function via preferential recognition of 3-fucosylated glycans. *FASEB J.* **2020**, *34*, 735–753. [CrossRef]

46. Thijssen, V.L. Galectins in Endothelial Cell Biology and Angiogenesis: The Basics. *Biomolecules* **2021**, *11*, 1386. [CrossRef]

47. Bueno-Sánchez, J.C.; Gómez-Gutiérrez, A.M.; Maldonado-Estrada, J.G.; Quintana-Castillo, J.C. Expression of placental glycans and its role in regulating peripheral blood NK cells during preeclampsia: A perspective. *Front. Endocrinol.* **2023**, *14*, 1087845. [CrossRef]

48. Scott, D.W.; Patel, R.P. Endothelial heterogeneity and adhesion molecules N-glycosylation: Implications in leukocyte trafficking in inflammation. *Glycobiology* **2013**, *23*, 622–633. [CrossRef]

49. Figueroa-Diesel, H.; Hernandez-Andrade, E.; Acosta-Rojas, R.; Cabero, L.; Gratacos, E. Doppler changes in the main fetal brain arteries at different stages of hemodynamic adaptation in severe intrauterine growth restriction. *Ultrasound Obstet. Gynecol.* **2007**, *30*, 297–302. [CrossRef]

50. Sauer, S.; Meissner, T.; Moehler, T.A. Furan-Based Lewis-Y-(CD174)-Saccharide Mimetic Inhibits Endothelial Functions and In Vitro Angiogenesis. *Adv. Clin. Exp. Med.* **2015**, *24*, 759–768. [CrossRef]

51. Moehler, T.M.; Sauer, S.; Witzel, M.; Andrulis, M.; Garcia-Vallejo, J.J.; Grobholz, R.; Willhauck-Fleckenstein, M.; Greiner, A.; Goldschmidt, H.; Schwartz-Albiez, R. Involvement of alpha 1-2-fucosyltransferase I (FUT1) and surface-expressed Lewis(y) (CD174) in first endothelial cell-cell contacts during angiogenesis. *J. Cell Physiol.* **2008**, *215*, 27–36. [CrossRef]

52. Gourvas, V.; Dalpa, E.; Konstantinidou, A.; Vrachnis, N.; Spandidos, D.A.; Sifakis, S. Angiogenic factors in placentas from pregnancies complicated by fetal growth restriction (review). *Mol. Med. Rep.* **2012**, *6*, 23–27. [CrossRef]

53. Schoots, M.H.; Gordijn, S.J.; Scherjon, S.A.; van Goor, H.; Hillebrands, J.L. Oxidative stress in placental pathology. *Placenta* **2018**, *69*, 153–161. [CrossRef] [PubMed]

54. Rapoport, E.M.; Matveeva, V.K.; Vokhmyanina, O.A.; Belyanchikov, I.M.; Gabius, H.J.; Bovin, N.V. Localization of Galectins within Glycocalyx. *Biochemistry* **2018**, *83*, 727–737. [CrossRef] [PubMed]

55. Pankiewicz, K.; Szczerba, E.; Fijalkowska, A.; Szamotulska, K.; Szewczyk, G.; Issat, T.; Maciejewski, T.M. The association between serum galectin-3 level and its placental production in patients with preeclampsia. *J. Physiol. Pharmacol.* **2020**, *71*, 845–856. [CrossRef]

56. de Boer, R.A.; Yu, L.; van Veldhuisen, D.J. Galectin-3 in cardiac remodeling and heart failure. *Curr. Heart Fail. Rep.* **2010**, *7*, 1–8. [CrossRef]

57. Blanda, V.; Bracale, U.M.; Di Taranto, M.D. Fortunato, Galectin-3 in Cardiovascular Diseases. *Int. J. Mol. Sci.* **2020**, *21*, 9232. [CrossRef]

58. Dorogova, I.V.; Panina, E.S. Comparison of the BPLab® sphygmomanometer for ambulatory blood pressure monitoring with mercury sphygmomanometry in pregnant women: Validation study according to the British Hypertension Society protocol. *Vasc. Health Risk Manag.* **2015**, *13*, 245–249. [CrossRef]

59. Knirel, Y.A.; Gabius, H.-J.; Blixt, O.; Rapoport, E.M.; Khasbiullina, N.R.; Shilova, N.V.; Bovin, N.V. Human tandem-repeat-type galectins bind bacterial non-βGal polysaccharides. *Glycoconj. J.* **2014**, *31*, 7–12. [CrossRef]

60. Khong, T.Y.; Mooney, E.E.; Ariel, I.; Balmus, N.C.; Boyd, T.K.; Brundler, M.A.; Derricott, H.; Evans, M.J.; Faye-Petersen, O.M.; Gillan, J.E.; et al. Sampling and Definitions of Placental Lesions: Amsterdam Placental Workshop Group Consensus Statement. *Arch. Pathol. Lab. Med.* **2016**, *140*, 698–713. [CrossRef]

61. Detection of Glycoproteins Using Lectins in Histochemistry, ELISA, and Western Blot Applications, Supplemental Protocol. 2022. Available online: https://vectorlabs.com/wp-content/uploads/2023/01/VL_LIT3055_Detect.Glycoproteins_SuppProtocol.LBL02552.pdf (accessed on 10 September 2023).

62. Vuskovic, M.I.; Xu, H.; Bovin, N.V.; Pass, H.I.; Huflejt, M.E. Processing and analysis of serum antibody binding signals from Printed Glycan Arrays for diagnostic and prognostic applications. *Int. J. Bioinform. Res Appl.* **2011**, *7*, 402–426. [CrossRef]

Sci 2023, 24, 15420

International Journal of
Molecular Sciences

Review

Events Leading to the Establishment of Pregnancy and Placental Formation: The Need to Fine-Tune the Nomenclature on Pregnancy and Gestation

Giuseppe Benagiano [1,2], Salvatore Mancuso [3], Sun-Wei Guo [4] and Gian Carlo Di Renzo [5,6,7,*]

1   Faculty of Medicine and Surgery, Sapienza University of Rome, 00185 Rome, Italy; pinoingeneva@bluewin.ch
2   Geneva Foundation for Medical Education and Research, 1206 Geneva, Switzerland
3   Faculty of Medicine and Surgery, Catholic University of the Sacred Heart, 00168 Rome, Italy; salmancuso71@gmail.com
4   Research Institute, Shanghai Obstetrics & Gynecology Hospital, Fudan University, Shanghai 200011, China; hoxa10@outlook.com
5   Center for Perinatal and Reproductive Medicine, University of Perugia, 06156 Perugia, Italy
6   Department of Obstetrics, Gynecology and Perinatology, I.M. Sechenov First Moscow State Medical University, 119146 Moscow, Russia
7   Department of Obstetrics & Gynecology, Wayne State University School of Medicine, Detroit, MI 48201, USA
*   Correspondence: giancarlo.direnzo@unipg.it

Citation: Benagiano, G.; Mancuso, S.; Guo, S.-W.; Di Renzo, G.C. Events Leading to the Establishment of Pregnancy and Placental Formation: The Need to Fine-Tune the Nomenclature on Pregnancy and Gestation. *Int. J. Mol. Sci.* **2023**, 24, 15420. https://doi.org/10.3390/ijms242015420

Academic Editor: Giovanni Tossetta

Received: 30 August 2023
Revised: 10 October 2023
Accepted: 18 October 2023
Published: 21 October 2023

Abstract: Today, there is strong and diversified evidence that in humans at least 50% of early embryos do not proceed beyond the pre-implantation period. This evidence comes from clinical investigations, demography, epidemiology, embryology, immunology, and molecular biology. The purpose of this article is to highlight the steps leading to the establishment of pregnancy and placenta formation. These early events document the existence of a clear distinction between embryonic losses during the first two weeks after conception and those occurring during the subsequent months. This review attempts to highlight the nature of the maternal–embryonic dialogue and the major mechanisms active during the pre-implantation period aimed at "selecting" embryos with the ability to proceed to the formation of the placenta and therefore to the completion of pregnancy. This intense molecular cross-talk between the early embryo and the endometrium starts even before the blastocyst reaches the uterine cavity, substantially initiating and conditioning the process of implantation and the formation of the placenta. Today, several factors involved in this dialogue have been identified, although the best-known and overall, the most important, still remains *Chorionic Gonadotrophin*, indispensable during the first 8 to 10 weeks after fertilization. In addition, there are other substances acting during the first days following fertilization, the *Early Pregnancy Factor*, believed to be involved in the suppression of the maternal response, thereby allowing the continued viability of the early embryo. The *Pre-Implantation Factor,* secreted between 2 and 4 days after fertilization. This linear peptide molecule exhibits a self-protective and antitoxic action, is present in maternal blood as early as 7 days after conception, and is absent in the presence of non-viable embryos. The *Embryo-Derived Platelet-activating Factor*, produced and released by embryos of all mammalian species studied seems to have a role in the ligand-mediated trophic support of the early embryo. The implantation process is also guided by signals from cells in the decidualized endometrium. Various types of cells are involved, among them epithelial, stromal, and trophoblastic, producing a number of cellular molecules, such as cytokines, chemokines, growth factors, and adhesion molecules. Immune cells are also involved, mainly uterine natural killer cells, macrophages, and T cells. In conclusion, events taking place during the first two weeks after fertilization determine whether pregnancy can proceed and therefore whether placenta's formation can proceed. These events represent the scientific basis for a clear distinction between the first two weeks following fertilization and the rest of gestation. For this reason, we propose that a new nomenclature be adopted specifically separating the two periods. In other words, the period from fertilization and birth should be named "gestation", whereas that from the completion of the process of implantation leading to the formation of the placenta, and birth should be named "pregnancy".

pregnancy could be detected clinically. Of relevance, most of the women with unrecognized early pregnancy losses had normal fertility, and, indeed, 95% of them became clinically pregnant within two years.

Demographic and epidemiological evidence also calls for the presence of a major early wastage. Indeed, human fecundity, i.e., the capacity to bear live children, defined as '*the probability to produce a vital term newborn per menstrual cycle during which there was normal sexual activity*' [19], rarely exceeds 35–40% [4]. In fact, Woods estimated that apparent fecundability varies from 0.14 to 0.31 (0.17–0.38 when adjusted for fetal loss) [20].

A recent investigation of Japanese couples trying to conceive their first child utilized the "time to conception" (TTP) as either "natural" or "total". Using TTP-total and TTP-natural, the sterile proportion of the whole sample was, respectively, 2% and 14%, and the interquartile range of fecundability (IQRF) was, respectively, 0.10 (CI: 0.04, 0.19) and 0.11 (CI: 0.05, 0.19). The IQRF was 0.18 (CI: 0.10, 0.29) for women aged 24 years or younger and 0.05 (CI: 0.02, 0.13) for 35–39 years old when TTP-all was used; the TTP-natural was 0.18 (CI: 0.10, 0.29) for women aged 24 years or younger and 0.06 (CI: 0.00, 0.15) for 35–39 years old. This investigation concluded that fecundability is overall lower at higher ages, while interquartile ranges are overlapping [21].

A summary of information on fecundability in the human available up to 2010 is presented in Table 1.

Table 1. Human fecundity over the last four centuries.

Population Studied	Publication	Fecundity Index
Canada (Québec), 18th century	Henrypin (1954) [22]	0.31
China 20th century	Wang et al (2003) [23]	0.40
France, 17th and 18th centuries	Charbonneaux (1970) [24]	0.21
Great Britain, 20th century	Vessey et al. (1976) [25]	0.21
Mexico, 20th century	Balakrishnan (1979) [26]	0.21
Peru, 20th century	Balakrishnan (1979) [26]	0.17
The Netherlands, 20th century *	van Noord Zaadstra et al. (1991) [27]	0.54
USA (Hutterite sect) 20th century	Sheps (1965) [28]	0.28
USA, 20th century	Zinamen et al. (1996) [29]	0.30

* After 2 months at age 31. Modified and reprinted with permission from Benagiano et al., 2010 [4].

Reasons for the reported wide variations in evaluating human fecundity have been critically examined by Smarr et al. [30] who pointed out that there is no 'population (bio)marker' that can be used and, as such, fecundity can only be assessed indirectly utilizing a variety of individual- or couple-based endpoints, defaulting both evaluation and monitoring to rely on rates of births (fertility) or adverse outcomes.

In terms of possible causes for the relatively low human fecundity, there is once again increasing evidence that it is due to pre-clinical pregnancy loss, suggestive of spontaneous failure of implantation. Therefore, the mechanisms underlying this phenomenon need to be critically analyzed.

In a 1996 clinical study, Zinaman et al. found a maximal fertility rate of approximately 30% per cycle in the first two cycles, with a pregnancy wastage of 31%. Importantly, 40% of these losses occurred so early after fertilization that the presence of gestation could be identified only through the measurement of urinary hCG [29]. Another clinical investigation followed a cohort of 217 women attempting to become pregnant, observing early loss rates ranging from a low estimate of 11.0% to a high one of 26.9% [31]. More recently, Leridon provided an average figure of 20–25% (at ages 20–30) and pointed out that, although human fetal mortality is high, amounting to 12–15% of confirmed pregnancies, "*an even higher proportion of 'products of conception' do not develop normally and are evacuated within a few weeks, before the woman becomes aware of her pregnancy*". He cites a figure of around 50% of all conceptions and states that in their great majority these failures are due to severe genetic abnormalities, concluding that *"human reproduction has a high error rate, but most of these errors are corrected by eliminating the products of conception"* [32].

A number of investigations have attempted an evaluation of pre-clinical embryonic losses in in vitro fertilization (IVF) cycles.

At the turn of the millennium, Simon et al. [33] recruited 145 subjects undergoing IVF and 92 undergoing oocyte donations. In subjects undergoing IVF, positive implantation was documented in 60.7% of embryo transfer (ET) cycles, but only 20.7% resulted in viable pregnancies, whereas the remaining miscarried; this occurred at an early pre-clinical stage in 72.4% of the failed ET cycles. In ovum donation cycles, positive implantation was recorded in 69.6% of ET cycles. Of these, 37.0% miscarried; among them, pre-clinical losses accounted for 70.6% of the total.

Boomsma et al. [34] utilized rates of rising urinary hCG (indicating initiation of implantation) that did not lead to a subsequent positive pregnancy test. They found that in over 50% of women undergoing ET, a rise in hCG could be documented, indicating an implanting embryo. However, in approximately one-third of these implanting embryos, a pre-clinical pregnancy loss occurred.

Calculations of early embryonic losses have been scrutinized by Jarvis [35], who expressed the opinion that: (i) the hypothesis by Roberts and Lowe [13] has no practical quantitative value; (ii) life-table analyses cannot evaluate losses at very early stages of development; (iii) measurement of hCG can only reveal losses occurring from the second week of gestation; and (iv) calculations by Hertig and his group [10–12] are highly imprecise and cast doubt on the validity of their analysis.

In support of his views, Jarvis quotes the large variations of losses published in a number of reputable scientific publications, before and during implantation (30–70% [36], >50% [37], and 75% [38]). He then mentions his re-analysis of the results of hCG investigations and concludes that approximately 40–60% of embryos may be lost between fertilization and birth. His critique of the work of Hertig's group is based on the fact that all estimates of early embryo mortality are subject to commensurate inaccuracy in the absence of reliable fertilization probabilities which, as stressed by Short [39] are *"surprisingly difficult to estimate"*.

Jarvis' critical reanalysis has merit, but—according to his own revision of published data—early pregnancy wastage would still reach at least 50%.

Finally, in 2020, Wilcox et al. [40], after combining data from epidemiologic, demographic, laboratory, and in vitro fertilization investigations, constructed an empirical framework aimed at producing plausible estimates of fecundability, sterility, transient anovulation, patterns of intercourse, and the proportion of ova fertilized in the presence of sperm. After combining all this information, they generated an estimation of preimplantation loss, to be considered an average for fertile couples, concluding that *"under a plausible range of assumptions. . . 40 to 50% of fertilized ova fail to implant"* even in normally fertile couples.

4. Mechanisms Involved in Early Embryo Selection

The first week of development is characterized by daily changes that can be summarized as follows:

Day 1: Fertilization takes place with the fusion of the sperm and egg to form the *zygote*.

Day 2: The division of the zygote begins, and two cells are formed.

Day 3: A solid spheric 'ball' of cells, the *morula*, is generated by the further division of the zygote.

Day 4: The morula continues to divide, while in its center a small cavity begins to form as the morula transforms into the *blastocyst*.

Day 5: The blastocyst begins to implant in the uterus.

Day 6: The implantation process continues, and the blastocyst begins to differentiate with the formation of an outer layer: the trophoblast.

Day 7: Implantation is complete, with the blastocyst fully embedded in the decidua; at this stage, the process of placental formation begins and is characterized by the formation of the two layers: cytotrophoblast and syncytiotrophoblast.

In 2017, Makrigiannakis et al. [41] summarized the steps involved in the process of implantation as follows:

1. The blastocyst moves towards the uterine cavity and contemporarily the decidualized endometrium evolves to a receptive phenotype, resulting in a biochemical cross-talk with the embryo.
2. The pre-implantation embryo begins to secrete factors capable of modulating the implantation site while, in turn, the decidua secretes cytokines and growth factors modulating embryonic differentiation and development.
3. In the presence of a proper biochemical environment, the embryo and the decidua jointly promote trophoblast invasion. This process can be altered in a number of ways resulting in early embryonic loss.

A schematic view of the major phases of the implantation process is presented in Figure 1.

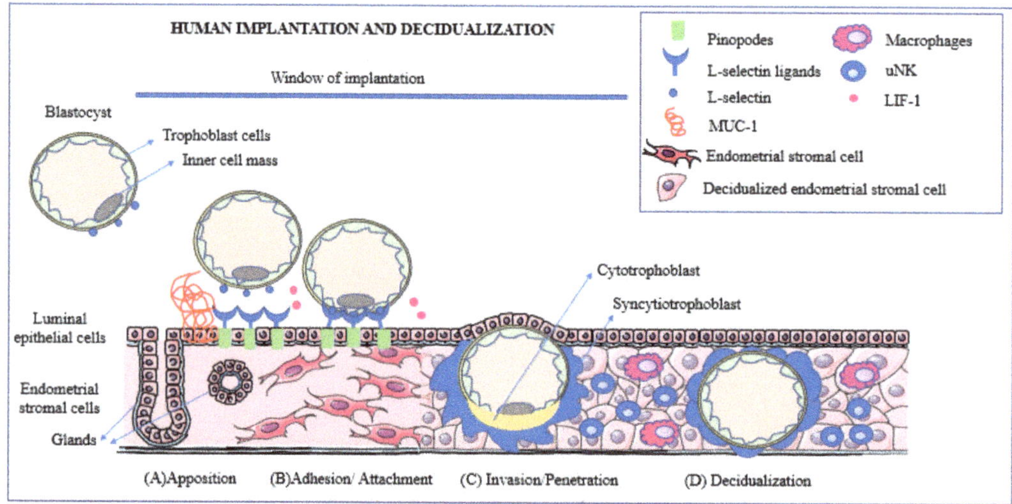

Figure 1. Schematic representation of the main phases of human implantation: apposition, adhesion/attachment, invasion/penetration, and decidualization. (**A**) (Apposition): the blastocyst expresses L-selectin that reacts with its ligands. The presence of Mucin-1 repels the blastocyst and prevents it from attaching outside of the window of receptivity. (**B**) (Adhesion): the blastocyst promotes cleavage of Mucin-1 at the implantation site to ensure successful attachment. (**C**) (Invasion): blastocyst trophoblast cells penetrate the endometrial epithelium and reach the stroma. As soon as implantation is initiated and the embryo reaches the stromal cells surrounding the embryo, these transform into decidualized cells. (**D**) (Decidualization): during the decidualization process, immune cells, such as macrophages and uterine natural killer (uNK) cells, play an important role promoting an environment conducive to successful implantation. Reprinted with permission from: Ochoa-Bernal and Fazleabas (2020) [42].

Over the last decades, information has been obtained concerning the mechanisms leading to this very early embryonic demise. These have been clearly summarized by Macklon and Brosens [43], who proposed that maternal systems are in place to prevent the implantation of poorly viable embryos. To achieve this, the decidualization of the endometrium acts as a biosensor through which signals from the early embryos are converted into a 'go' or 'no-go' endometrial response and therefore successful implantation.

Brosens et al. [44] have further elaborated the path that led to new insights into the processes that govern maternal selection of human embryos in early gestation, pointing out the major challenge for the maternal organism (and specifically for the decidualized endometrium) to eliminate defective embryos without at the same time preventing implantation of normal blastocysts. Two distinct mechanisms seem to be at work: the first implies that a proportion of embryos simply fail to implant; the second is that they are rejected soon after the beginning of the process of implantation in the endometrial luminal epithelium [45].

4.1. Factors Secreted by the Pre-Implantation Embryo

Fundamental new knowledge has been gained thanks to assisted reproduction techniques. An early investigation by Shutt and Lopata [46] of embryos cultured in vitro over a period of 3–4 days, found that the corona cells surrounding the fertilized ovum could secrete daily a mean amount of 50 ng of progesterone and approximately 100 pg each of estradiol, prostaglandin-E_2 (PGE2), and PGF2α, respectively. This experiment provided the first evidence of how the early embryo is supplied with vital substances for its survival before the maternal organism intervenes.

After the blastocyst reaches the uterine cavity, several substances of embryonic origin become involved in the intense cross-talk between the early embryo and endometrium, initiating before and conditioning the process of implantation and of the placenta's formation.

4.1.1. The Early Pregnancy Factor

In the 1970s, Morton et al. [47], starting from the observation that human lymphocytes showed a depression in their activity when incubated in serum from pregnant women, identified the above-mentioned EPF. Biochemically, EPF is a homolog of chaperonin 10 and belongs to the heat shock family of proteins with immunosuppressive and growth factor properties [48] and seems involved in the suppression of the maternal response, thereby allowing the continued viability of the early embryo. Nahhas and Barnea [49] believe that EPF represents a link between fertilization and immunomodulation and that during the pre-implantation period, EPF is of maternal origin, whereas after nidation it becomes of embryonic origin.

4.1.2. The Pre-Implantation Factor

A second early-secreted substance, coined *Pre-Implantation Factor* (PIF), was identified by Barnea et al. [50] in 1994. Starting from the observation that in men and non-pregnant women, the proportion of lymphocytes bound by platelets is significantly different from that of pregnant women ($p < 0.0001$), they developed an assay to measure the presence of PIF. This was detected in subjects who had successfully undergone IVF-ET, followed by a normal pregnancy by 4 days after transfer. PIF is a linear peptide molecule consisting of 15 amino acids, which exhibits a self-protective and antitoxic action and is present in maternal blood as early as 7 days after conception and absent in the presence of non-viable embryos [51].

Yang et al. [52] identified 21 proteins capable of interacting with PIF. Of particular interest is myosin heavy chain 10 (MYH10), since silencing its expression in human endometrial adenocarcinoma (HEC-1-B) cell culture, significantly attenuates cell migration and invasion capacities. It seems therefore that MYH10-mediated cell migration and invasion act in conjunction with PIF in promoting trophoblast invasion. PIF has different biological functions in mammalian species and plays a major role in the embryo's neural system development and neuroprotection; updated information obtained from evidence-based studies as of 2020 is available [53].

Finally, it has been recently suggested that PIF accentuates the decidualization process and the production of endometrial factors that limit trophoblast invasion. Therefore, by controlling both trophoblast and endometrial cells, PIF seems to play a major role in the process of human embryo implantation [54].

4.1.3. The Embryo-Derived Platelet-Activating Factor

A third early factor, the *embryo-derived platelet-activating factor* (PAF) (1-o-alkyl-2-acetyl-sn-glycero-3-phosphocholine) was identified by O'Neill et al. [55]. Subsequently, Roudebush et al. [56] proved a correlation between PAF levels in human embryo culture media and pregnancy outcome: as PAF levels increase, so does the corresponding pregnancy rate.

The activity of PAF is regulated at the level of its synthesis and degradation as well as by the expression of a specific cell surface receptor (PAFr); PAF is eventually degraded by the enzyme PAF-acetyl-hydrolase (PAF-AH). It has been suggested that PAF may constitute the embryonic signal controlling embryo transport to the uterus. Velasquez et al. [57] have documented the co-localization of both PAFr and PAF-AH in the epithelium and stromal cells of the fallopian tubes, granting credibility to this theory.

In 2005, O'Neill summarized knowledge of the factor [58], noting that PAF is produced and released by the embryos of all mammalian species studied to date and is considered the best-described embryotrophin, with a role in the ligand-mediated trophic support of the early embryo.

Sato et al. [59] histologically investigated the aggregates of trophoblasts invading spiral arteries and observed deposition of maternal platelets in them. Furthermore, there is evidence that these platelets are activated and, in an in vitro system, they are capable of increasing the ability of trophoblast to invade spiral arteries.

Recently, PAF has been identified as a fetus-derived mediator repressing placental progesterone receptor A (PR-A) in the human placenta leading to the activation of pro-labor signaling. It seems therefore that PAF represents a novel fetal-derived candidate for initiation of labor [60].

4.1.4. The Human Chorionic Gonadotrophin

In spite of the progress made in the biology of these pre-implantation factors, the best known and the most important among them still remains hCG, indispensable for maintaining the corpus luteum. Fishel et al. [61] were the first to detect its presence in the medium surrounding two embryos cultured for more than 7 days after IVF. This places hCG among the pre-implantation factors. In fact, its mRNA is transcribed as early as the 8-cell stage [62].

4.2. Factors Produced by the Endometrium

Successful implantation requires the exquisitely coordinated migration and invasion of trophoblast cells from the outer capsule of the blastocyst into the endometrium [42]. Over and above the described substances produced by the early embryo, the process is also guided by additional signals from cells in the decidualized endometrium. There are numerous recent descriptions of the mechanisms involved in successful implantation, e.g., [3,32,34,36,39,42,43,56,57,63]. For this reason, a detailed analysis of the implantation process is beyond the scope of this review.

Briefly, however, for the implantation process to be successfully achieved, a well-orchestrated interplay of various cell types is necessary. These consist of epithelial, stromal, and immune cells, and trophoblasts, producing a number of cellular molecules, such as cytokines, chemokines, growth factors, and adhesion molecules [42,63]. In addition, endometrial stroma cells have been shown to promote trophoblast invasion through the generation of an inflammatory environment modulated by TNF-α (tumor necrosis factor α) [64]. Furthermore, through the production of IL-17 (interleukine-17), stromal cells promote trophoblast migration [65]. Finally, adequate glycolysis also appears to be necessary to provide all the energy and macromolecules needed for implantation and early pregnancy [66].

4.3. The Role of the Immune System in Early Pregnancy Wastage

Some twenty years ago, Clark, starting from the fact that embryos bear paternal and embryonic antigens foreign to the maternal immune system, asked the question: could some otherwise normal embryos be 'rejected'? [67]. In a subsequent review [68], he searched and critically analyzed the evidence in animals and humans, finding that various treatments may improve the live birth rate. Unfortunately, not enough evidence exists to indicate when in gestation such mechanisms may be active.

Early investigations by Haddad et al. [69,70] provided some experimental evidence in a murine model of the presence of activated macrophages at implantation sites before overt embryo damage occurs. In terms of mechanism, they showed that increased nitric oxide production by decidual macrophages is involved in early murine embryo loss.

Vento-Tormo et al. [71] have determined the cellular composition of human decidua, documenting the existence of subsets of perivascular and stromal cells located in distinct decidual layers. These include immune cells, principally uNK cells (~70%), macrophages (~20%), and T cells (~10%). In a mouse model, the influx and expansion of these immune cells are controlled by decidual cells, since effector T cells cannot accumulate within the decidua, thanks to the epigenetic silencing of key T cell-attracting inflammatory chemokine genes in decidual stromal cells [72]. According to Brosens et al. [44], cooperation between different innate immune cell populations is essential for optimal decidualization. Therefore, an adverse impact of impaired decidualization on local immune populations and vice versa, hampering the distinction between cause and effect.

A recent review of endometrial immune dysfunction in recurrent pregnancy loss [73] found experimental and clinical evidence suggesting that derangement of the endometrial immune environment can be involved in recurrent implantation failure. Indeed, both changes and modulation of the activity of the immune cells at the level of the endometrium occur early in gestation and the maternal immune system involvement is extended to the endometrial tissue breakdown, vascular remodeling, and placentation. What is not clear is whether the involvement of the immune system occurs at the very early pre-implantation stage, although there are data in support of the hypothesis that the process of nidation evokes an initial, early, inflammatory reaction which is promptly followed by the establishment of an anti-inflammatory decidual environment, allowing the survival of the conceptus and the progression of the pregnancy.

In addition, regulatory T cells (Tregs) also seem to play a vital role in implantation and sustaining pregnancy [74]. The fact that a healthy woman can successfully carry her genetically disparate fetus to term without immune rejection strongly indicates that immune cells are actively involved in pregnancy. Tregs are thymus-derived and can be recruited to non-lymphoid tissues to curtail inflammation and maintain immunological self-tolerance and homeostasis [75]. Tregs may also differentiate and proliferate from CD4+ naive T cells when stimulated with immunosuppressive cytokines such as TGF-β and IL-10 [76]. The potent immunosuppressive activity of Tregs comes from their ability to influence the polarization, expansion, and effector function of T effector cells [77]. Tregs can inhibit effector immunity, restrict inflammation, and support maternal vascular adaptations, and, as such, facilitate trophoblast invasion and placental access to the maternal blood supply. Ample data have shown that insufficient Treg numbers or inadequate functional competence are involved in idiopathic infertility and recurrent miscarriage, as well as later-onset pregnancy complications, ranging from preeclampsia and fetal growth restriction [78].

5. Factors Predictive of Embryonic Loss

The Embryo–Maternal Dialogue

In reviewing the situation with PIF and other early factors, Barnea [79] pointed out that the embryo–maternal dialogue starts shortly after fertilization and is exerted through both local and systemic signaling. First comes the PIF, secreted in humans already at the 2–4-cell stage, that initiates the modulation of cellular immunity. Then comes a class of novel proteins/peptides coined *developmental proteins* (DPs), present in the embryo before a mature immune system has developed. They promote normal proliferation while controlling any abnormal one and seemingly acting through specific receptors. When the development of an embryo becomes incompatible with life, DPs may lead to growth arrest, a decline of pre-implantation factors, reactivationof the immune system, and, ultimately, pregnancy rejection. Drawing attention to the fact that, despite the use of adjuvant therapies, the cumulative rates of live births following IVF-ET remains at ~40%, Yang et al. [52] stressed that the low pregnancy rates, even in the presence of high fertility rates, are due to implantation failure. As reported above, they tried to construct a profile of the DPs reacting with PIF in an attempt to clarify the molecular mechanisms by which PIF promotes trophoblast invasion (see Section 4.1.2).

Brosens et al. [44] have identified multiple decidual checkpoints conditioning the implantation of the blastocyst. The first is appropriate timing since a delayed rise in hCG levels beyond the putative implantation window is strongly associated with the interruption of gestation in the first two weeks of pregnancy. The second is the action of migratory decidual cells encapsulating the implanting embryo, serving as biosensors, and acting in both negative and positive selection. In the presence of low-quality embryos, these cells engage in a stress response inhibiting the secretion of implantation factors and hindering embryo encapsulation, while normal embryos secrete factors enhancing the expression of maternal implantation and metabolic genes, thus actively promoting the completion of the implantation process. The third is represented by a sufficient secretion of hCG to rescue ovarian progesterone production until around 8 to 10 gestation weeks when the placenta takes over progesterone production [80].

Brosens et al. [44] concluded that at the time of implantation, both positive and negative selection mechanisms are active in a process that is intrinsically dynamic and adaptable (Figure 2).

Information is also available on the mechanisms through which the early embryo identifies its own defects, such as, first and foremost, apoptosis which begins to appear at the blastocyst level [81,82]. The starting point has been the observation that many embryos grown in vitro contain unequal-sized blastomeres and multiple cellular fragments and, when fragmentation becomes excessive, their developmental potential both in vitro and in vivo is severely limited [82]. Hardy [83] reported that embryos cultured in vitro, when evaluated after their developmental arrest, often show features characteristic of apoptosis, whereas embryos that seem to be developing normally show no such features before compaction.

A recent review indicates that the most studied mechanisms of embryo fragmentation are apoptotic cell death, membrane compartmentalization of altered DNA, cytoskeletal disorders, and vesicle formation. These phenomena may result in the extrusion of entire blastomeres, the release of apoptotic bodies and other vesicles, and the formation of micronuclei [82]. Indeed, back in 1996, Jurisicova et al. [81] evaluated arrested, fragmented human embryos and were able to detect extensive condensation and degradation of chromatin, which is suggestive of apoptosis. Of importance, no such abnormalities were observed in embryos with regular-sized blastomeres and absence of fragmentation; these findings provided evidence for the existence of the mechanism coined *programmed cell death*. In human embryos, such a mechanism seems to be triggered at a stage prior to blastocyst formation, leading to preimplantation embryo death.

According to Leidenfrost et al. [84], in the bovine model, errors and even failure of the first few cleavage divisions frequently cause immediate embryo death, which constitutes the main source of developmental heterogeneity. This seems to be due to a systemically and developmentally controlled elimination of cells, while the nature and mechanisms of the inner cell mass cell death are unclear.

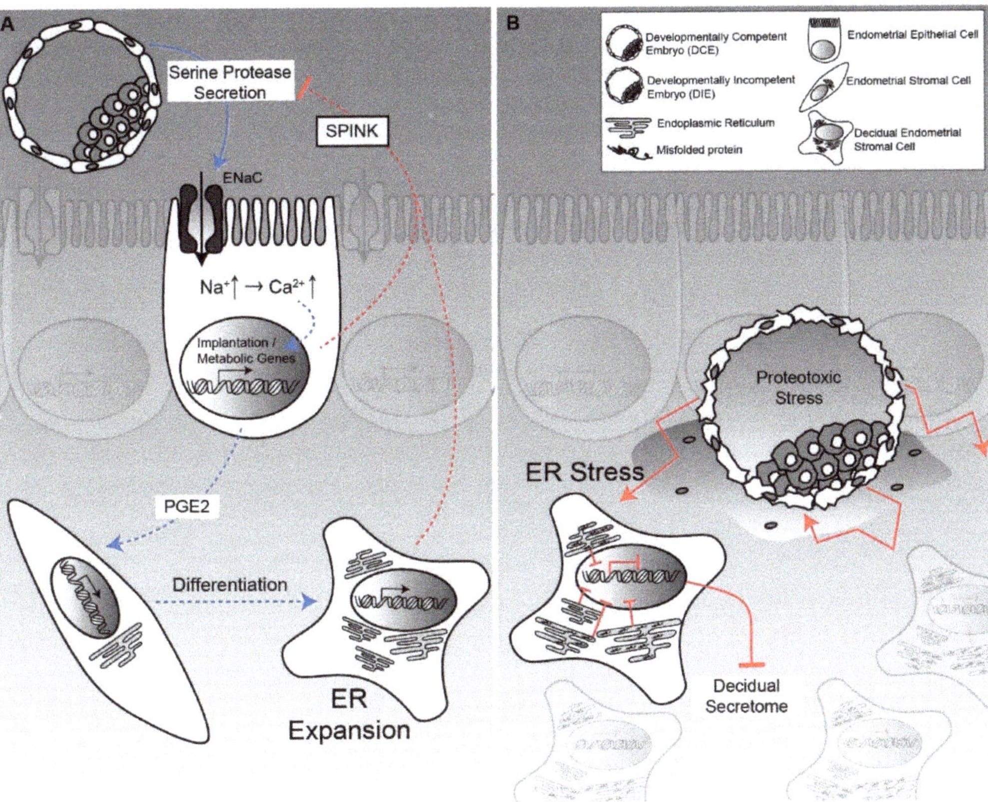

Figure 2. Positive and negative mechanisms contribute actively to the selection of human embryos at implantation. (**A**) Developmentally competent human embryos secrete serine proteases that activate epithelial Na1 channel (ENaC) that induce genes involved in implantation and post-implantation embryo development. (**B**) Developmentally compromised embryos By contrast, induce excessive protease activity, leading to accumulation of misfolded proteins and ER stress, compromising decidual cell functions and triggering early maternal rejection. From: Brosens JJ et al. (2014) [44].

An early investigation by Liu et al. [85] failed to find a significant difference in the expression frequency of all studied genes between viable and nonviable or arrested embryos, but with one exception: *BCL-2* was only detected in viable embryos. Shortly after, Jurisicova et al. [86] reported that human oocytes and preimplantation embryos possess abundant levels of transcripts encoding cell death suppressors and apoptosis inducer genes, *Bax* and *Caspase-2*. They also conducted an analysis of gene expression in single human embryos exhibiting various degrees of fragmentation at the 2-, 4-, and 8-cell stages and reported that at the 4-cell stage, embryos displaying $30 \pm 50\%$ fragmentation showed a significant increase in the *Harakiri* gene and *Caspase-3*, suggesting that cellular fragmentation could be regulated by certain components of a genetic program of cell death. They stressed that cell fate (i.e., survival/differentiation or death) is determined by the outcome of specific intracellular interactions between pro- and anti-apoptotic proteins, many of which are expressed during oocyte and preimplantation embryo development [87].

The same group [88] also examined changes in mitochondrial membrane potential over the preimplantation stages of mouse and human embryos. They found that mouse zygotes and early embryos contain highly-polarized mitochondria and observed a transient increase in the ratio of high to low $\Delta\Psi m$ (delta psi m, a measure of the mitochondrial membrane potential) with increasing cleavage. In human 8-cell embryos, there was an increased ratio of high- to low-polarized mitochondria, as well as increasing degrees of embryo fragmentation, leading to the conclusion that an aberrant shift in $\Delta\Psi$m is either associated with or can contribute to a decreased developmental potential. Komatsu et al. [89] measured the $\Delta\Psi$m of the in vivo-fertilized 1- and 2-cell stage, and of IVF embryos and found that the $\Delta\Psi$m of in vivo-fertilized embryos was highly upregulated, whereas in a number of IVF embryos remained unchanged. In addition, the development of low-$\Delta\Psi$m 2-cell stage IVF embryos tended to be arrested after the 2-cell stage. Zhao et al. [90] tried to profile the role of the mitofusin2 gene *Mfn2* (a key player in many mitochondrial activities, such as fusion, trafficking, turnover, and contacts with other organelles [91]) in mouse embryos and determine the underlying mechanism of *Mfn2* function in embryo development. They concluded that low in vitro expression of *Mfn2* causes mitochondrial dysfunction and attenuates blastocyst formation rate. These findings indicate that *Mfn2* could affect preimplantation embryo development through mitochondrial function and cellular apoptosis [90].

Recently, a study by Haouzi et al. [92] of genes involved in the regulation of the apoptotic and survival pathways of mouse and human embryos found that components of the major apoptotic and survival signaling pathways were expressed during early human and mouse embryonic development in a species-specific manner.

According to Brosens et al. [44], genome-wide screening of blastomeres in IVF cycles seems to harbor cells with complex large-scale structural chromosomal imbalances, mostly by mitotic non-disjunction. In addition, a vast array of chromosomal errors has been detected in human embryos throughout all stages of pre-implantation development [44].

In an attempt to identify visual markers of early pregnancy loss in women undergoing IVF, Amitai et al. assembled an expansive set of 314 morphological, morphokinetic, and dynamic features derived from measurable static and dynamic properties of preimplantation embryo development [93]. They identified a subset of six non-redundant morphodynamical features possessing high predictive capacity. Among them, of particular interest were features that account for the distribution of the nucleolus precursor bodies within the small pronucleus and pronuclei dynamics. Using these features, they developed a "decision-support tool" for prioritizing embryos for transfer based on their predicted implantation potential.

The challenge now is to find a way to apply the identification of these markers to natural pregnancies.

6. The Need for New Terminology

As summarized here, human preimplantation embryos exhibit in vitro high levels of apoptotic and high rates of developmental arrest during the first week [82]. In fact, in vitro and in vivo, errors and failures of the first and the next three cleavage divisions frequently cause immediate embryo death or lead to aberrant subsequent development. At the same time, the ability of human embryos to eliminate/expel abnormal blastomeres as cell debris/fragments and carry out—whenever possible—self-correction or quality control has been also documented [94]. Finally, Monsivais et al. [95] have shown that for the correct regulation of uterine receptivity, a convergence is necessary of bone morphogenetic proteins (BMPs, members of the TGF-β family that regulate the post-implantation and mid-gestation stages of pregnancy) and steroid hormone signaling pathways.

For a number of years, in assisted reproduction, the establishment of a viable pregnancy has been defined by a rise in circulating hCG. The use of this marker shows that in IVF the successful implantation of an embryo represents the major milestone in determining the success of gestation since it has been shown that only 50% of transferred embryos implant and that half of these embryos are subsequently lost [34].

The documented occurrence of massive early/preimplantation embryo loss provides the scientific basis to necessitate a clear distinction between the first 1 or 2 weeks and the subsequent 9 months of gestation. In the first phase, a physiological loss of fertilized oocytes/early embryos, calculated at around 50%, occurs, whereas during the second phase, the pathologic wastage of embryos/fetuses has been estimated to be around 15%, although increasing with maternal age. Given this stark contrast, it seems appropriate to distinguish these two periods also using a different nomenclature.

This is why we fully concur with the idea of, on the one hand, employing the word **gestation** to identify the period from fertilization (whether in vitro or in utero) to birth, and, on the other, to utilize the word **pregnancy** when referring to the period after implantation is completed [1]. The idea is certainly not new in view of the above-mentioned definition of pregnancy in the IVF clinic as the rise of hCG upon embryo implantation [96]. Extending this concept to all pregnancies would create a simple, clear nomenclature.

This seems the position taken by several organizations, which—in practice—affirm that there is no pregnancy before implantation: among them is the World Health Organization [97], which, referring to "Medicines for Reproductive Health and Perinatal Care" (Item 22), under 'Oral Hormonal Contraceptives' (22.1.1), lists hormonal emergency contraceptives levonorgestrel and ulipristal; and the other is the American College of Obstetricians and Gynecologists [98].

7. Conclusions

The continuing progress in our understanding of the complexity of interactions between the maternal organism and the early embryo is changing our overall outlook on the initial steps in establishing a pregnancy through placental formation. The first two weeks after fertilization must today be viewed as the critical period during which a major embryo selection process takes place in which a proportion that may surpass 50% of them is physiologically eliminated because they are unfit to progress toward birth.

The new knowledge, along with a refined nomenclature to distinguish gestation and pregnancy, should help us in improving the effectiveness of the various assisted reproduction technologies, as well as providing the scientific basis for a clear distinction between the first two weeks following fertilization and the rest of gestation.

Funding: No funding was obtained for this work.

Institutional Review Board Statement: No Review Board approval is necessary for this review.

Informed Consent Statement: No human subjects are involved and therefore no informed content form was prepared.

Data Availability Statement: Not applicable.

Conflicts of Interest: The authors declare no conflict of interest.

References

1. Benagiano, G.; Mancuso, S.; Gianaroli, L.; Di Renzo, G.C. Gestation vs. Pregnancy. *Am. J. Obstet. Gynecol.* **2023**, *229*, 91–92. [CrossRef] [PubMed]
2. Green, R.M. Embryo as epiphenomenon: Some cultural, social and economic forces driving the stem cell debate. *J. Med. Ethics* **2008**, *34*, 840–844. [CrossRef] [PubMed]
3. Macklon, N.S.; Geraedts, J.P.; Fauser, B.C. Conception to ongoing pregnancy: The 'black box' of early pregnancy loss. *Hum. Reprod. Update* **2002**, *8*, 333–343. [CrossRef] [PubMed]
4. Benagiano, G.; Farris, M.; Grudzinskas, G. Fate of fertilized human oocytes. *Reprod. Biomed. Online* **2010**, *21*, 732–741. [CrossRef] [PubMed]
5. The Annual Capri Workshop Group. Early pregnancy loss: The default outcome for fertilized human oocytes. *J. Assist. Reprod. Genet.* **2020**, *37*, 1057–1063. [CrossRef]
6. Steiner, A.Z.; Pritchard, D.A.; Young, S.L.; Herring, A.H. Peri-implantation intercourse lowers fecundability. *Fertil. Steril.* **2014**, *102*, 178–182. [CrossRef]
7. World Health Organisation. Congenital Anomalies. Available online: https://www.who.int/health-topics/congenital-anomalies#tab=tab_1 (accessed on 22 July 2022).
8. Nybo Andersen, A.M.; Wohlfahrt, J.; Christens, P.; Olsen, J.; Melbye, M. Maternal age and fetal loss: Population-based register linkage study. *BMJ* **2000**, *320*, 1708–1712. [CrossRef]

9.   Du Fosse, N.A.; van der Hoorn, M.P.; van Lith, J.M.M.; le Cessie, S.; Lashley, E. Advanced paternal age is associated with an increased risk of spontaneous miscarriage: A systematic review and meta-analysis. *Hum. Reprod. Update* **2020**, *26*, 650–669. [CrossRef]

10. Hertig, A.T.; Rock, J.; Adams, E.C. A description of 34 human ova within the first 17 days of development. *Am. J. Anat.* **1956**, *98*, 435–493. [CrossRef]

11. Hertig, A.T.; Rock, J.; Adams, E.C.; Menkin, M.C. Thirty-four fertilized human ova, good, bad and indifferent, recovered from 210 women of known fertility: A study of biologic wastage in early human pregnancy. *Pediatrics* **1959**, *23 Pt 2*, 202–211. [CrossRef]

12. Hertig, A.T.; Rock, J. Searching for early fertilized human ova. *Gynecol. Investig.* **1973**, *4*, 121–139. [CrossRef] [PubMed]

13. Roberts, C.J.; Lowe, C.R. Where have all the conceptions gone? *Lancet* **1975**, *305*, 498–499. [CrossRef]

14. Shepard, T.H.; Fantel, A.G. Embryonic and early fetal loss. *Clin. Perinatol.* **1979**, *6*, 219–243. [CrossRef]

15. Morton, H.; Tinneberg, H.R.; Rolfe, B.; Wolf, M.; Mettler, L. Rosette inhibition test: A multicentre investigation of early pregnancy factor in humans. *J. Reprod. Immunol.* **1982**, *4*, 251–261. [CrossRef]

16. Rolfe, B.E. Detection of fetal wastage. *Fertil. Steril.* **1982**, *37*, 655–660. [CrossRef] [PubMed]

17. Smart, Y.C.; Fraser, I.S.; Roberts, T.K.; Clancy, R.L.; Cripps, A.W. Fertilization and early pregnancy loss in healthy women attempting conception. *Clin. Reprod. Fertil.* **1982**, *1*, 177–184.

18. Wilcox, A.J.; Weinberg, C.R.; O'Connor, J.F.; Baird, D.D.; Schlatterer, J.P.; Canfield, R.E.; Armstrong, E.G.; Nisula, B.C. Incidence of early loss of pregnancy. *N. Engl. J. Med.* **1988**, *319*, 189–194. [CrossRef]

19. Olsen, J.; Rachootin, P. Invited commentary: Monitoring fecundity over time—If we do it, then let's do it right. *Am. J. Epidemiol.* **2003**, *157*, 94–97. [CrossRef]

20. Wood, J.W. Fecundity and natural fertility in humans. *Oxf. Rev. Reprod. Biol.* **1989**, *11*, 61–109.

21. Konishi, S.; Kariya, F.; Hamasaki, K.; Takayasu, L.; Ohtsuki, H. Fecundability and Sterility by Age: Estimates Using Time to Pregnancy Data of Japanese Couples Trying to Conceive Their First Child with and without Fertility Treatment. *Int. J. Environ. Res. Public. Health.* **2021**, *18*, 5486. [CrossRef]

22. Henripin, J. *La Population Canadienne au début du XVIIIème Siècle [Canadian Population at the Beginning of the XVIII Century]*; Presses Universitaires de France: Paris, France, 1954.

23. Wang, X.; Chen, C.; Wang, L.; Chen, D.; Guang, W.; French, J. Conception, early pregnancy loss, and time to clinical pregnancy: A population-based prospective study. *Fertil. Steril.* **2003**, *79*, 1517–1521. [CrossRef] [PubMed]

24. Charbonneaux, P. *Tourouvre-au-Perche au XVIIème et XVIIIème Siècle [Tourouvre-au-Perche during the XVII and XVIII Centuries]*; Presses Universitaires de France: Paris, France, 1970.

25. Vessey, M.; Doll, R.; Peto, R.; Johnson, B.; Wiggins, P. A long-term follow-up study of women using different methods of contraception. An interim Report. *J. Biosoc. Sci.* **1976**, *8*, 373–427. [CrossRef] [PubMed]

26. Balakrishnan, T.R. Probability of conception, conception delay and estimates of fecundability in rural and semi-urban areas of certain Latin-American countries. *Soc. Biol.* **1979**, *26*, 226–231. [CrossRef] [PubMed]

27. van Noord-Zaadstra, B.M.; Looman, C.W.; Alsbach, H.; Habbema, J.D.; te Velde, E.R.; Karbaat, J. Delaying childbearing: Effect of age on fecundity and outcome of pregnancy. *Br. Med. J.* **1991**, *302*, 1361–1365. [CrossRef] [PubMed]

28. Sheps, M.C. An analysis of reproductive patterns in an American isolate. *Popul. Stud.* **1965**, *19*, 65–77. [CrossRef]

29. Zinaman, M.J.; Clegg, E.D.; Brown, C.C.; O'Connor, J.; Selevan, S.G. Estimates of human fertility and pregnancy loss. *Fertil. Steril.* **1996**, *65*, 503–509. [CrossRef]

30. Smarr, M.M.; Sapra, K.J.; Gemmill, A.; Kahn, L.G.; Wise, L.A.; Lynch, C.D.; Factor-Litvak, P.; Mumford, S.L.; Skakkebaek, N.E.; Slama, R.; et al. Is human fecundity changing? A discussion of research and data gaps precluding us from having an answer. *Hum. Reprod.* **2017**, *32*, 499–504. [CrossRef]

31. Ellish, N.J.; Saboda, K.; O'Connor, J.; Nasca, P.C.; Stanek, E.J.; Boyle, C. A prospective study of early pregnancy loss. *Hum. Reprod.* **1996**, *11*, 406–412. [CrossRef]

32. Leridon, H. Human fecundity: Situation and outlook. *Popul. Soc.* **2010**, *471*, 1–4.

33. Simon, C.; Landeras, J.; Zuzuarregui, J.L.; Martin, J.C.; Remohi, J.; Pellicer, A. Early pregnancy losses in in vitro fertilization and oocyte donation. *Fertil. Steril.* **1999**, *72*, 1061–1065. [CrossRef]

34. Boomsma, C.M.; Kavelaars, A.; Eijkemans, M.J.; Lentjes, E.G.; Fauser, B.C.; Heijnen, C.J.; Macklon, N.S. Endometrial secretion analysis identifies a cytokine profile predictive of pregnancy in IVF. *Hum. Reprod.* **2009**, *24*, 1427–1435. [CrossRef] [PubMed]

35. Jarvis, G.E. Misjudging early embryo mortality in natural human reproduction. *F1000Res.* **2020**, *9*, 702. [CrossRef] [PubMed]

36. Kennedy, T.G. Physiology of implantation. In *In Vitro Fertilization and Assisted Reproduction*; Gomel, V., Leung, P.C.K., Eds.; Monduzzi Editore: Bologna, Italy, 1997; pp. 729–735.

37. Jones, R.E.; Lopez, K.H. Pregnancy. In *Human Reproductive Biology*, 4th ed.; Jones, R.E., Lopez, K.H., Eds.; Elsevier: Amsterdam, The Netherlands, 2014; pp. 175–204.

38. Sharkey, A.M.; Macklon, N.S. The science of implantation emerges blinking into the light. *Reprod. Biomed. Online* **2013**, *27*, 453–460. [CrossRef]

39. Short, R.V. When a conception fails to become a pregnancy. *Ciba Found. Symp.* **1978**, *64*, 377–394.

40. Wilcox, A.J.; Harmon, Q.; Doody, K.; Wolf, D.P.; Adashi, E.Y. Preimplantation loss of fertilized human ova: Estimating the unobservable. *Hum. Reprod.* **2020**, *35*, 743–750. [CrossRef] [PubMed]
41. Makrigiannakis, A.; Vrekoussis, T.; Zoumakis, E.; Kalantaridou, S.N.; Jeschke, U. The Role of HCG in Implantation: A Mini-Review of Molecular and Clinical Evidence. *Int. J. Mol. Sci.* **2017**, *18*, 1305. [CrossRef]
42. Kim, S.M.; Kim, J.S. A Review of Mechanisms of Implantation. *Dev. Reprod.* **2017**, *21*, 351–359. [CrossRef]
43. Macklon, N.S.; Brosens, J.J. The human endometrium as a sensor of embryo quality. *Biol. Reprod.* **2014**, *91*, 98. [CrossRef]
44. Brosens, J.J.; Bennett, P.R.; Abrahams, V.M.; Ramhorst, R.; Coomarasamy, A.; Quenby, S.; Lucas, E.S.; McCoy, R.C. Maternal selection of human embryos in early gestation: Insights from recurrent miscarriage. *Semin. Cell Dev. Biol.* **2022**, *131*, 14–24. [CrossRef]
45. Brosens, J.J.; Salker, M.S.; Teklenburg, G.; Nautiyal, J.; Salter, S.; Lucas, E.S.; Steel, J.H.; Christian, M.; Chan, Y.-W.; Boomsma, C.M.; et al. Uterine selection of human embryos at implantation. *Sci. Rep.* **2014**, *4*, 3894. [CrossRef]
46. Shutt, D.A.; Lopata, A. The secretion of hormones during the culture of human preimplantation embryos with corona cells. *Fertil. Steril.* **1981**, *35*, 413–416. [CrossRef]
47. Morton, H.; Rolfe, B.; Clunie, G.J. An early pregnancy factor detected in human serum by the rosette inhibition test. *Lancet* **1977**, *1*, 394–397. [CrossRef]
48. Morton, H. Early pregnancy factor: An extracellular chaperonin 10 homologue. *Immunol. Cell Biol.* **1998**, *76*, 483–496. [CrossRef]
49. Nahhas, F.; Barnea, E. Human embryonic origin early pregnancy factor before and after implantation. *Am. J. Reprod. Immunol.* **1990**, *22*, 105–108. [CrossRef]
50. Barnea, E.R.; Lahijani, K.I.; Roussev, R.; Barnea, J.D.; Coulam, C.B. Use of lymphocyte platelet bindingassay for detecting a preimplantation factor: A quantitative assay. *Am. J. Reprod. Immunol.* **1994**, *32*, 133–138. [CrossRef]
51. Barnea, E.R.; Simon, J.; Levine, S.P.; Coulam, C.B.; Taliadouros, G.S.; Leavis, P.C. Progress in characterization of pre-implantation factor in embryo cultures and in vivo. *Am. J. Reprod. Immunol.* **1999**, *42*, 95–99.
52. Yang, M.; Yang, Y.; She, S.; Li, S. Proteomic investigation of the effects of preimplantation factor on human embryo implantation. *Mol. Med. Rep.* **2018**, *17*, 3481–3488. [CrossRef]
53. Zare, F.; Seifati, S.M.; Mahdi Dehghan-Manshadi, M.; Fesahat, F. Preimplantation Factor (PIF): A peptide with various functions. *JBRA Assist. Reprod.* **2020**, *24*, 214–218. [CrossRef]
54. Dos Santos, E.; Moindjie, H.; Sérazin, V.; Arnould, L.; Rodriguez, Y.; Fathallah, K.; Barnea, E.R.; Vialard, F.; Dieudonné, M.-N. Preimplantation factor modulates trophoblastic invasion throughout the decidualization of human endometrial stromal cells. *Reprod. Biol. Endocrinol.* **2021**, *19*, 96. [CrossRef]
55. O'Neill, C.; Collier, M.; Ryan, J.P.; Spinks, N.R. Embryo-derived platelet-activating factor. *J. Reprod. Fertil. Suppl.* **1989**, *37*, 19–27.
56. Roudebush, W.E.; Wininger, J.D.; Jones, A.E.; Wright, G.; Toledo, A.A.; Kort, H.I.; Massey, J.B.; Shapiro, D.B. Embryonic platelet-activating factor: An indicator of embryo viability. *Hum. Reprod.* **2002**, *17*, 1306–1310. [CrossRef] [PubMed]
57. Velasquez, L.A.; Maisey, K.; Fernandez, R.; Valdes, D.; Cardenas, H.; Imarai, M.; Delgado, J.; Aguilera, J.; Croxatto, H.B. PAF receptor and PAF acetylhydrolase expression in the endosalpinx of the human Fallopian tube: Possible role of embryo-derived PAF in the control of embryo transport to the uterus. *Hum. Reprod.* **2001**, *16*, 1583–1587. [CrossRef] [PubMed]
58. O'Neill, C. The role of PAF in embryo physiology. *Hum. Reprod. Update.* **2005**, *11*, 215–228. [CrossRef] [PubMed]
59. Sato, Y.; Fujiwara, H.; Konishi, I. Role of platelets in placentation. *Med. Mol. Morphol.* **2010**, *43*, 129–133. [CrossRef]
60. Palomares, K.T.; Parobchak, N.; Ithier, M.C.; Aleksunes, L.M.; Castaño, P.M.; So, M.; Faro, R.; Heller, D.; Wang, B.; Rosen, T. Fetal Exosomal Platelet-activating Factor Triggers Functional Progesterone Withdrawal in Human Placenta. *Reprod. Sci.* **2021**, *28*, 252–262. [CrossRef]
61. Fishel, S.B.; Edwards, R.G.; Evans, C.J. Human chorionic gonadotropin secreted by preimplantation embryos cultured in vitro. *Science* **1984**, *223*, 816–818. [CrossRef]
62. Bonduelle, M.L.; Dodd, R.; Liebaers, I.; Van Steirteghem, A.; Williamson, R.; Akhurst, R. Chorionic gonadotrophin-beta mRNA, a trophoblast marker, is expressed in human 8-cell embryos derived from tripronucleate zygotes. *Hum. Reprod.* **1988**, *3*, 909–914. [CrossRef]
63. Ochoa-Bernal, M.A.; Fazleabas, A.T. Physiologic Events of Embryo Implantation and Decidualization in Human and Non-Human Primates. *Int. J. Mol. Sci.* **2020**, *21*, 1973. [CrossRef]
64. You, Y.; Stelzl, P.; Joseph, D.N.; Aldo, P.B.; Maxwell, A.J.; Dekel, N.; Liao, A.; Whirledge, S.; Mor, G. TNF-alpha Regulated Endometrial Stroma Secretome Promotes Trophoblast Invasion. *Front. Immunol.* **2021**, *12*, 737401. [CrossRef]
65. Zhang, Z.; Yang, Y.; Lv, X.; Liu, H. Interleukin-17 promotes proliferation, migration, and invasion of trophoblasts via regulating PPAR-γ/RXR-α/Wnt signaling. *Bioengineered* **2022**, *13*, 1224–1234. [CrossRef]
66. Chi, F.; Sharpley, M.S.; Nagaraj, R.; Sen Roy, S.; Banerjee, U. Glycolysis-Independent Glucose Metabolism Distinguishes TE from ICM Fate during Mammalian Embryogenesis. *Dev. Cell* **2020**, *53*, 9–26.e4. [CrossRef] [PubMed]
67. Clark, D.A. Is there any evidence for immunologically mediated or immunologically modifiable early pregnancy failure? *J. Assist. Reprod. Genet.* **2003**, *20*, 63–72. [CrossRef] [PubMed]
68. Clark, D.A. Immunological factors in pregnancy wastage: Fact or fiction. *Am. J. Reprod. Immunol.* **2008**, *59*, 277–300. [CrossRef] [PubMed]

69. Haddad, E.K.; Duclos, A.J.; Antecka, E.; Lapp, W.S.; Baines, M.G. Role of interferon-gamma in the priming of decidual macrophages for nitric oxide production and early pregnancy loss. *Cell Immunol.* **1997**, *181*, 68–75. [CrossRef] [PubMed]

70. Haddad, E.K.; Duclos, A.J.; Lapp, W.S.; Baines, M.G. Early embryo loss is associated with the prior expression of macrophage activation markers in the decidua. *J. Immunol.* **1997**, *158*, 4886–4892. [CrossRef] [PubMed]

71. Vento-Tormo, R.; Efremova, M.; Botting, R.A.; Turco, M.Y.; Vento-Tormo, M.; Meyer, K.B.; Park, J.-E.; Stephenson, E.; Polański, K.; Goncalves, A.; et al. Single-cell reconstruction of the early maternal-fetal interface in humans. *Nature* **2018**, *563*, 347–353. [CrossRef] [PubMed]

72. Nancy, P.; Tagliani, E.; Tay, C.S.; Asp, P.; Levy, D.E.; Erlebacher, A. Chemokine gene silencing in decidual stromal cells limits T cell access to the maternal-fetal interface. *Science* **2012**, *336*, 1317–1321. [CrossRef]

73. Ticconi, C.; Pietropolli, A.; Di Simone, N.; Piccione, E.; Fazleabas, A. Endometrial Immune Dysfunction in Recurrent Pregnancy Loss. *Int. J. Mol. Sci.* **2019**, *20*, 5332. [CrossRef]

74. Robertson, S.A.; Care, A.S.; Moldenhauer, L.M. Regulatory T cells in embryo implantation and the immune response to pregnancy. *J. Clin. Investig.* **2018**, *128*, 4224–4235. [CrossRef]

75. Kanamori, M.; Nakatsukasa, H.; Okada, M.; Lu, Q.; Yoshimura, A. Induced Regulatory T Cells: Their Development, Stability, and Applications. *Trends Immunol.* **2016**, *37*, 803–811. [CrossRef]

76. Josefowicz, S.Z.; Rudensky, A. Control of regulatory T cell lineage commitment and maintenance. *Immunity* **2009**, *30*, 616–625. [CrossRef] [PubMed]

77. Vent-Schmidt, J.; Han, J.M.; MacDonald, K.G.; Levings, M.K. The role of FOXP3 in regulating immune responses. *Int. Rev. Immunol.* **2014**, *33*, 110–128. [CrossRef] [PubMed]

78. Sadlon, T.; Brown, C.Y.; Bandara, V.; Hope, C.M.; Schjenken, J.E.; Pederson, S.M.; Breen, J.; Forrest, A.; Beyer, M.; Robertson, S.; et al. Unravelling the molecular basis for regulatory T-cell plasticity and loss of function in disease. *Clin. Transl. Immunol.* **2018**, *7*, e1011. [CrossRef] [PubMed]

79. Barnea, E.R. Embryo maternal dialogue: From pregnancy recognition to proliferation control. *Early Pregnancy* **2001**, *5*, 65–66.

80. Csapo, A.I.; Pulkkinen, M.O.; Ruttner, B.; Sauvage, J.P.; Wiest, W.G. The significance of the human corpus luteum in pregnancy maintenance. I. Preliminary studies. *Am. J. Obstet. Gynecol.* **1972**, *112*, 1061–1067. [CrossRef]

81. Jurisicova, A.; Varmuza, S.; Casper, R.F. Programmed cell death and human embryo fragmentation. *Mol. Hum. Reprod.* **1996**, *2*, 93–98. [CrossRef]

82. Cecchele, A.; Cermisoni, G.C.; Giacomini, E.; Pinna, M.; Vigano, P. Cellular and Molecular Nature of Fragmentation of Human Embryos. *Int. J. Mol. Sci.* **2022**, *23*, 1349. [CrossRef]

83. Hardy, K. Apoptosis in the human embryo. *Rev. Reprod.* **1999**, *4*, 125–134. [CrossRef]

84. Leidenfrost, S.; Boelhauve, M.; Reichenbach, M.; Gungor, T.; Reichenbach, H.D.; Sinowatz, F.; Wolf, E.; Habermann, F.A. Cell arrest and cell death in mammalian preimplantation development: Lessons from the bovine model. *PLoS ONE* **2011**, *6*, e22121. [CrossRef]

85. Liu, H.C.; He, Z.Y.; Mele, C.A.; Veeck, L.L.; Davis, O.; Rosenwaks, Z. Expression of apoptosis-related genes in human oocytes and embryos. *J. Assist. Reprod. Genet.* **2000**, *17*, 521–533. [CrossRef]

86. Jurisicova, A.; Antenos, M.; Varmuza, S.; Tilly, J.L.; Casper, R.F. Expression of apoptosis-related genes during human preimplantation embryo development: Potential roles for the Harakiri gene product and Caspase-3 in blastomere fragmentation. *Mol. Hum. Reprod.* **2003**, *9*, 133–141. [CrossRef] [PubMed]

87. Jurisicova, A.; Acton, B.M. Deadly decisions: The role of genes regulating programmed cell death in human preimplantation embryo development. *Reproduction* **2004**, *128*, 281–291. [CrossRef] [PubMed]

88. Acton, B.M.; Jurisicova, A.; Jurisica, I.; Casper, R.F. Alterations in mitochondrial membrane potential during preimplantation stages of mouse and human embryo development. *Mol. Hum. Reprod.* **2004**, *10*, 23–32. [CrossRef] [PubMed]

89. Komatsu, K.; Iwase, A.; Mawatari, M.; Wang, J.; Yamashita, M.; Kikkawa, F. Mitochondrial membrane potential in 2-cell stage embryos correlates with the success of preimplantation development. *Reproduction* **2014**, *147*, 627–638. [CrossRef]

90. Zhao, N.; Zhang, Y.; Liu, Q.; Xiang, W. Mfn2 Affects Embryo Development via Mitochondrial Dysfunction and Apoptosis. *PLoS ONE* **2015**, *10*, e0125680. [CrossRef]

91. Filadi, R.; Pendin, D.; Pizzo, P. Mitofusin 2: From functions to disease. *Cell Death Dis.* **2018**, *9*, 330. [CrossRef]

92. Haouzi, D.; Boumela, I.; Chebli, K.; Hamamah, S. Global, Survival, and Apoptotic Transcriptome during Mouse and Human Early Embryonic Development. *Biomed. Res. Int.* **2018**, *2018*, 5895628. [CrossRef]

93. Amitai, T.; Kan-Tor, Y.; Or, Y.; Shoham, Z.; Shofaro, Y.; Richter, D.; Har-Vardi, I.; Ben-Meir, A.; Srebnik, N.; Buxboim, A. Embryo classification beyond pregnancy: Early prediction of first trimester miscarriage using machine learning. *J. Assist. Reprod. Genet.* **2023**, *40*, 309–322. [CrossRef]

94. Orvieto, R.; Shimon, C.; Rienstein, S.; Jonish-Grossman, A.; Shani, H.; Aizer, A. Do human embryos have the ability of self-correction? *Reprod. Biol. Endocrinol.* **2020**, *18*, 98. [CrossRef]

95. Monsivais, D.; Clementi, C.; Peng, J.; Titus, M.M.; Barrish, J.P.; Creighton, C.J.; Lydon, J.P.; DeMayo, F.J.; Matzuk, M.M. Uterine ALK3 is essential during the window of implantation. *Proc. Natl. Acad. Sci. USA* **2016**, *113*, E387–E395. [CrossRef]

96. Chung, K.; Sammel, M.D.; Coutifaris, C.; Chalian, R.; Lin, K.; Castelbaum, A.J.; Freedman, M.F.; Barnhart, K.T. Defining the rise of serum HCG in viable pregnancies achieved through use of IVF. *Hum. Reprod.* **2006**, *21*, 823–828. [CrossRef] [PubMed]

97. World Health Organization. WHO Model List of Essential Medicines—22nd List. 2021. Available online: https://www.who.int/publications/i/item/WHO-MHP-HPS-EML-2021.02 (accessed on 15 September 2023).

98. American College of Obstetricians, Gynecologists. Committee on Practice B-G. ACOG Practice Bulletin No. 200: Early Pregnancy Loss. *Obstet. Gynecol.* **2018**, *132*, e197–e207. [CrossRef] [PubMed]

International Journal of
Molecular Sciences

Review

Thrombotic Alterations under Perinatal Hypoxic Conditions: HIF and Other Hypoxic Markers

Alejandro Berna-Erro [1], María Purificacion Granados [2], Juan Antonio Rosado [1,*] and Pedro Cosme Redondo [1]

1 Department of Physiology (Phycell), University of Extremadura, Avd de la Universidad s/n, 10003 Caceres, Spain; alejandrobe@unex.es (A.B.-E.); pcr@unex.es (P.C.R.)
2 Pharmacy Unit of Extremadura County Health Service, Health Center of Talayuela, 10310 Talayuela, Spain; mpc77@gmail.com
* Correspondence: jarosado@unex.es; Tel.: +34-927257106 (ext. 551376)

Abstract: Hypoxia is considered to be a stressful physiological condition, which may occur during labor and the later stages of pregnancy as a result of, among other reasons, an aged placenta. Therefore, when gestation or labor is prolonged, low oxygen supply to the tissues may last for minutes, and newborns may present breathing problems and may require resuscitation maneuvers. As a result, poor oxygen supply to tissues and to circulating cells may last for longer periods of time, leading to life-threatening conditions. In contrast to the well-known platelet activation that occurs after reperfusion of the tissues due to an ischemia/reperfusion episode, platelet alterations in response to reduced oxygen exposition following labor have been less frequently investigated. Newborns overcome temporal hypoxic conditions by changing their organ functions or by adaptation of the intracellular molecular pathways. In the present review, we aim to analyze the main platelet modifications that appear at the protein level during hypoxia in order to highlight new platelet markers linked to complications arising from temporal hypoxic conditions during labor. Thus, we demonstrate that hypoxia modifies the expression and activity of hypoxic-response proteins (HRPs), including hypoxia-induced factor (HIF-1), endoplasmic reticulum oxidase 1 (Ero1), and carbonic anhydrase (CIX). Finally, we provide updates on research related to the regulation of platelet function due to HRP activation, as well as the role of HRPs in intracellular Ca^{2+} homeostasis.

Keywords: Hypoxia; HIF-1; calcium; platelets; neonates

Citation: Berna-Erro, A.; Granados, M.P.; Rosado, J.A.; Redondo, P.C. Thrombotic Alterations under Perinatal Hypoxic Conditions: HIF and Other Hypoxic Markers. *Int. J. Mol. Sci.* **2023**, *24*, 14541. https://doi.org/10.3390/ijms241914541

Academic Editor: Giovanni Tossetta

Received: 28 July 2023
Revised: 7 September 2023
Accepted: 8 September 2023
Published: 26 September 2023

1. Placental Alteration Leading to Fetal Hypoxemia

The specialization of throphoblast cells allows for the generation of a complex functional tissue that has two main roles during the development of the fetus: (1) oxygen supply; and (2) the generation of a physical barrier that controls the interchange of particles and cells between the mother and the fetus. Throphoblastic cells interact with two vascular beds, avoiding direct contact between maternal and neonatal blood: the uterus vascular bed and extraembryonic mesenchyme of the allantois [1]. Uterine-linked trophoblast cells are responsible for supplying maternal blood to placenta, while trophoblast cells from the allantoic vasculature efficiently extract nutrients that end in the fetal vasculature [1]. Low oxygen supply to the embryo has been demonstrated before the implantation phase, and this stressful situation drastically changes upon placenta formation. Interestingly, at up to 12 weeks of gestation, a difference in the partial pressure of oxygen between the placenta (17.9 mm Hg) and endometrium (39.6 mm Hg) can be seen, providing evidence of hypoxemia during this stage of embryo development [2].

Placental alterations due to hypertension or preeclampsia are present in 3–7% of pregnant women worldwide. Often, defective trophoblastic invasion and an impaired development of the uterine spiral arteries lead to abnormal placentation, the main cause of hypoxia/reoxygenation phenomena; further, it may result in fluctuations of the gradient of oxygen, altered antioxidant capacity, placenta oxidative stress; and a reduction in nitric oxide (NO) [3]. On the other hand, low oxygen supply to the fetus may occur due to early senescence of the placenta. Senescence of the placenta is characterized by the increased expression of p53, p21, and p16 transcriptional factors, as well as premature senescence of the extravillous trophoblasts. The latter is associated with abnormal placental development, that in turn leads to blood hypoperfusion and, subsequently, develops into fetal growth restriction, preterm birth, and stillbirth [4].

2. Hypoxia during Labor

Oxygen saturation (SpO_2) blood tests determine the percentage of hemoglobin bound to oxygen. It is an indirect indicator of tissue oxygenation. The average arterial SpO_2 value usually found in healthy adults is around 98%, and this indirectly indicates an adequate oxygen supply to the whole organism. The threshold considered as "below normal" oxygen saturation (hypoxemia) is when SpO_2 values are below 90% [5]. Hypoxemia is not a dangerous condition as long as tissues are adequately prefunded and oxygenated. Conversely, severe hypoxemia can lead to defective tissue oxygenation or to hypoxia that comprises tissue damage and leads to subsequent deleterious effects in physiology and metabolism [6].

2.1. Risk of Perinatal Hypoxemia and Hypoxia during Gestation

The fetus must face hypoxemia during labor, since oxygen saturation in arterial blood drops considerably, lasting several minutes after parturition [7]. The fetus grows in a hypoxemic environment in utero that, surprisingly, seems to be optimal for its development. For instance, SpO_2 values of around 70% are frequently found in the umbilical vein as compared with that of 98% found in mothers [8]. Thus, the fetus should have a particular metabolism that is adapted to mild hypoxemic conditions; for instance, fetuses express increased levels of fetal hemoglobin, which is much more efficient at capturing oxygen than adult hemoglobin [9]. Immediately before parturition, fetal oxygen saturation drops drastically after each uterine contraction [10]. Next, the fetus must face an additional downward trend in oxygen saturation and a marked transient hypoxemia during delivery, so that the mean preductal SpO_2 decreases to 50–73% around 10 min before parturition, re-marking the hypoxic temporal conditions. However, SpO_2 values quickly increase up to 90% after 5 min upon the start of breathing, thereafter achieving the normal values (95%) found in adults after 12 min, or up to 97.6% during the first week after birth [7,11]. Each labor varies in time and in the level of fetal stress. These can be prolonged under certain situations or complications, further increasing the length and degree of neonatal hypoxemia. For instance, SpO_2 values in arterial blood decrease by 3% during cesarean delivery as compared to vaginal parturition [12]. This transient hypoxemia during normal labor does not pose a threat, but newborns must adapt themselves quickly and increase oxygen saturation in the blood by oxygen intake from a new, rich oxygen atmosphere. If the transition fails, neonates suffer prolonged hypoxemia that can develop into life-threating asphyxia and hypoxia, a prevailing complication in labor as compared to other stages of life [13]. The appearance of neonatal hypoxia as a consequence of prolonged hypoxemia indicates the existence of a transient adaptive mechanism against neonatal damage, since newborns are not indefinitely "protected" from prolonged hypoxemia. Complications that can lead to fetal and neonatal hypoxia include maternal heart, lung and kidney diseases, but also decreased maternal breathing from anesthesia, anemia, maternal diabetes, or low maternal blood pressure. Additionally, pathologic intrauterine hypoxia also arises as a consequence of maternal hypertension and preeclampsia; fetal intrauterine growth restriction; asphyxia due to compression of the umbilical cord; premature birth; traumatic delivery; intravascular volume contraction after birth; immature mucosal and skin barriers; the failure of lung expansion; leftover fluid in the lungs; and meconium aspiration [13,14].

2.2. Risk of Hypoxia after Birth

Two out of every thousand newborns suffer from asphyxia after birth in developed countries, a number that rises tenfold in the absence of healthcare in non-developed countries [15]. A common disorder that triggers neonatal hypoxia is hyaline membrane disease, or neonatal respiratory distress syndrome (NRDS). NRDS is a failure in pulmonary surfactant maturation that impedes adequate lung and alveoli expansion within 24 h after birth, hindering correct pulmonary contraction and normal breathing. The prevalence of NRDS is higher in preterm newborns because their lungs are not ready to produce enough surfactants. Inflammation, due to meconium aspiration or previous obstructive processes, also leads to NRDS, worsening hypoxemia conditions during labor [16]. Uncontrolled subsequent hypoxia as a consequence of asphyxia, together with increased inflammation derived from NRDS, often leads to additional complications, such as pulmonary hypertension in newborns (PHN), which increases hypoxia and further damages tissue mostly localized in the pulmonary vascular tree. Uncontrolled severe respiratory distress syndrome can impede complete lung development and it can progress to permanent damage over time, frequently evolving towards bronchopulmonary dysplasia (BPD), among other chronic breathing problems [17]. Common complications of neonatal hypoxic disease include sepsis, asthma, defective vision, or a delay in learning body movements. Serious complications include bleeding or thrombotic ischemia in the brain, with further consequences such as delayed cognitive development leading to learning or behavioral disabilities, and even cerebral palsy, multiorgan failure, and neonatal death. Among others, disorders in neonatal blood coagulation, usually referred to as thrombotic and hemostatic disorders, are the most common complications associated with neonatal hypoxic pathologies. Hypoxia/asphyxia causes altered coagulation in the immature neonatal hemostatic system, greatly complicating the course of the disease, and increasing the risk of perinatal death. Thrombotic disorders may appear localized, for instance, in the pulmonary vasculature, or scattered all over the vascular tree. Further, respiratory distress syndrome leads to endothelial activation and decreased numbers of circulating platelets (thrombocytopenia) due to increased platelet adhesion [16]. Untreated hypoxia often leads to intraventricular hemorrhage, a common cause of death in preterm infants. The complete mechanism underlying this pathology is not clear, but an immature neonatal hemostatic system contributes to the appearance of the disease [18].

2.3. Hypoxic-Evoked Thrombotic Alterations during Labor

Platelets are the main cellular components of the hemostatic system. Platelets rise from megakaryocytic fragmentation within the bone marrow in response to thrombopoietin [19]. Thrombopoietin, combined with interleukins and others, induces megakaryocyte transformation into giant multinuclear cells, leading to cell cytoplasmic fragmentation and, finally, evoking the generation of pro-platelets that subsequently mature into platelets due to fragmentation into pulmonary vasculature [19]. Platelets participate in the coagulation cascade that comprises a wide range of proteins generated in the liver and released to the blood stream, which act as agonists (pro-thrombotic) or inhibitors (anti-thrombotic) of blood coagulation. Platelets work together with the coagulation cascade to stop excessive blood loss during traumatic hemorrhage (hemostasis). Bleeding arrest is achieved by plugging broken vessels with a clot of aggregated platelets and blood cells (thrombus). Pathological uncontrolled thrombi formation may lead to obstructive complications (thrombosis), leading to vessel occlusion. Severe vessel occlusion may compromise blood supply and organ function. Platelets also participate in inflammation and immunity [20].

Increasing evidence indicates the maturation of fetal hemostatic system in childhood [21]. Thus, studies in murine models suggest the existence of megakaryocytic and hematopoietic transcriptional networks controlling developmental changes in platelets [22]. This implies significant differences between newborns and adults. For instance, neonatal megakaryocytes are smaller but hyperproliferative compared to those of adults [23]. Neonatal platelets are less reactive or hyporeactive in response to principle physiologic agonists such as ADP, thrombin or collagen [24–26]. Key surface proteins involved in platelet activation are also decreased [27]. These differences are much more evident in preterm newborns, who present longer bleeding times [28]. Mechanisms underlying neonatal hyporeactivity are not completely understood, but the proposed hypotheses are decreased receptor expression, changes in signaling pathways. or decreased intracellular Ca^{2+} mobilization [29]. However, decreased neonatal platelet activity is counterbalanced by a more robust response by the coagulation cascade [30], increasing, for instance, the expression and activity of the pro-thrombotic von Willebrand factor (vWF) [31].

Conversely, evidence suggests that the concentrations of some pro-thrombotic factors of the coagulation cascade are decreased, although this requires further investigation [32]. Interestingly, anti-thrombotic factors of the coagulation cascade, such as protein C, protein S, antithrombin III, heparin cofactor II and plasminogen, are also decreased, showing less inhibitory and fibrinolytic activity, thereby compensating the minor presence of pro-thrombotic factors [33]. This decrease in anti-thrombotic factors reinforces pro-thrombotic activity, indicating that neonatal hemostasis mostly relies on the coagulation cascade, as compared to adults. The consequence of the reinforced role of the coagulation cascade in newborns is an increased prevalence of thromboembolic complications when the coagulation cascade is pathologically altered. Additionally, increasing evidence suggest that therapies developed for adults but given to newborns should be adjusted. In this sense, platelet transfusion often solves the problem of neonatal thrombocytopenia, but this procedure is not without risks and it is under evaluation [34]. Transfusion of adult platelet plasma to thrombocytopenic newborns shortens aggregation rates and increases the risk of thrombosis due to the reinforced neonatal coagulation cascade [35]. Additionally, thrombocytopenic neonates must face other adverse effects associated with transfusion [36].

The incidence of thromboembolic complications during labor is higher than in other stages of life because of the traumatic nature of this event. One fourth of neonates, for instance, suffer from thrombocytopenia. Severe thrombocytopenia often leads to prolonged bleeding times, increasing the incidence of hemorrhage [37]. Neonatal thrombocytopenia may result from complications during fetal development, during labor or immediately after parturition, hence representing a serious risk factor for mortality in newborns. As we have noted, the fetus grows normally in a hypoxemic environment. However, severe intrauterine hypoxia as a consequence of maternal pathologies may lead to neonatal thrombocytopenia. In this case, the cause of thrombocytopenia arises from impaired platelet production by defective megakaryocytic function [37]. Megakaryopoiesis significantly increases in asphyxiated infants, but decreases as platelet counts reach normal numbers, indicating a possible compensation mechanism for low platelet production [38].

3. Altered Hemostasis during Exposition to Perinatal Hypoxia

3.1. Changes in the Coagulation System

Asphyxiated newborns, as a consequence of complications after birth, develop mild or severe thrombosis among other disorders, followed by secondary thrombocytopenia, which drastically increases the risk of morbidity [37]. The molecular mechanism is under investigation, but the relevant role of the coagulation cascade in their immature hemostatic systems seems to sensitize them to these coagulation disorders. Together with an associated inflammation, hypoxia damages and activates the endothelial cells of the vascular wall. These cells react to the injury, increasing the production of the vasoconstrictor endothelin-1 (ET-1), pro-thrombotic reactive oxygen species (ROS), thromboxane A_2 (TxA_2) and tissue factor, but conversely decreasing the production of anti-thrombotic and vasodilators such as nitric oxide (NO) and prostacyclin (PGI_2). Antithrombotic factors of the coagulation cascade, such as antithrombin III, protein C, and protein S, also decrease dramatically during asphyxia, reducing the fibrinolytic activity necessary for the prevention of thrombus formation and clearance (thrombolysis) [39]. Activated endothelial cells, together with increased pro-thrombotic ROS and TxA_2, activate circulating platelets in turn [40], promoting their adhesion to the vessel wall and generating thrombocytopenia due to this excessive platelet consumption. Decreased anti-thrombotic factors cannot counter-balance this process.

3.2. Changes in Platelet Reactivity

In the absence of neonatal evidence, studies in adults suggest that hypoxic stress increases the reactivity of platelets by changing their protein expression profile [41]. For instance, stressed platelets express hypoxia-inducible factor (HIF)-2, which increases the expression and release of pro-thrombotic factors such as plasminogen-activator inhibitor-1 (PAI-1). This increase in activation and granule secretion is mainly achieved by calcium (Ca^{2+}) entry through tstore-operated calcium entry (SOCE) into the platelets [42–44]. Rodent models also point to an increase in cyclooxygenase-2 (COX-2) in platelets, as well as its main product TxA_2, which greatly amplifies their activation and aggregation [43,45]. As a consequence, severe hypoxia often leads to excessive platelet activation and scattered adhesion to the endothelium of the vascular tree, referred to as disseminated intravascular coagulation (DIC). The prevalence of DIC is particularly high in neonates, which may result in life-threatening conditions. Commonly targeted organs during the development of DIC are the lungs; liver; and kidneys, frequently as a result of renal venous thrombosis [46]; as well as the brain, due to cerebral sinovenous thrombosis, which increases the risk of neonatal encephalopathy [47]. The result is that 50% of neonates who develop DIC finally die [48].

3.3. Complications in Newborns as a Consequence of Hypoxia

Moreover, excessive consumption of platelets during DIC results in thrombocytopenia, increasing the risk of excessive bleeding and hemorrhage [49]. In summary, 7.1% of neonates suffering from respiratory distress syndrome, a common cause of asphyxia, develop DIC, 50% develop moderate to severe thrombocytopenia, and 14.3% develop hemorrhagic episodes [50].

On the other hand, released factors by the activated endothelium as a consequence of asphyxia also act as vasodilators and vasoconstrictors, and their imbalance promotes the appearance of pulmonary hypertension of the newborn (PHN) as a result of increased vasoconstriction in the pulmonary vessels. Vasoconstriction leads to restricted blood flow, triggering hypoxia or worsening an initial hypoxic condition. PHN occurs in 2 per 1000 newborns, normally as a complication of a previous existing disease. For instance, 10% of newborns with failed cardiopulmonary transition develop PHN. Other causes favoring PHN are maternal drug intake, asphyxia, meconium aspiration syndrome, respiratory distress syndrome, or pneumonia. PHN is resolved in most cases, but 5% of newborns develop severe persistent PHN (PPHN), an important risk factor for mortality. Twenty-five percent of survivors develop neurological defects. PHN is also associated with thrombocytopenia, among other complications [51,52]. While platelet activation has been reported in adults with pulmonary hypertension, it has not been explored in neonates. Studies in animal models also suggest that PPHN itself increases platelet activation [53]. Therefore, the increased TxA_2 generated by endothelial cells and adhered platelets [40], together with ET-1, stimulates contraction of the smooth muscle cells (SMCs), which are located in the vessel wall. Animal models suggest that another secreted factor, such as serotonin, released by the activated platelets might contribute to vasoconstriction and PHN [53].

The release of these platelet agonists are mediated by intracellular Ca^{2+} mobilization and Ca^{2+} entry through SOCE [42]. The decrease in NO and PGI_2 cannot counter-balance vasoconstriction and, therefore, vasoconstrictors, such as thromboxane and endothelin-1, activate receptors coupled to G-proteins (Gq/11R) linked with, among others, an increased inositol triphosphate (IP_3) and inhibition of the production of cyclic guanosine monophosphate (cGMP) in neonatal SMCs, and, subsequently, evoking enhanced SOCE activation. As result, hypoxia promotes contraction of SMCs and the maintenance of PHN [54,55]. Thus, PHN contributes to the maintenance of hypoxia among other cardiovascular alterations, aggravating the condition and increasing the risk of mortality. Increased fibrin/fibrinogen deposition in the pulmonary vasculature under hypoxia, and the absence of fibrinolytic activity, amplifies this positive feedback loop, and triggers further thrombosis in the lungs [56]. PPHN and hypoxia also induce pulmonary vascular remodeling through increased SOCE, by proliferation in SMCs, and by the expression of hypoxia-sensitive genes triggered by the master hypoxia regulator, HIF-1, thereby leading to pulmonary arterial wall-thickening and further prolonging the pulmonary hypoxia [55]. Neonatal hypoxemia and asphyxia are often treated with extracorporeal oxygenation or by hypothermia to prevent severe brain damage. Unfortunately, a common complication of these treatments is an increased risk of thrombosis, which is solved by anticoagulant administration [57]. For instance, a swine model of asphyxia revealed that therapeutic re-oxygenation triggers unwanted transient platelet activation, eliciting deranged coagulation [58]. The decreased temperature during therapeutic hypothermia slows the hemostatic system down even further, increasing the risk of uncontrolled bleeding and hemorrhage [59].

4. Cell Markers of Hypoxia

4.1. HIF-1 Is the Keyhole during Hypoxia

Among other proteins, we aim to explore here the most relevant cellular markers of hypoxia, being HIF-1 identified in most tissues investigated under hypoxic conditions. Hypoxia inducible factor 1 is a large protein (HIF-1; 92 kDa) formed by the association of two subunits, alpha and beta. Codified by a gene located at the chromosome 14 (14q23.2) and, despite that it also responds to endoplasmic reticulum stress, both the expression and function of HIF-1 drastically increases in response to hypoxic conditions; therefore, it has been highlighted as the main signaling-regulator during hypoxia. Although the HIF-1β subunit is very stable, independently of the hypoxic conditions, HIF-1α mRNA is constantly expressed by several mammalian cells [60]. However, under normal oxygenation (normoxia), HIF-1α is highly unstable at the protein level. Conversely, under hypoxic conditions, this subunit becomes very stable [60]. As a transcription factor, the former HIF-1 complex associates to DNA sequence 5$'$-TACGTG-3$'$ of its targeted genes, which requires a previous interaction with the HIF-1 partner Arnt (DNA-binding factor) that in turn stabilizes the HIF-1α subunit [61].

Other members of the HIF family are HIF-2α and HIF-3α [62] and, despite all of them sharing similarities, it has been shown that they do not present redundant functions. All of them regulate gene transcription in different cell types in response to hypoxia [63]. In fact, under certain circumstances, HIF-1α differently controls to HIF-2α. For instance, in clear cell renal carcinoma, HIF-2α promotes cell survival, while HIF-1α activation leads to cell death [63]. In line with this observation, the hematopoietic growth factor, erythropoietin (EPO), was demonstrated to be upregulated by HIF-2α under hypoxic conditions in an in vitro model of hypoxia–ischemia [64]. Conversely, EPO was shown to downregulate HIF-1α through promoting the activity of the prolyl hydroxylase domain 2 (PHD-2) and MMP-9 [64]. On the other hand, the role of HIF-3α has not yet been well described, but it is known that transcription of the HIF-3α locus results in at least five splice variants [65]. In samples collected from patients suffering hepatic carcinoma, HIF-3α was shown to be upregulated in 46% of the samples; nonetheless, it was also observed to be downregulated in another 42% of cases [66]. On the other hand, the expression of HIF-3α directly correlated with HIF-2α in these samples, but not with HIF-1α. Therefore, considering that HIF-3α only presents a transactivation domain (TADs) that is sensitive to hypoxia, in contraposition with HIF-1α and HIF-2α that contain 2 TADs [67], it has been concluded that HIF-3α may be a hypoxic gene that can be targeted by HIF-1α or HIF-2α. Interestingly, in hepatocellular carcinoma, the overexpression of HIF-2α promoted HIF-3α expression, while HIF-1α remained unmodified [66]. In contrast, HIF-3α was largely found to co-localize with HIF-1α in human renal carcinoma cells, but the regulation of HIF-3α remains elusive. It has been clearly demonstrated in the literature that HIF-3α function, once it becomes active, downregulates the function of HIF-1α and HIF-2α [68]. A recent study indicates that HIF-3α can be regulated by methylation in two gene regions, and that methylation in the second region could be linked with pre-eclampsia, as demonstrated by analyzing HIF-3α in mononuclear cells isolated from umbilical cord blood [69]; however, this study was largely descriptive and the authors did not delve into the molecular mechanism behind the altered methylation of the HIF-3α gene.

4.2. Regulation of HIF-1 Function

Regarding the regulation of HIF-1α, several proteins have been proposed, such as MUC-1, Pin1, and the tumor suppressor gene von Hippel–Lindau protein (VHL), among others, as presented in Figure 1 [70]. In normoxia, the presence of oxygen favors HIF-1α inactivation due to hydroxylation at the two proline residues (Pro402 and Pro564) that belong to the alpha subunit; as a result, HIF-1α interacts with VHL and the E3 ubiquitin ligase complex to favor the degradation of HIF-1α [70]. Furthermore, hydroxylation at the asparagine residue within the carboxy-terminal domain blocks interaction with the coactivator of HIF-1α, the p300 [71]. The proposed mechanism works as follows: prolyl hydroxylase (PHD) detects the O_2 concentration at cellular level and then evokes the hydroxylation at the ODD domain of HIF-1α [72]. In hypoxia, since the hydroxylation enzymes, PHD1-3, require O_2 and α-ketoglutarate as substrates, the privation of O_2 should be able to reduce hydroxylation at the prolyl residues and, therefore, impair the ubiquitination and degradation of HIF-1α. The accumulation of HIF-1α subunits and their binding to HIF-1β resembles the active HIF-1α factor that associates with the Arnt cofactor and facilitates the recognition and association with the hypoxia-response elements (HER) 3'-flanking regions of the targeted genes, favoring its translation [70,73].

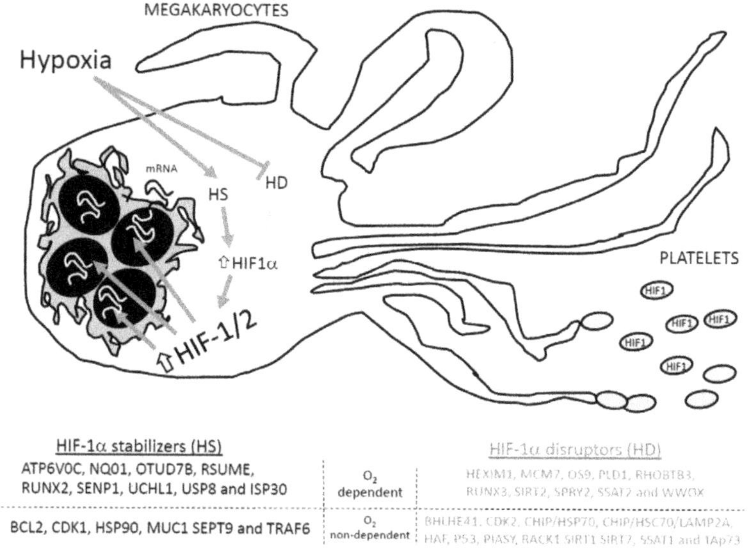

Figure 1. Proteins involved in HIF-1a stabilization and degradation. Adapted from Semanza GL, 201855. HD: HIF-1 disruptors; HS: HIF-1 stabilizers. Hypoxia activates hypoxia stabilizers (HS) while blocking hypoxia disrupters (HD), which increases HIF-1α conforming to the HIF-1/2 complex that induces translation of the targeted genes, and promotes the arresting of HIF1 in the generated platelets.

HIF-1α could be degraded by SUMOylation, proteasomal degradation machinery and/or chaperone mediated autophagy [70]. Therefore, regulation of HIF-1α relies on a delicate balance between the activity of those proteins that favors its degradation and those that avoid it. Among others, proteins like HEXIM1, MCM7, OS9, PLD1, RHOBTB3, RUNX3, SIRT2, SPRY2, SSAT2 and WWOX promote the normoxia-dependent and PHD/VHL-dependent degradation of HIF-1. Conversely, ATP6V0C, NQ01, OTUD7B, RSUME, RUNX2,

SENP1, UCHL1, USP8 and ISP30 interact with HIF-1α to avoid its O_2-dependent degradation [71]. In addition, O_2-independent degradation of HIF-1α was reported to occur upon its interaction with the following proteins: BHLHE41, CDK2, CHIP/HSP70, CHIP/HSC70/LAMP2A, HAF, P53, PIASY, RACK1 SIRT1 SIRT7, SSAT1, and TAp73. Meanwhile, HIF-1α interaction with BCL2, CDK1, HSP90, MUC1 SEPT9 and TRAF6 seems to avoid O_2-independent degradation [71,72]. In addition, in silico analysis revealed that many other proteins may interact with HIF-1α, but the relevance of these interactions in cell physiology are not defined yet [73].

The HIF-1 targeted genes list includes: TIMP-1 (matrix and barriers reorganizing function) [74], CD18 (inflammation) [75]; EPO (erythropoietin) [76]; TF and TFRC (iron metabolism) [77]; the VEGF signaling pathway that regulates angiogenesis (including VEGF, EGF, Flt-1, PAI-1, ANGPT, Tie-2 and TIMP-1) [78]; EDN1, iNOs, eNOS, HMOX1 and ANP that are involved in vascular tone [79]; Glut1, which promotes anaerobic metabolism by increasing glycolysis; PDK-1, an inhibitor of tricarboxylic acid cycle metabolism; hexokinase, PFKL, GAPDH, ALDOA, ENO1, PGK1, PFK3 and LDHA that are involved in the promotion of the anaerobic metabolism [80]; and Bcl-2 and p21/p27, as regulators of proliferation and apoptosis [81]. Finally, a novel actor has been described downstream of HIF-1, stanniocalcin-2, whose role remains largely unknown but its expression in cancer cells was upregulated and it seems to protect cells from apoptosis [82]. In addition, stanniocalcin-2 was demonstrated to participate in the regulation of Ca^{2+} homeostasis in human platelets by acting as an antagonist of extracellular non-capacitive calcium entry driven by Orai3 [83]. In line with this observation, intermittent hypoxia favors ROS production that activates PLCγ, generating IP_3 that mobilizes Ca^{2+} and evokes the activation of calmodulin kinase, CamK [84]. CamK activation phosphorylates and recruits p300/CBP to the HIF-1α/β complex, favoring the activation of the complex and, subsequently, favoring the translation of the downstream genes [84,85]. The expression of the STC2 gene may require the recruitment of p300 and HDAC7, according to the researchers' observations [84]. Similarly, the STC2 paralogue, STC1, was reported to also be upregulated by hypoxia in an HIF-1 dependent mechanism in rat alveolar type II cells [86]. Finally, the activity of HIF-1α was reported to be blocked by drugs like PX-478 2HCl, which was used in in vitro studies to induce apoptosis in tumoral cells, and is in phase 1 of a clinical trial with the reference number, NCT00522652; physicians are now testing its efficacy in treating advanced solid tumors and lymphoma.

4.3. Alternative Hypoxic-Sensitive Signaling Pathways to HIF-1

The activation of alternative signaling pathways, independent to HIF-1, have been described in cells under hypoxic conditions (see Table 1, which represents those actually expressed in human platelets). In the following section, we describe relevant alternative markers and second messengers involved in cell survival under hypoxic conditions, including the Carbonic anhydrase IX (CAIX), ERO1a, KDM4B, KDM3A, PDK1, GLUT1, osteopontin, and BNIP-3.

Table 1. Hypoxic-response proteins (HRPs) with a possible role in platelet physiology.

Hypoxia Marker	Platelets *	Expression in Other Cells	References
KDM4B & KDM3A	Yes	U2OS, MCF7, HeLa, IMR32 and HL60 cell lines	Van Oorschot RV et al. [87]
GLUT1	Yes	Universally expressed	Filder TP et al. [88]
Osteopontin	Yes	Widely expressed	Chanzu et al. [89]
BNIP-3	No	Megakaryocytes, other stem cells, breast cancer cells, etc.	Malherbe JAJ et al. [90]
HSP-70	Yes	Caco-2 & HT29 cell lines	Jackson et al. [91]

* Corroborated at protein level in platelets.

Carbonic anhydrase family (CA) groups contain fifteen members, the most relevant of which are the CAIX and CAXII members. As zinc metalloenzymes, they regulate pH within the cells and in their surrounding medium by transforming the CO_2 to bicarbonate [92]. Tumor cells maintain an acidic pH in the surrounding microenvironment due to CAIX activity, thus avoiding the activation of a hypoxia-related cascade [92]. However, in most of the tumor samples investigated, CAIX seems to be upregulated by HIF-1; CAIX($-$)/HIF-1α (+) and CAIX (+)/HIF-1α (–) can sometimes be detected in cervix cancer samples [93]. Interestingly, CAIX overexpression was also observed in cells grown in high densities, where its activation seems to relay in PI3K activity instead of on HIF-1 activation [93].

Endoplasmic reticulum oxidase 1 (ERO1) is a glycosylated flavonase mainly located at the mitochondria-associated membrane (MAMs), where it regulates the ER redox and Ca^{2+} homeostasis [94,95]. ERO1α and β isoforms have been described, and their distribution among tissue could be different [96]. Interestingly, unlike ERO1β, ERO1α was shown to be stimulated in an in vivo hypoxic rat model and, further, in mouse and human culture cells [96]. Conversely, the ER stressor inductor, tunicamycin, was shown to evoke an unfolded protein response (UPR) that upregulated ERO-1β, but not ERO-1α [96], thereby demonstrating two different regulatory mechanisms for each ERO1 subtype in response to different ER stressors. Regarding the physiological relevance of ERO1, it has been shown that VEGF activation, and the subsequent angiogenesis, seems to be regulated by ERO1-α downstream of HIF-1α, in response to hypoxia [97].

Histone lysine demethylases (KDMs) have been shown to be upregulated by HIF-1α due to hypoxia, like KDM3A, KDM4B, KDM4C and KDM6B [98]. The physiological relevance of this signaling pathway is based on the fact that DNA and histone methylation are the main epigenetic mechanisms for regulating gene expression. Thus, histone methylation is regulated by the balance between histone methyltransferases and histone demethylases; therefore, the activation of these proteins promotes gene expression by removing the methylation-induced break to gene translation. Interestingly, KDM3A and KDM4B genes contain three putative hypoxia response elements, which confer to this protein a unique regulatory mechanism that is driven by hypoxia as compared to other KDM family members who lack these domains [99]. In fact, the expression of KDM3A and KDM4B, but not other KDMs, was reported in U2OS, MCF7, HeLa, IMR32 and HL60 cells cultured at 0.5% O_2 pressure [87,100].

PDK-1 and Glut-1 cancer cells exposed to hypoxic conditions often reprogram their metabolism to a pure glycolytic metabolism, which may work even in anoxic conditions. As result of this adaptation, PDK-1 and Glut-1 activities are exacerbated in order to impair pyruvate processing in the mitochondria, enhancing the glucose transport and subsequently triggering an exacerbated glycolysis. It has been shown that these metabolic changes may be supported by the mTORC1-HIF-1 axes in CD8+ T cells [101]. An early observation concluded that HIF-1 activation favored selective GLUT regulation [88,102]; thus, while low-affinity/high capacity GLUT2 activity resulted in repression, other high-affinity/low capacity isoforms, like GLUT1 and GLUT3, were upregulated during hypoxia [102]. Similarly, other glycolytic enzymes may also be overexpressed in hypoxic conditions, as with PFK-1, aldolase A, PKM, triose-phosphate isomerase and LDH-A. However, increased glycolysis would not be sufficient under conditions where cells still consume O_2 in the tricarboxylic acid cycle within the mitochondria as a result of pyruvate degradation. Therefore, cells subjected to hypoxia often block the mitochondrial tricarboxylic cycle by enhancing phosphorylation of the pyruvate dehydrogenase due to the activation of PDK-1 [103]. Both metabolic regulatory mechanisms were observed in several cancer cells.

Osteopontin (OPN) was initially described as a regulator of osteoclast function during bone remodeling, since it was secreted and found complexed in hydroxyapatite deposits within the mineralized bone matrix. The vitronectin receptor on the surfaces of osteoclasts recognize OPN and directs these cells to the bone locations that must be remodeled [104]. Additionally, the cytokine-related role of OPN was also shown to promote interferon-gamma and interleukin-12 generation. Overexpression and upregulation of

OPN by hypoxia and radiotherapy were reported in certain tumors [105]. Interestingly, in breast cancer biopsies, a positive feedback loop between hypoxia and OPN was demonstrated. The authors claim that OPN favors HIF-1 induction during hypoxia, but not HIF-2, and that subsequently, VEGF activation leads to angiogenesis. This signaling pathway would require the activation of integrin-linked kinase (ILK)/AKT-mediated nuclear factor (NF)-κB p65 [106]. The latest studies consider OPN as a good cancer molecular marker of poor prognosis and a potential plasma marker in patients suffering from different types of cancers [107–109].

BNIP-3. HIF-1α stabilization is critical for cell survival under hypoxia. However, a close look at the two transactivation domains, (N-TAD and C-TAD) of HIF-1, reveals a dual function of HIF-1, which was revealed as a result of the use of the C-TAD inhibitor [110]. Under these experimental conditions, the N-TAD-targeted genes included the Bcl-2/adenovirus E1B 19-kDa interacting protein 3 (*bnip3*) among others, the maximum expression of which resulted in severe hypoxia in areas close to the necrotic parts of the solid tumors [110]. BNIP-3 belongs to the BH3-only subfamily of the Bcl-2 protein family that, upon dimerization, antagonized the pro-survival activity of proteins such as Bcl-2 and Bcl-X$_L$ [111]. It has been demonstrated that the activation of BNIP-3 evokes cell death [112]; however, exogenous overexpression of this protein in several cultured cell types, such as MEFs, MCf7, PC3 and LS174 cells, failed to reproduce previous results found in cardiomyocytes [110]. Interestingly, BNIP3 was reported to be involved in mitochondrial autophagy (mitophagy), a mechanism that may play a protective role against hypoxia-induced cell death due to the disruption of those altered mitochondria that release elevated amounts of ROS [113]. Later on, it was shown that in nucleus pulposus cells locatedwithin avascularized areas within the intervertebral discs that are constantly supporting hypoxic conditions, HIF-1 should be activated and that, subsequently, autophagy must be constantly modulated [114]. Interestingly, these cells respond to hypoxia by increasing the LC3 content, an autophagy cell marker that is involved in the generation of autophagosomes that were HIF-1/BNIP3 independent [114]. In addition, some cancer cell types overcome the hypoxia-evoked cell death by inducing aberrant methylation and the silencing of BNIP3, as was observed in colorectal and gastric cancers [115].

All these HRE elements are now being subjected to intense investigations in order to ascertain the hypoxia-related effect in cell physiology and in pathologic conditions; thus, the knowledge regarding these HRE will increase in the following years. Interestingly, most of these molecular markers of hypoxia have been described in platelets, with the exception of BNIP-3 [63–66]. Nonetheless, GLUT1 and HSP70 are currently being investigated in platelet. HSP70 is particularly relevant in diabetic patients [116], who have also been also described as suffering ROS overproduction and altered glucose metabolism; hence, both mechanisms seem to be highly intricate in platelets.

5. Platelet Function Alteration in Response to Hypoxia: HIF-1 in Platelet Function

Wide bodies of evidence regarding the hypoxia effects on platelet function have arisen from studies conducted on subjects exposed to high altitudes Thrombotic events derived from hypoxemic conditions at high altitudes rely on the generation of a pro-thrombotic-like status, which, according to the literature, is similar to that found in patients suffering from venous thromboembolism [117]. Two recent publications detailed the molecular alterations that take place in platelets exposed to high altitudes and hypoxia. The authors concluded that, in addition to an enhanced HIF-1α expression, platelets also presented alterations in many other molecular pathways and proteins, such as elevated vWF, CD40, P-seletin and PF4, which facilitate the appearance of the coagulation and thrombophilia profiles characteristic of these patients [117].

These results were later corroborated by other studies using hypobaric and hypoxic conditions which demonstrated alterations in several platelet proteins, including the membrane glycoproteins, GP4, GP6 and GP9; integrin subunits (ITGA2B); and chemokines located in alpha-granules (SELP, PF4V1) [118]. In line with this, a previous publication demonstrated alterations in HIF-1α expression in subjects living at high altitude in comparison to subjects living at sea-level; another 137 genes were also differentially expressed under these conditions. It is worth mentioning that among these, the authors clearly differentiated between proteins belonging to the oxidative stress-related protein, focal adhesion, and those proteins involved in the complement and coagulation cascades, respectively [119]. Considering the latter, the authors reported an altered expression of F11, HRG, SERPINF2, SERPIND1, the expression of which was reduced at high altitudes. Meanwhile, PF4 and vWF were overexpressed. These molecular alterations resulted in a significant increase in prothrombin time and elevated partial thromboplastin time [119].

Platelet exposition to hypoxia activates HRPs; therefore, it is likely that HIF-1 regulates platelet function. Surprisingly, not many papers can be found in the literature regarding the role of HIF-1 on hypoxia-dependent alterations in coagulation; few studies directly analyze the role of HIF-1 in platelet function. This contradiction is due to cancer cells and endothelial cells, particularly those located in the lung vasculature, activates HIF-1 under hypoxic conditions, which in turn activates the secretion of PAI-1, therefore recruiting platelets to the tumor location. This aberrant platelet function avoids the detection of cancer cells by the immune system [120]. Therefore, those studies focus the interested of researches that analyze the molecules and alteration derived or affecting to endothelial cells, meanwhile those affecting directly to platelets remains elusive.

A deep search in the literature about HIF-1 role in platelets leads us to conclude that this protein participates in platelet generation and function, as it is following described. Regarding the role of hypoxia in platelet generation, it has been reported that both HIF-1 and HIF-2 were detected in megakaryocytes. In line with this observation, those patients diagnosed of immune thrombocytopenia (ITP) that are suffering from a high rate of platelet destruction, presented low HIF-1α expression in the bone marrow [121]. Additionally, administration of HIF-1α activator, IOX-2, restored both megakaryocyte maturation and platelet counts in a murine model of ITP [121]. Other authors have reported that iron deficiency leads to pathologic excessive number of platelets known as thrombocytosis, which occurred independently of the megakaryopoietic growth factors trombopoietin, IL-5 or IL-11 [122]. Interestingly, iron deficiency in cord blood cell cultures evoked the expression of megakaryopoietic markers that lead to pro-platelet formation and subsequent molecular analysis revealed enhanced HIF-2α and VEGFA. These data indicate that HIF-2 role is relevant to platelet production in response to iron deficiency [122].

Further evidence can be found in patients suffering the Chuvash polycythemia disease. These patients are homozygous for C598T mutation in the von Hippel-Lindau protein (VHL) that as mentioned above is the protein responsible for HIF-1α ubiquitination [123]. Chuvash polycythemia patients often die of thrombotic events. Authors claimed that these patients have elevated amounts of reduced glutathione (GSH), a natural cell antioxidant found in these patients that resulted in elevated glutamate cysteine ligase (GCL). In line with this experiments done using a murine model expressing VHL mutants, resulted in a decreased HIF-1 expression that positively correlates with GCL levels, then HIF-1α may impair the cellular balance of reactive oxygen species by altering the GSH generation [123]. GHS has been described to contribute to the appearance of vascular diseases due to increasing platelet function [124], in fact, GHS intracellular production in response to platelet agonists like Thr or arachidonic acid, is involved in thromboxane B_2 and, subsequently, contributes with platelets aggregation [125].

Finally, later evidence regarding the role of hypoxia in platelet function turn up of the increased number of patients suffering sleep apnea and chronic obstructive pulmonary diseases [43]. In normoxic conditions circulating platelets express HIF-2α that is exacerbated in hypoxic conditions associated with chronic obstructive pulmonary disease, while mRNA of HIF-1α and HIF-2α was found in circulating platelets [43]. Thrombin and other platelet agonists also evoke increasing HIF-2α expression which leads to PAI-1 expression and secretion that in turn activates platelets, closing the hypoxia platelet activation cycle [43]. Recently, it has been published a list of microRNA associated to hypoxia and coagulation after an in silico extensive study [126]. Authors claimed that hsa-mir-4433a-3p, hsa-mir-4667-5p, hsa-mir-6735-5p, hsa-mir-6777-3p and hsa-mir-6815-3p regulates simultaneously hypoxia related genes (among them HIF-1α, HIF-2α and HIF-3α, Arnt and Arnt2) and genes involved in coagulation, which may be the start point for interesting future investigations, but without the appropriate experimental confirmation these data lack of physiological relevance. Finally, HIF prolyl-hydrolases inhibition using IOX-2, a dual PHD and HIF antagonist, was evidenced to impair platelet aggregation in response to CRP and thrombin [19]. Authors claimed that HIF-1α expression was enhanced in platelets due to the antagonist treatment that is also linked with a reduction in ROS production [19].

Finally, as shown in Figure 2, HIF-1α plays a crucial role in chronic hypoxia by inducing the TRPC-SOCE signaling axis. Numerous studies from different groups have reported on the overexpression of, for example, TRPC6 and TRPC1 in pulmonary arterial smooth muscle cell (PASMC) [127]; RPC3, TRPC6 and TRPC1 in cultured neonatal rat cardiac myocytes [128]; and TRPC6 with enhanced HIF-1α/ZEB2 axis in a middle cerebral artery occlusion (MCAO) model [129]. Nonetheless, there is not a unique association between HIF-1α and SOCE members. HIF-1α directly controls STIM1 transcription and is required for STIM1-mediated SOCE via the activation of Ca^{2+}/calmodulin-dependent protein kinase II and p300 in hepatocarcinoma cells [130]. In ovary carcinoma cells exposed to placental growth factor (PlGF), Orai1/STIM1 expression is enhanced, contributing to the upregulation of HIF-1α [131]. Conversely, Wang et al., 2017 reported HIF-1α-dependent upregulation of Orai2, but not Orai1, in pulmonary arterial smooth muscle cells (PASMCs) and mice models [132]. Orai3 is induced by hypoxic environments in some cancer cells, for example, the induction of Orai3 under hypoxia is mediated by HIF-1α in the TNBC MDA-MB-468 cell line, increasing the TRPC1 ion channel induced by HIF-1α [133]. Curiously, in spite of the fact that TRPC1 silencing suppresses HIF-1α induction by hypoxia, Orai3 does not show the same results [134].

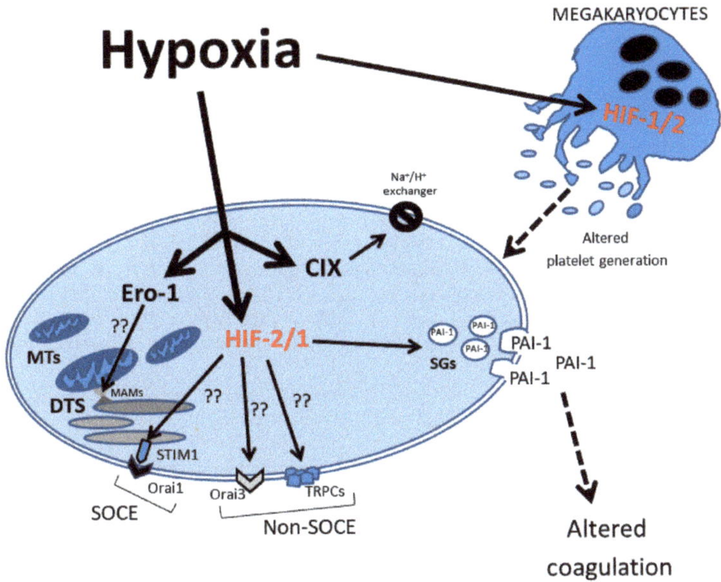

Figure 2. Schematic representation of hypoxia effects on the signaling pathways in neonatal platelets. The HIF-1/2 regulatory effect on several proteins that control intracellular Ca^{2+} homeostasis is represented. DTS: dense tubular system; MTs: mitochondria; SGs: secretory granules; PAI-1: plasminogen activator inhibitor 1; SOCE: store-operated calcium entry; MAMs: mitochondria-associated membranes.

6. Conclusions

Platelet function is modified during perinatal hypoxia. The main HRPs, HIF-1 and HIF-2, were reported to regulate platelet production and function, as further confirmed by the deregulated platelet function shown in patients suffering from Chuvash polycythemia disease. Hypoxia modifies, among others, HIF-1 function, which might in turn modify Ca^{2+} homeostasis in platelets, leading to altered platelets function during labor. Therefore, further research is required to demonstrate the presence of the hypoxia-regulated mechanism discussed in this review.

Author Contributions: A.B.-E. and P.C.R. are responsible for the conceptualization of the manuscripts, and for the writing of different sections of the manuscript. M.P.G. wrote part of the manuscript and drew the images. J.A.R. wrote the rest of the manuscript and revised the final text. All authors have read and agreed to the published version of the manuscript.

Funding: This review was supported by the Junta de Extremadura-FEDER (IB18020 and GR18061) and MICIN (PID2019-104084GB-C21 and PIB2022-136279-NB-C21). CC was awarded with a Predoctoral fellowship from Junta de Extremadura-FEDER (PD16072).

Institutional Review Board Statement: Not applicable.

Informed Consent Statement: Not applicable.

Data Availability Statement: Not applicable.

Conflicts of Interest: The authors declare no conflict of interest.

References

1. Soares, M.J.; Iqbal, K.; Kozai, K. Hypoxia and placental development. *Birth Defects Res.* **2017**, *109*, 1309. [CrossRef]
2. James, J.L.; Stone, P.R.; Chamley, L.W. The regulation of trophoblast differentiation by oxygen in the first trimester of pregnancy. *Hum. Reprod. Update* **2006**, *12*, 137–144. [CrossRef]
3. Guerby, P.; Tasta, O.; Swiader, A.; Pont, F.; Bujold, E.; Parant, O.; Vayssiere, C.; Salvayre, R.; Negre-Salvayre, A. Role of oxidative stress in the dysfunction of the placental endothelial nitric oxide synthase in preeclampsia. *Redox Biol.* **2021**, *40*, 101861. [CrossRef]
4. Qi, H.; Xiong, L.; Tong, C. Aging of the placenta. *Aging* **2022**, *14*, 5294. [CrossRef]
5. Lam, F.; Subhi, R.; Houdek, J.; Schroder, K.; Battu, A.; Graham, H. The prevalence of hypoxemia among pediatric and adult patients presenting to healthcare facilities in low- and middle-income countries: Protocol for a systematic review and meta-analysis. *Syst. Rev.* **2020**, *9*, 67. [CrossRef]
6. Bhutta, B.S.; Alghoula, F.; Berim, I. *Hypoxia*; StatPearls: Tampa, FL, USA, 2022.
7. Kamlin, C.O.F.; O'Donnell, C.P.F.; Davis, P.G.; Morley, C.J. Oxygen saturation in healthy infants immediately after birth. *J. Pediatr.* **2006**, *148*, 585–589. [CrossRef]
8. Cetin, I.; Taricco, E.; Mandò, C.; Radaelli, T.; Boito, S.; Nuzzo, A.M.; Giussani, D.A. Fetal Oxygen and Glucose Consumption in Human Pregnancy Complicated by Fetal Growth Restriction. *Hypertension* **2020**, *75*, 748–754. [CrossRef]
9. Manning, J.M.; Manning, L.R.; Dumoulin, A.; Padovan, J.C.; Chait, B. Embryonic and Fetal Human Hemoglobins: Structures, Oxygen Binding, and Physiological Roles. *Sub-Cell. Biochem.* **2020**, *94*, 275–296. [CrossRef]
10. McNamara, H.; Johnson, N. The effect of uterine contractions on fetal oxygen saturation. *BJOG: Int. J. Obstet. Gynaecol.* **1995**, *102*, 644–647. [CrossRef]
11. Dildy, G.A.; van den Berg, P.P.; Katz, M.; Clark, S.L.; Jongsma, H.W.; Nijhuis, J.G.; Loucks, C.A. Intrapartum fetal pulse oximetry: Fetal oxygen saturation trends during labor and relation to delivery outcome. *Am. J. Obstet. Gynecol.* **1994**, *171*, 679–684. [CrossRef]
12. Rabi, Y.; Yee, W.; Chen, S.Y.; Singhal, N. Oxygen saturation trends immediately after birth. *J. Pediatr.* **2006**, *148*, 590–594. [CrossRef]
13. Yadav, N.; Damke, S. Study of risk factors in children with birth asphyxia. *Int. J. Contemp. Pediatr.* **2017**, *4*, 518–526. [CrossRef]
14. Thompson, L.; Crimmins, S.; Telugu, B.; Turan, S. Intrauterine hypoxia: Clinical consequences and therapeutic perspectives. *Res. Rep. Neonatol.* **2015**, *5*, 79. [CrossRef]
15. Gillam-Krakauer, M.; Gowen, C.W., Jr. *Birth Asphyxia*; StatPearls Publishing: St. Petersburg, FL, USA, 2020.
16. Tuder, R.M.; Yun, J.H.; Bhunia, A.; Fijalkowska, I. Hypoxia and chronic lung disease. *J. Mol. Med.* **2007**, *85*, 1317–1324. [CrossRef]
17. Zysman-Colman, Z.; Tremblay, G.M.; Bandeali, S.; Landry, J.S. Bronchopulmonary dysplasia—Trends over three decades. *Paediatr. Child Health* **2013**, *18*, 86–90. [CrossRef]
18. Chessells, J.M.; Wigglesworth, J.S. Coagulation studies in preterm infants with respiratory distress and intracranial haemorrhage. *Arch. Dis. Child.* **1972**, *47*, 564–570. [CrossRef]
19. Gu, W.; Qi, J.; Zhang, S.; Ding, Y.; Qiao, J.; Han, Y. Inhibition of Hypoxia-Inducible Factor Prolyl-Hydroxylase Modulates Platelet Function. *Thromb. Haemost.* **2022**, *122*, 1693–1705. [CrossRef]

20. Berna-Erro, A.; Redondo, P.; Lopez, E.; Albarran, L.; Rosado, J. Molecular Interplay between Platelets and the Vascular Wall in Thrombosis and Hemostasis. *Curr. Vasc. Pharmacol.* **2013**, *11*, 409–430. [CrossRef]
21. Nowak-Göttl, U.; Limperger, V.; Kenet, G.; Degenhardt, F.; Arlt, R.; Domschikowski, J.; Clausnizer, H.; Liebsch, J.; Junker, R.; Steppat, D. Developmental hemostasis: A lifespan from neonates and pregnancy to the young and elderly adult in a European white population. *Blood Cells Mol. Dis.* **2017**, *67*, 2–13. [CrossRef]
22. Stolla, M.C.; Catherman, S.C.; Kingsley, P.D.; Grant Rowe, R.; Koniski, A.D.; Fegan, K.; Vit, L.; McGrath, K.E.; Daley, G.Q.; Palis, J. Lin28b regulates age-dependent differences in murine platelet function. *Blood Adv.* **2019**, *3*, 72–82. [CrossRef]
23. Nishihira, H.; Toyoda, Y.; Miyazaki, H.; Kigasawa, H.; Ohsaki, E. Growth of macroscopic human megakaryocyte colonies from cord blood in culture with recombinant human thrombopoietin (c-mpl ligand) and the effects of gestational age on frequency of colonies. *Br. J. Haematol.* **1996**, *92*, 23–28. [CrossRef] [PubMed]
24. Muli, M.M.; Hathaway, W.E. Altered platelet function in newborns. *Pediatr. Res.* **1970**, *4*, 229–237. [CrossRef] [PubMed]
25. Davenport, P.; Sola-Visner, M. Platelets in the neonate: Not just a small adult. *Res. Pract. Thromb. Haemost.* **2022**, *6*, e12719. [CrossRef] [PubMed]
26. Cowman, J.; Quinn, N.; Geoghegan, S.; Müllers, S.; Oglesby, I.; Byrne, B.; Somers, M.; Ralph, A.; Voisin, B.; Ricco, A.J.; et al. Dynamic platelet function on von Willebrand factor is different in preterm neonates and full-term neonates: Changes in neonatal platelet function. *J. Thromb. Haemostasis JTH* **2016**, *14*, 2027–2035. [CrossRef] [PubMed]
27. Šimák, J.; Holada, K.; Janota, J.; Straňák, Z. Surface expression of major membrane glycoproteins on resting and TRAP- activated neonatal platelets. *Pediatr. Res.* **1999**, *46*, 445–449. [CrossRef]
28. Del Vecchio, A.; Latini, G.; Henry, E.; Christensen, R.D. Template bleeding times of 240 neonates born at 24 to 41 weeks gestation. *J. Perinatol.* **2008**, *28*, 427–431. [CrossRef]
29. Rajasekhar, D.; Barnard, M.R.; Bednarek, F.J.; Michelson, A.D. Platelet hyporeactivity in very low birth weight neonates. *Thromb. Haemost.* **1997**, *77*, 1002–1007. [CrossRef]
30. Kettner, S.C.; Pollak, A.; Zimpfer, M.; Seybold, T.; Prusa, A.R.; Herkner, K.; Kuhle, S. Heparinase-modified thrombelastography in term and preterm neonates. *Anesth. Analg.* **2004**, *98*, 1650–1652. [CrossRef]
31. Saxonhouse, M.A.; Sola, M.C. Platelet function in term and preterm neonates. *Clin. Perinatol.* **2004**, *31*, 15–28. [CrossRef]
32. Attard, C.; van der Straaten, T.; Karlaftis, V.; Monagle, P.; Ignjatovic, V. Developmental hemostasis: Age-specific differences in the levels of hemostatic proteins. *J. Thromb. Haemost.* **2013**, *11*, 1850–1854. [CrossRef]
33. Andrew, M.; Castle, V.; Saigal, S.; Carter, C.; Kelton, J.G. Clinical impact of neonatal thrombocytopenia. *J. Pediatr.* **1987**, *110*, 457–464. [CrossRef] [PubMed]
34. Sola-Visner, M.; Bercovitz, R.S. Neonatal Platelet Transfusions and Future Areas of Research. *Transfus. Med. Rev.* **2016**, *30*, 183–188. [CrossRef] [PubMed]
35. Ferrer-Marin, F.; Chavda, C.; Lampa, M.; Michelson, A.D.; Frelinger, A.L.; Sola-Visner, M. Effects of in vitro adult platelet transfusions on neonatal hemostasis. *J. Thromb. Haemost.* **2011**, *9*, 1020–1028. [CrossRef] [PubMed]
36. Margraf, A.; Nussbaum, C.; Sperandio, M. Ontogeny of platelet function. *Blood Adv.* **2019**, *3*, 692–703. [CrossRef] [PubMed]
37. Roberts, I.; Murray, N.A. Neonatal thrombocytopenia: Causes and management. *Arch. Dis. Child. Fetal Neonatal Ed.* **2003**, *88*, F359–F364. [CrossRef]
38. Aly, H.; El Beshlawy, A.; Badrawi, N.; Mohsen, L.; Mansour, E.; Ramy, N.; Patel, K. Thrombopoietin level is increased in the serum of asphyxiated neonates: A prospective controlled study. *J. Perinatol.* **2005**, *25*, 320–324. [CrossRef]
39. El Beshlawy, A.; Hussein, H.A.; Abou-Elew, H.H.; Kader, M.S.E.M.A. Study of protein C, protein S, and antithrombin III in hypoxic newborns*. *Pediatr. Crit. Care Med.* **2004**, *5*, 163–166. [CrossRef]
40. Pönicke, K.; Sternitzky, R.; Mest, H.J. Stimulation of aggregation and thromboxane A2 formation of human platelets by hypoxia. *Prostaglandins Leukot. Med.* **1987**, *29*, 49–59. [CrossRef]
41. Cameron, S.J.; Mix, D.S.; Ture, S.K.; Schmidt, R.A.; Mohan, A.; Pariser, D.; Stoner, M.C.; Shah, P.; Chen, L.; Zhang, H.; et al. Hypoxia and ischemia promote a maladaptive platelet phenotype. *Arterioscler. Thromb. Vasc. Biol.* **2018**, *38*, 1594–1606. [CrossRef]
42. Lopez, E.; Bermejo, N.; Berna-Erro, A.; Alonso, N.; Salido, G.M.; Redondo, P.C.; Rosado, J.A. Relationship between calcium mobilization and platelet α- And δ-granule secretion. A role for TRPC6 in thrombin-evoked δ-granule exocytosis. *Arch. Biochem. Biophys.* **2015**, *585*, 75–81. [CrossRef]
43. Chaurasia, S.N.; Kushwaha, G.; Kulkarni, P.P.; Mallick, R.L.; Latheef, N.A.; Mishra, J.K.; Dash, D. Platelet HIF-2α promotes thrombogenicity through PAI-1 synthesis and extracellular vesicle release. *Haematologica* **2019**, *104*, 2482–2492. [CrossRef]
44. Berna-Erro, A.; Ramesh, G.; Delgado, E.; Corbacho, A.J.; Ferrer-Marín, F.; Teruel, R.; Granados, M.P.; Rosado, J.A.; Redondo, P.C. CAPN1 (Calpain1)-Dependent Cleavage of STIM1 (Stromal Interaction Molecule 1) Results in an Enhanced SOCE (Store-Operated Calcium Entry) in Human Neonatal Platelets. *Arterioscler. Thromb. Vasc. Biol.* **2023**, *43*, e151–e170. [CrossRef] [PubMed]
45. Pidgeon, G.P.; Tamosiuniene, R.; Chen, G.; Leonard, I.; Belton, O.; Bradford, A.; Fitzgerald, D.J. Intravascular thrombosis after hypoxia-induced pulmonary hypertension: Regulation by cyclooxygenase-2. *Circulation* **2004**, *110*, 2701–2707. [CrossRef] [PubMed]
46. Saracco, P.; Parodi, E.; Fabris, C.; Cecinati, V.; Molinari, A.C.; Giordano, P. Management and investigation of neonatal thromboembolic events: Genetic and acquired risk factors. *Thromb. Res.* **2009**, *123*, 805–809. [CrossRef]
47. Sweetman, D.; Kelly, L.A.; Zareen, Z.; Nolan, B.; Murphy, J.; Boylan, G.; Donoghue, V.; Molloy, E.J. Coagulation Profiles Are Associated with Early Clinical Outcomes in Neonatal Encephalopathy. *Front. Pediatr.* **2019**, *7*, 399. [CrossRef]

48. Nowak-Göttl, U.; Von Kries, R.; Göbel, U. Neonatal symptomatic thromboembolism in Germany: Two year survey. *Arch. Dis. Child. Fetal Neonatal Ed.* **1997**, *76*, 163–167. [CrossRef]

49. Suzuki, S.; Morishita, S. Hypercoagulability and dic in high-risk infants. In Proceedings of the Seminars in Thrombosis and Hemostasis; Thieme Medical Publishers, Inc.: New York, NY, USA, 1998; Volume 24, pp. 463–466. [CrossRef]

50. Aronis, S.; Platokouki, H.; Photopoulos, S.; Adamtziki, E.; Xanthou, M. Indications of Coagulation and/or Fibrinolytic System Activation in Healthy and Sick Very-Low-Birth-Weight Neonates. *Neonatology* **1998**, *74*, 337–344. [CrossRef]

51. Abman, S.H. Persistent Pulmonary Hypertension of the Newborn: Pathophysiology and Treatment. *Adv. Pulm. Hypertens.* **2006**, *5*, 22–30. [CrossRef]

52. Mota, R.; Rocha, G.; Flor-de-Lima, F.; Guimarães, H. Persistent pulmonary hypertension—The neonatal period and evaluation at 2 years of age. *J. Pediatr. Neonatal Individ. Med.* **2016**, *5*, 50119. [CrossRef]

53. Davizon-Castillo, P.; Allawzi, A.; Sorrells, M.; Fisher, S.; Baltrunaite, K.; Neeves, K.; Nozik-Grayck, E.; DiPaola, J.; Delaney, C. Platelet activation in experimental murine neonatal pulmonary hypertension. *Physiol. Rep.* **2020**, *8*, e14386. [CrossRef]

54. Steinhorn, R.H. Neonatal pulmonary hypertension. In Proceedings of the Pediatric Critical Care Medicine; Lippincott Williams and Wilkins: Philadelphia, PA, USA, 2010; Volume 11, pp. S79–S84. [CrossRef]

55. Dakshinamurti, S. Pathophysiologic mechanisms of persistent pulmonary hypertension of the newborn. *Pediatr. Pulmonol.* **2005**, *39*, 492–503. [CrossRef] [PubMed]

56. Lawson, C.A.; Yan, S.D.; Yan, S.F.; Liao, H.; Zhou, Y.S.; Sobel, J.; Kisiel, W.; Stern, D.M.; Pinsky, D.J. Monocytes and tissue factor promote thrombosis in a murine model of oxygen deprivation. *J. Clin. Investig.* **1997**, *99*, 1729–1738. [CrossRef] [PubMed]

57. Rais Bahrami, K.; Van Meurs, K.P. ECMO for neonatal respiratory failure. *Semin. Perinatol.* **2005**, *29*, 15–23. [CrossRef] [PubMed]

58. Postma, S.; Emara, M.; Obaid, L.; Johnson, S.T.; Bigam, D.L.; Cheung, P.-Y. Temporal platelet aggregatory function in hypoxic newborn piglets reoxygenated with 18%, 21%, and 100% oxygen. *Shock* **2007**, *27*, 448–454. [CrossRef] [PubMed]

59. Forman, K.R.; Diab, Y.; Wong, E.C.C.; Baumgart, S.; Luban, N.L.C.; Massaro, A.N. Coagulopathy in newborns with hypoxic ischemic encephalopathy (HIE) treated with therapeutic hypothermia: A retrospective case-control study. *BMC Pediatr.* **2014**, *14*, 277. [CrossRef] [PubMed]

60. Kallio, P.J.; Pongratz, I.; Gradin, K.; McGuire, J.; Poellinger, L. Activation of hypoxia-inducible factor 1α: Posttranscriptional regulation and conformational change by recruitment of the Arnt transcription factor. *Proc. Natl. Acad. Sci. USA* **1997**, *94*, 5667–5672. [CrossRef]

61. Wenger, R.H.; Gassmann, M. Oxygen(es) and the hypoxia-inducible factor-1. *Biol. Chem.* **1997**, *378*, 609–616.

62. Holmquist-Mengelbier, L.; Fredlund, E.; Löfstedt, T.; Noguera, R.; Navarro, S.; Nilsson, H.; Pietras, A.; Vallon-Christersson, J.; Borg, Å.; Gradin, K.; et al. Recruitment of HIF-1α and HIF-2α to common target genes is differentially regulated in neuroblastoma: HIF-2α promotes an aggressive phenotype. *Cancer Cell* **2006**, *10*, 413–423. [CrossRef]

63. Ratcliffe, P.J. HIF-1 and HIF-2: Working alone or together in hypoxia? *J. Clin. Investig.* **2007**, *117*, 862–865. [CrossRef]

64. Souvenir, R.; Flores, J.J.; Ostrowski, R.P.; Manaenko, A.; Duris, K.; Tang, J. Erythropoietin Inhibits HIF-1α Expression via Upregulation of PHD-2 Transcription and Translation in an In Vitro Model of Hypoxia-Ischemia. *Transl. Stroke Res.* **2014**, *5*, 118–127. [CrossRef]

65. Tanaka, T.; Wiesener, M.; Bernhardt, W.; Eckardt, K.U.; Warnecke, C. The human HIF (hypoxia-inducible factor)-3α gene is a HIF-1 target gene and may modulate hypoxic gene induction. *Biochem. J.* **2009**, *424*, 143–151. [CrossRef]

66. Liu, P.; Fang, X.; Song, Y.; Jiang, J.X.; He, Q.J.; Liu, X.J. Expression of hypoxia-inducible factor 3α in hepatocellular carcinoma and its association with other hypoxia-inducible factors. *Exp. Ther. Med.* **2016**, *11*, 2470–2476. [CrossRef]

67. Tian, H.; McKnight, S.L.; Russell, D.W. Endothelial PAS domain protein 1 (EPAS1), a transcription factor selectively expressed in endothelial cells. *Genes Dev.* **1997**, *11*, 72–82. [CrossRef]

68. Maynard, M.A.; Evans, A.J.; Hosomi, T.; Hara, S.; Jewett, M.A.; Ohh, M. Human HIF-3alpha4 is a dominant-negative regulator of HIF-1 and is down-regulated in renal cell carcinoma. *FASEB J.* **2005**, *19*, 1396–1406. [CrossRef]

69. Mansell, T.; Ponsonby, A.L.; Januar, V.; Novakovic, B.; Collier, F.; Burgner, D.; Vuillermin, P.; Ryan, J.; Saffery, R.; Carlin, J.; et al. Early-life determinants of hypoxia-inducible factor 3A gene (HIF3A) methylation: A birth cohort study. *Clin. Epigenetics* **2019**, *11*, 96. [CrossRef]

70. Semenza, G.L. A compendium of proteins that interact with HIF-1α. *Exp. Cell Res.* **2017**, *356*, 128–135. [CrossRef] [PubMed]

71. Schofield, C.J.; Ratcliffe, P.J. Oxygen sensing by HIF hydroxylases. *Nat. Rev. Mol. Cell Biol.* **2004**, *5*, 343–354. [CrossRef] [PubMed]

72. Yang, S.L.; Wu, C.; Xiong, Z.F.; Fang, X. Progress on hypoxia-inducible factor-3: Its structure, gene regulation and biological function (Review). *Mol. Med. Rep.* **2015**, *12*, 2411–2416. [CrossRef] [PubMed]

73. Semenza, G.L. Regulation of physiological responses to continuous and intermittent hypoxia by hypoxia-inducible factor 1. In Proceedings of the Experimental Physiology; 2006; Volume 91, pp. 803–806. [CrossRef]

74. Ilie, M.I.; Lassalle, S.; Long-Mira, E.; Hofman, V.; Zangari, J.; Bénaim, G.; Bozec, A.; Guevara, N.; Haudebourg, J.; Birtwisle-Peyrottes, I.; et al. In papillary thyroid carcinoma, TIMP-1 expression correlates with BRAF V600E mutation status and together with hypoxia-related proteins predicts aggressive behavior. *Virchows Arch.* **2013**, *463*, 437–444. [CrossRef] [PubMed]

75. Kong, T.; Eltzschig, H.K.; Karhausen, J.; Colgan, S.P.; Shelley, C.S. Leukocyte adhesion during hypoxia is mediated by HIF-1-dependent induction of β2 integrin gene expression. *Proc. Natl. Acad. Sci. USA* **2004**, *101*, 10440–10445. [CrossRef]

76. Yu, Y.; Ma, L.; Zhang, H.; Sun, W.; Zheng, L.; Liu, C.; Miao, L. EPO could be regulated by HIF-1 and promote osteogenesis and accelerate bone repair. *Artif. Cells Nanomed. Biotechnol.* **2020**, *48*, 206–217. [CrossRef] [PubMed]

77. Ding, H.; Yan, C.Z.; Shi, H.; Zhao, Y.S.; Chang, S.Y.; Yu, P.; Wu, W.S.; Zhao, C.Y.; Chang, Y.Z.; Duan, X.L. Hepcidin is involved in iron regulation in the ischemic brain. *PLoS ONE* **2011**, *6*, e25324. [CrossRef]
78. Sasagawa, T.; Nakamura, T. Traumatic spondyloptosis at the thoracolumbar junction in a patient with diffuse idiopathic skeletal hyperostosis: A case report. *J. Orthop. Sci.* **2018**, *25*, 926–930. [CrossRef]
79. Liu, Z.; Huang, J.; Zhong, Q.; She, Y.; Ou, R.; Li, C.; Chen, R.; Yao, M.; Zhang, Q.; Liu, S. Network-based analysis of the molecular mechanisms of multiple myeloma and monoclonal gammopathy of undetermined significance. *Oncol. Lett.* **2017**, *14*, 4167–4175. [CrossRef] [PubMed]
80. Lai, J.H.; Jan, H.J.; Liu, L.W.; Lee, C.C.; Wang, S.G.; Hueng, D.Y.; Cheng, Y.Y.; Lee, H.M.; Ma, H.I. Nodal regulates energy metabolism in glioma cells by inducing expression of hypoxia-inducible factor 1. *Neuro-Oncology* **2013**, *15*, 1330–1341. [CrossRef] [PubMed]
81. Yin, J.; Ni, B.; Liao, W.G.; Gao, Y.Q. Hypoxia-induced apoptosis of mouse spermatocytes is mediated by HIF-1α through a death receptor pathway and a mitochondrial pathway. *J. Cell. Physiol.* **2018**, *233*, 1146–1155. [CrossRef]
82. Law, A.Y.S.; Wong, C.K.C. Stanniocalcin-2 is a HIF-1 target gene that promotes cell proliferation in hypoxia. *Exp. Cell Res.* **2010**, *316*, 466–476. [CrossRef]
83. López, E.; Gómez-Gordo, L.; Cantonero, C.; Bermejo, N.; Pérez-Gómez, J.; Granados, M.P.; Salido, G.M.; Dionisio, J.A.R.; Liberal, P.C.R. Stanniocalcin 2 regulates non-capacitative Ca²⁺ entry and aggregation in mouse platelets. *Front. Physiol.* **2018**, *9*, 266. [CrossRef]
84. Shi, L.; Xu, H.; Wei, J.; Ma, X.; Zhang, J. Caffeine induces cardiomyocyte hypertrophy via p300 and CaMKII pathways. *Chem. -Biol. Interact.* **2014**, *221*, 35–41. [CrossRef]
85. Hui, A.S.; Bauer, A.L.; Striet, J.B.; Schnell, P.O.; Czyzyk-Krzeska, M.F. Calcium signaling stimulates translation of HIF-α during hypoxia. *FASEB J.* **2006**, *20*, 466–475. [CrossRef]
86. Ito, Y.; Zemans, R.; Correll, K.; Yang, I.V.; Ahmad, A.; Gao, B.; Mason, R.J. Stanniocalcin-1 is induced by hypoxia inducible factor in rat alveolar epithelial cells. *Biochem. Biophys. Res. Commun.* **2014**, *452*, 1091–1097. [CrossRef]
87. Van Oorschot, R.; Hansen, M.; Koornneef, J.M.; Marneth, A.E.; Bergevoet, S.M.; Van Bergen, M.G.J.M.; Van Alphen, F.P.J.; Van Der Zwaan, C.; Martens, J.H.A.; Vermeulen, M.; et al. Molecular mechanisms of bleeding disorderassociated GFI1BQ287* mutation and its affected pathways in megakaryocytes and platelets. *Haematologica* **2019**, *104*, 1460–1472. [CrossRef]
88. Fidler, T.P.; Campbell, R.A.; Funari, T.; Dunne, N.; Balderas, A.E.; Middleton, E.A.; Chaudhuri, D.; Weyrich, A.S.; Abel, D.E. Deletion of GLUT1 and GLUT3 Reveals Multiple Roles for Glucose Metabolism in Platelet and Megakaryocyte Function. *Cell Rep.* **2017**, *20*, 881–894. [CrossRef]
89. Chanzu, H.; Lykins, J.; Wigna-Kumar, S.; Joshi, S.; Pokrovskaya, I.; Storrie, B.; Pejler, G.; Wood, J.P.; Whiteheart, S.W.; Preston, R. Platelet α-granule cargo packaging and release are affected by the luminal proteoglycan, serglycin. *J. Thromb. Haemost.* **2021**, *19*, 1082–1095. [CrossRef]
90. Malherbe, J.A.J.; Fuller, K.A.; Arshad, A.; Nangalia, J.; Romeo, G.; Hall, S.L.; Meehan, K.S.; Guo, B.; Howman, R.; Erber, W.N. Megakaryocytic hyperplasia in myeloproliferative neoplasms is driven by disordered proliferative, apoptotic and epigenetic mechanisms. *J. Clin. Pathol.* **2016**, *69*, 155–163. [CrossRef]
91. Jackson, J.W.; Rivera-Marquez, G.M.; Beebe, K.; Tran, A.D.; Trepel, J.B.; Gestwicki, J.E.; Blagg, B.S.J.; Ohkubo, S.; Neckers, M. Pharmacologic dissection of the overlapping impact of heat shock protein family members on platelet function. *J. Thromb. Haemost.* **2020**, *18*, 1197–1209. [CrossRef]
92. Švastová, E.; Hulíková, A.; Rafajová, M.; Zaťovičová, M.; Gibadulinová, A.; Casini, A.; Cecchi, A.; Scozzafava, A.; Supuran, C.T.; Pastorek, J.; et al. Hypoxia activates the capacity of tumor-associated carbonic anhydrase IX to acidify extracellular pH. *FEBS Lett.* **2004**, *577*, 439–445. [CrossRef]
93. Kaluz, S.; Kaluzová, M.; Liao, S.Y.; Lerman, M.; Stanbridge, E.J. Transcriptional control of the tumor- and hypoxia-marker carbonic anhydrase 9: A one transcription factor (HIF-1) show? *Biochim. Et Biophys. Acta Rev. Cancer* **2009**, *1795*, 162–172. [CrossRef]
94. Sevier, C.S.; Kaiser, C.A. Ero1 and redox homeostasis in the endoplasmic reticulum. *Biochim. Et Biophys. Acta Mol. Cell Res.* **2008**, *1783*, 549–556. [CrossRef]
95. Seervi, M.; Sobhan, P.K.; Joseph, J.; Ann Mathew, K.; Santhoshkumar, T.R. ERO1α-dependent endoplasmic reticulummitochondrial calcium flux contributes to ER stress and mitochondrial permeabilization by procaspase-activating compound-1 (PAC-1). *Cell Death Dis.* **2013**, *4*, e968. [CrossRef]
96. Gess, B.; Hofbauer, K.H.; Wenger, R.H.; Lohaus, C.; Meyer, H.E.; Kurtz, A. The cellular oxygen tension regulates expression of the endoplasmic oxidoreductase ERO1-Lα. *Eur. J. Biochem.* **2003**, *270*, 2228–2235. [CrossRef]
97. May, D.; Itin, A.; Gal, O.; Kalinski, H.; Feinstein, E.; Keshet, E. Ero1-Lα plays a key role in a HIF-1-mediated pathway to improve disulfide bond formation and VEGF secretion under hypoxia: Implication for cancer. *Oncogene* **2005**, *24*, 1011–1020. [CrossRef]
98. Salminen, A.; Kaarniranta, K.; Kauppinen, A. Hypoxia-inducible histone lysine demethylases: Impact on the aging process and age-related diseases. *Aging Dis.* **2016**, *7*, 180–200. [CrossRef]
99. Beyer, S.; Kristensen, M.M.; Jensen, K.S.; Johansen, J.V.; Staller, P. The histone demethylases JMJD1A and JMJD2B are transcriptional targets of hypoxia-inducible factor HIF. *J. Biol. Chem.* **2008**, *283*, 36542–36552. [CrossRef]
100. Pollard, P.J.; Loenarz, C.; Mole, D.R.; McDonough, M.A.; Gleadle, J.M.; Schofield, C.J.; Ratcliffe, P.J. Regulation of Jumonji-domain-containing histone demethylases by hypoxia-inducible factor (HIF)-1α. *Biochem. J.* **2008**, *416*, 387–394. [CrossRef]

101. Finlay, D.K.; Rosenzweig, E.; Sinclair, L.V.; Carmen, F.C.; Hukelmann, J.L.; Rolf, J.; Panteleyev, A.A.; Okkenhaug, K.; Cantrell, D.A. PDK1 regulation of mTOR and hypoxia-inducible factor 1 integrate metabolism and migration of CD8+ T cells. *J. Exp. Med.* **2012**, *209*, 2441–2453. [CrossRef]

102. Ebert, B.L.; Firth, J.D.; Ratcliffe, P.J. Hypoxia and mitochondrial inhibitors regulate expression of glucose transporter-1 via distinct cis-acting sequences. *J. Biol. Chem.* **1995**, *270*, 29083–29089. [CrossRef]

103. Papandreou, I.; Cairns, R.A.; Fontana, L.; Lim, A.L.; Denko, N.C. HIF-1 mediates adaptation to hypoxia by actively downregulating mitochondrial oxygen consumption. *Cell Metab.* **2006**, *3*, 187–197. [CrossRef]

104. Singh, A.; Gill, G.; Kaur, H.; Amhmed, M.; Jakhu, H. Role of osteopontin in bone remodeling and orthodontic tooth movement: A review. *Prog. Orthod.* **2018**, *19*, 18. [CrossRef]

105. Chong, H.C.; Tan, C.K.; Huang, R.L.; Tan, N.S. Matricellular proteins: A sticky affair with cancers. *J. Oncol.* **2012**, *2012*, 351089. [CrossRef]

106. Raja, R.; Kale, S.; Thorat, D.; Soundararajan, G.; Lohite, K.; Mane, A.; Karnik, S.; Kundu, G.C. Hypoxia-driven osteopontin contributes to breast tumor growth through modulation of HIF1α-mediated VEGF-dependent angiogenesis. *Oncogene* **2014**, *33*, 2053–2064. [CrossRef]

107. Wohlleben, G.; Hauff, K.; Gasser, M.; Waaga-Gasser, A.M.; Grimmig, T.; Flentje, M.; Polat, B. Hypoxia induces differential expression patterns of osteopontin and CD44 in colorectal carcinoma. *Oncol. Rep.* **2018**, *39*, 442–448. [CrossRef]

108. Gu, X.; Gao, X.S.; Ma, M.; Qin, S.; Qi, X.; Li, X.; Sun, S.; Yu, H.; Wang, W.; Zhou, D. Prognostic significance of osteopontin expression in gastric cancer: A meta-analysis. *Oncotarget* **2016**, *7*, 69666. [CrossRef]

109. Duarte-Salles, T.; Misra, S.; Stepien, M.; Plymoth, A.; Muller, D.; Overvad, K.; Olsen, A.; Tjønneland, A.; Baglietto, L.; Severi, G.; et al. Circulating Osteopontin and Prediction of Hepatocellular Carcinoma Development in a Large European Population. *Cancer Prev. Res.* **2016**, *9*, 758–765. [CrossRef]

110. Bellot, G.; Garcia-Medina, R.; Gounon, P.; Chiche, J.; Roux, D.; Pouysségur, J.; Mazure, N.M. Hypoxia-Induced Autophagy Is Mediated through Hypoxia-Inducible Factor Induction of BNIP3 and BNIP3L via Their BH3 Domains. *Mol. Cell. Biol.* **2009**, *29*, 2570–2581. [CrossRef]

111. Chen, G.; Cizeau, J.; Velde, C.V.; Park, J.H.; Bozek, G.; Bolton, J.; Shi, L.; Dubik, D.; Greenberg, A. Nix and Nip3 form a subfamily of pro-apoptotic mitochondrial proteins. *J. Biol. Chem.* **1999**, *274*, 7–10. [CrossRef]

112. Webster, K.A.; Graham, R.M.; Bishopric, N.H. BNip3 and signal-specific programmed death in the heart. *J. Mol. Cell. Cardiol.* **2005**, *38*, 35–45. [CrossRef]

113. Zhang, H.; Bosch-Marce, M.; Shimoda, L.A.; Yee, S.T.; Jin, H.B.; Wesley, J.B.; Gonzalez, F.J.; Semenza, G.L. Mitochondrial autophagy is an HIF-1-dependent adaptive metabolic response to hypoxia. *J. Biol. Chem.* **2008**, *283*, 10892–10903. [CrossRef]

114. Choi, H.; Merceron, C.; Mangiavini, L.; Seifert, E.L.; Schipani, E.; Shapiro, I.M.; Risbud, M.V. Hypoxia promotes noncanonical autophagy in nucleus pulposus cells independent of MTOR and HIF1A signaling. *Autophagy* **2016**, *12*, 1631–1646. [CrossRef]

115. Murai, M.; Toyota, M.; Suzuki, H.; Satoh, A.; Sasaki, Y.; Akino, K.; Ueno, M.; Takahashi, F.; Kusano, M.; Mita, H.; et al. Aberrant methylation and silencing of the BNIP3 gene in colorectal and gastric cancer. *Clin. Cancer Res.* **2005**, *11*, 1021–1027. [CrossRef]

116. Chiva-Blanch, G.; Peña, E.; Cubedo, J.; García-Arguinzonis, M.; Pané, A.; Gil, P.A.; Perez, A.; Ortega, E.; Padró, T.; Badimon, L. Molecular mapping of platelet hyperreactivity in diabetes: The stress proteins complex HSPA8/Hsp90/CSK2α and platelet aggregation in diabetic and normal platelets. *Transl. Res. J. Lab. Clin. Med.* **2021**, *235*, 1–14. [CrossRef]

117. Prabhakar, A.; Chatterjee, T.; Bajaj, N.; Tyagi, T.; Sahu, A.; Gupta, N.; Kumari, B.; Nair, V.; Kumar, B.; Ashraf, M.Z. Venous thrombosis at altitude presents with distinct biochemical profiles: A comparative study from the Himalayas to the plains. *Blood Adv.* **2019**, *3*, 3713–3723. [CrossRef]

118. Shang, C.; Wuren, T.; Ga, Q.; Bai, Z.; Guo, L.; Eustes, A.S.; McComas, K.N.; Rondina, M.T.; Ge, R. The human platelet transcriptome and proteome is altered and pro-thrombotic functional responses are increased during prolonged hypoxia exposure at high altitude. *Platelets* **2020**, *31*, 33–42. [CrossRef]

119. Du, X.; Zhang, R.; Ye, S.; Liu, F.; Jiang, P.; Yu, X.; Xu, J.; Ma, L.; Cao, H.; Shen, Y.; et al. Alterations of Human Plasma Proteome Profile on Adaptation to High-Altitude Hypobaric Hypoxia. *J. Proteome Res.* **2019**, *18*, 2021–2031. [CrossRef]

120. Maurer, S.; Kropp, K.N.; Klein, G.; Steinle, A.; Haen, S.P.; Walz, J.S.; Hinterleitner, C.; Märklin, M.; Kopp, H.G.; Salih, H.R. Platelet-mediated shedding of NKG2D ligands impairs NK cell immune-surveillance of tumor cells. *OncoImmunology* **2018**, *7*, e1364827. [CrossRef]

121. Qi, J.; You, T.; Pan, T.; Wang, Q.; Zhu, L.; Han, Y. Downregulation of hypoxia-inducible factor-1α contributes to impaired megakaryopoiesis in immune thrombocytopenia. *Thromb. Haemost.* **2017**, *117*, 1875–1886. [CrossRef]

122. Jimenez, K.; Khare, V.; Evstatiev, R.; Kulnigg-Dabsch, S.; Jambrich, M.; Strobl, H.; Gasche, C. Increased expression of HIF2α during iron deficiency-associated megakaryocytic differentiation. *J. Thromb. Haemost.* **2015**, *13*, 1113–1127. [CrossRef]

123. Cario, H.; Schwarz, K.; Jorch, N.; Kyank, U.; Petrides, P.E.; Schneider, D.T.; Uhle, R.; Debatin, K.M.; Kohne, E. Mutations in the von Hippel-Lindau (VHL) tumor suppressor gene and VHL-haplotype analysis in patients with presumable congenital erythrocytosis. *Haematologica* **2005**, *90*, 19–24.

124. Wang, L.; Wu, Y.; Zhou, J.; Ahmad, S.S.; Mutus, B.; Garbi, N.; Hämmerling, G.; Liu, J.; Essex, D.W. Platelet-derived ERp57 mediates platelet incorporation into a growing thrombus by regulation of the αIIbβ3 integrin. *Blood* **2013**, *122*, 3642–3650. [CrossRef]

125. Burch, J.W.; Services, P.T. Glutathione disulfide production during arachidonic acid oxygenation in human platelets. *Prostaglandins* **1990**, *39*, 123–134. [CrossRef]

126. Hembrom, A.A.; Srivastava, S.; Garg, I.; Kumar, B. Identification of regulatory microRNAs for hypoxia induced coagulation mechanism by In-silico analysis. *bioRxiv* **2020**, *2020*, 173112. [CrossRef]

127. Yang, K.; Lu, W.; Jia, J.; Xu, L.; Zhao, M.; Wang, S.; Jiang, H.; Xu, L.; Wang, J. Noggin inhibits hypoxia-induced proliferation by targeting store-operated calcium entry and transient receptor potential cation channels. *Am. J. Physiol.-Cell Physiol.* **2015**, *308*, C869–C878. [CrossRef] [PubMed]

128. Chu, W.; Wan, L.; Zhao, D.; Qu, X.; Cai, F.; Huo, R.; Wang, N.; Zhu, J.; Zhang, C.; Zheng, F.; et al. Mild hypoxia-induced cardiomyocyte hypertrophy via up-regulation of HIF-1α-mediated TRPC signaling. *J. Cell. Mol. Med.* **2012**, *16*, 2022–2034. [CrossRef] [PubMed]

129. Nakuluri, K.; Nishad, R.; Mukhi, D.; Kumar, S.; Nakka, V.P.; Kolligundla, L.P.; Narne, P.; Natuva, S.S.K.; Phanithi, P.B.; Pasupulati, A.K. Cerebral ischemia induces TRPC6 via HIF1α/ZEB2 axis in the glomerular podocytes and contributes to proteinuria. *Sci. Rep.* **2019**, *9*, 17897. [CrossRef]

130. Li, Y.; Guo, B.; Xie, Q.; Ye, D.; Zhang, D.; Zhu, Y.; Chen, H.; Zhu, B. STIM1 Mediates Hypoxia-Driven Hepatocarcinogenesis via Interaction with HIF-1. *Cell Rep.* **2015**, *12*, 388–395. [CrossRef]

131. Abdelazeem, K.N.M.; Droppova, B.; Sukkar, B.; al-Maghout, T.; Pelzl, L.; Zacharopoulou, N.; Ali Hassan, N.H.; Abdel-Fattah, K.I.; Stournaras, C.; Lang, F. Upregulation of Orai1 and STIM1 expression as well as store-operated Ca^{2+} entry in ovary carcinoma cells by placental growth factor. *Biochem. Biophys. Res. Commun.* **2019**, *512*, 467–472. [CrossRef]

132. Wang, J.; Xu, C.; Zheng, Q.; Yang, K.; Lai, N.; Wang, T.; Tang, H.; Lu, W. Orai1, 2, 3 and STIM1 promote store-operated calcium entry in pulmonary arterial smooth muscle cells. *Cell Death Discov.* **2017**, *3*, 1–11. [CrossRef]

133. Azimi, I.; Milevskiy, M.J.G.; Kaemmerer, E.; Turner, D.; Yapa, K.T.D.S.; Brown, M.A.; Thompson, E.W.; Roberts-Thomson, S.J.; Monteith, G.R. TRPC1 is a differential regulator of hypoxia-mediated events and Akt signalling in PTEN-deficient breast cancer cells. *J. Cell Sci.* **2017**, *130*, 2292–2305. [CrossRef]

134. Azimi, I.; Milevskiy, M.J.G.; Chalmers, S.B.; Yapa, K.T.D.S.; Robitaille, M.; Henry, C.; Baillie, G.J.; Thompson, E.W.; Roberts-Thomson, S.J.; Monteith, G.R. ORAI1 and ORAI3 in breast cancer molecular subtypes and the identification of ORAI3 as a hypoxia sensitive gene and a regulator of hypoxia responses. *Cancers* **2019**, *11*, 208. [CrossRef]

International Journal of
Molecular Sciences

Article

Fetal Oxygenation from the 23rd to the 36th Week of Gestation Evaluated through the Umbilical Cord Blood Gas Analysis

Luca Filippi [1,2,*], Francesca Pascarella [2], Alessandro Pini [3], Maurizio Cammalleri [4], Paola Bagnoli [4], Riccardo Morganti [5], Francesca Innocenti [2], Nicola Castagnini [2], Alice Melosi [2] and Rosa Teresa Scaramuzzo [2]

[1] Department of Clinical and Experimental Medicine, University of Pisa, 56126 Pisa, Italy
[2] Neonatology Unit, Azienda Ospedaliero-Universitaria Pisana, 56126 Pisa, Italy; fro221295@gmail.com (F.P.); f.innocenti17@gmail.com (F.I.); n.castagnini@studenti.unipi.it (N.C.); alicemelosi@icloud.com (A.M.); r.scaramuzzo@ao-pisa.toscana.it (R.T.S.)
[3] Department of Experimental and Clinical Medicine, University of Florence, 50121 Florence, Italy
[4] Unit of General Physiology, Department of Biology, University of Pisa, 56126 Pisa, Italy; maurizio.cammalleri@unipi.it (M.C.); paola.bagnoli@unipi.it (P.B.)
[5] Section of Statistics, Azienda Ospedaliero-Universitaria Pisana, 56126 Pisa, Italy; riccardo-morganti@ao-pisa.toscana.it
* Correspondence: luca.filippi@unipi.it; Tel.: +39-50-993677

Abstract: The embryo and fetus grow in a hypoxic environment. Intrauterine oxygen levels fluctuate throughout the pregnancy, allowing the oxygen to modulate apparently contradictory functions, such as the expansion of stemness but also differentiation. We have recently demonstrated that in the last weeks of pregnancy, oxygenation progressively increases, but the trend of oxygen levels during the previous weeks remains to be clarified. In the present retrospective study, umbilical venous and arterial oxygen levels, fetal oxygen extraction, oxygen content, CO_2, and lactate were evaluated in a cohort of healthy newborns with gestational age < 37 weeks. A progressive decrease in pO_2 levels associated with a concomitant increase in pCO_2 and reduction in pH has been observed starting from the 23rd week until approximately the 33–34th week of gestation. Over this period, despite the increased hypoxemia, oxygen content remains stable thanks to increasing hemoglobin concentration, which allows the fetus to become more hypoxemic but not more hypoxic. Starting from the 33–34th week, fetal oxygenation increases and ideally continues following the trend recently described in term fetuses. The present study confirms that oxygenation during intrauterine life continues to vary even after placenta development, showing a clear biphasic trend. Fetuses, in fact, from mid-gestation to near-term, become progressively more hypoxemic. However, starting from the 33–34th week, oxygenation progressively increases until birth. In this regard, our data suggest that the placenta is the hub that ensures this variable oxygen availability to the fetus, and we speculate that this biphasic trend is functional for the promotion, in specific tissues and at specific times, of stemness and intrauterine differentiation.

Keywords: newborn; intrauterine hypoxia; fetal hypoxia; differentiation

Citation: Filippi, L.; Pascarella, F.; Pini, A.; Cammalleri, M.; Bagnoli, P.; Morganti, R.; Innocenti, F.; Castagnini, N.; Melosi, A.; Scaramuzzo, R.T. Fetal Oxygenation from the 23rd to the 36th Week of Gestation Evaluated through the Umbilical Cord Blood Gas Analysis. *Int. J. Mol. Sci.* **2023**, *24*, 12487. https://doi.org/10.3390/ijms241512487

Academic Editor: Giovanni Tossetta

Received: 4 July 2023
Revised: 2 August 2023
Accepted: 4 August 2023
Published: 6 August 2023

1. Introduction

The embryo and the fetus live in a physiologically hypoxic environment (the so-called Everest in utero) [1], although the intrauterine environment is characterized by a variable level of hypoxia throughout pregnancy.

During the first stages of pregnancy, the concentrations of oxygen are very low, similar to those measurable within the non-pregnant uterus [2,3]. Over the first weeks of pregnancy, placental development favors the increase in oxygen availability to the fetoplacental unit, and, at the beginning of the second trimester of gestation, placental oxygenation triples the oxygen availability, reaching a maximum around the 16th week of gestation [3]. From this week onwards, a slow, gradual reduction in placental oxygen levels is observed, and this trend appears to be consistent with a progressive reduction in fetal oxygenation status [3]. In fact, a series of studies carried out by analyzing human umbilical cord venous and arterial blood samples obtained via cordocentesis demonstrated decreasing fetal partial pressure of oxygen (pO_2) and saturation of oxygen (SaO_2), which start from the 16–18th week of gestation and progress with advancing gestation [4–7].

However, some considerations led us to imagine a reversion of the trend of oxygenation during the final stages of pregnancy. This hypothesis was inspired by the observation that in many animals, the vasculature is still immature at birth and that its maturation occurs after birth. This is particularly evident in the brain of rodents, whose appropriate glial–vascular interaction is instituted after birth [8,9] and in the retina of mice, which is avascular at birth but begins to vascularize during the first week after birth [10] or even later [11], demonstrating, at least chronologically, a strict relationship between vessel maturation and an increase in oxygen exposure. If vascular maturity requires an increase in oxygen levels, it is legitimate to imagine that in humans, whose vascularization is completed during the last weeks of intrauterine life [12], there is a progressive increase in oxygen tension in the more advanced stages of pregnancy.

To verify if oxygenation increased in the last weeks of pregnancy, we recently performed an observational study that assessed the umbilical gas analysis collected from a cohort of healthy newborns with gestational ages $\geq$ 37 weeks. Results demonstrated a progressive increase in fetal oxygenation from the 37th to the 41st weeks of gestation (approximately an increase of about 1 mmHg per week) [13]. These findings are in accordance with a previous study that reported a slight increase in cord venous pO_2 between near-term and term newborns [14]. The pooled analysis of studies evaluating fetal oxygen levels suggests that oxygenation during intrauterine life continues to vary after placenta development, taking on a biphasic trend. Fetuses, in fact, from mid-gestation to near-term grow in an environment progressively more hypoxic, while, approaching the term of pregnancy, oxygenation reverses its trend. Knowledge of this dynamic assumes a particular importance because it helps us understand the physiological mechanisms that guarantee the maturation of vascularization, whether it takes place after birth, as in rodents, or in utero, as in humans.

The main objective of this study is to reconstruct, with the same methodology used to analyze the blood gas variations in term infants, the fetal oxygenation status from the 23rd to the 36th weeks of gestation. Our objectives are to verify that healthy fetuses become more hypoxemic over the weeks, to confirm that this trend at a certain point is reverted, and to identify the time when this reversal occurs.

2. Results

Out of 12,544 newborns born between 2016 and 2022, 11,375 newborns with GA $\geq$ 37 weeks were excluded from the analysis, and 1169 preterm newborns were eligible for this study. Umbilical cord gas analyses of 560 newborns were not performed, incomplete, or unreliable. Out of 609 samples, 5 with acidosis at birth were excluded (Figure 1).

Of the 604 newborns enrolled, males were prevalent (310/604, 51.2%). Neonates born by vaginal delivery were a minority (148/604, 24.5%). Table 1 shows the cumulative demographic and gas analytical parameters of all newborns.

According to what was recently observed in term newborns [13] and also in preterm neonates, the modality of delivery influences the blood gas analysis. Neonates born by vaginal delivery showed higher values of pO_2 and lactate both in venous and arterial cord samples. Lower levels of pH, partial pressure of carbon dioxide (pCO_2), and bicarbonate were evident only in venous cord samples. Neonates born by cesarean section displayed an oxygen extraction significantly higher than neonates born by spontaneous delivery, in line with what was demonstrated in term neonates [13]. To verify whether the incidence of vaginal or caesarean delivery substantially changed as gestation progressed, data from blood gas analysis were arbitrarily grouped into six newborn groups: a group of preterm newborns from 23 to 26^{+6} weeks of gestation, 27–28^{+6}, 29–30^{+6}, 31–33^{+6}, 34–35^{+6}, and

$36–36^{+6}$ (Figure 1). The percentage of vaginal or cesarean delivery was not statistically different among the groups, allowing us to evaluate the blood samples as a whole, regardless of the type of delivery, without further stratification.

12,544 preterm newborns born between 2016-2022 at Pisa University hospital

Exclusion:
- 11,375 newborns born at gestational age $\geq$ 37 weeks
- 560 newborns with analyses not performed, incomplete, or not reliable
- 5 newborns with pH $\leq$ 7.00 and/or base excess $\leq$ -12 mmol/L

604 umbilical cord samples analyzed

Gestational Age (weeks)	Newborns n=604	Vaginal delivery n=148 (24.5)	Cesarean section n=456 (75.5)
$23-26^{+6}$, n (%)	61	21 (34.4)	40 (65.6)
$27-28^{+6}$, n (%)	39	12 (30.8)	27 (69.2)
$29-30^{+6}$, n (%)	61	12 (19.7)	49 (80.3)
$31-33^{+6}$, n (%)	118	26 (22.0)	92 (78.0)
$34-35^{+6}$, n (%)	191	42 (22.0)	149 (78.0)
$36-36^{+6}$, n (%)	134	35 (26.1)	99 (73.9)

Figure 1. Flow chart illustrating patient enrollment of this retrospective observational cohort study.

Table 1. Umbilical cord blood gas analysis in all enrolled preterm newborns and separately analyzed by the type of delivery.

	All Preterm Newborns $n = 604$	Vaginal Delivery $n = 148$	Cesarean Section $n = 456$	p Value
GA, days, mean (SD)	230 (24)	228 (28)	231 (23)	0.127
Birth weight, g, mean (SD)	1855 (687)	1926 (728)	1832 (672)	0.155
Male, n (%)	310 (51.2)	80 (54.0)	230 (50.4)	0.474
Apgar Score at 5 min, mean (SD)	7.6 (1.2)	7.6 (1.4)	7.7 (1.1)	0.781
Umbilical venous cord sampling				
pH, mean (SD)	7.315 (0.08)	7.334 (0.08)	7.309 (0.08)	<0.001
pCO_2, mmHg, mean (SD)	43.2 (9.4)	39.9 (9.5)	44.3 (9.2)	<0.001
pO_2, mmHg, mean (SD)	22.8 (7.9)	25.3 (8.3)	22.0 (7.6)	<0.001
Bicarbonate, mmol/L, mean (SD)	21.7 (3.1)	20.9 (2.7)	22.0 (3.1)	<0.001
BE(B), mmol/L, mean (SD)	−4.3 (3.6)	−4.6 (3.5)	−4.3 (3.6)	0.250
Lactate, mmol/L, mean (SD)	2.8 (1.4)	3.3 (1.4)	2.7 (1.3)	<0.001
Hemoglobin, g/dL, mean (SD)	14.96 (2.3)	15.5 (1.9)	14.8 (2.4)	0.001
SaO_2, %, mean (SD)	52.9 (22.1)	57.4 (21.4)	51.3 (22.2)	0.011
Umbilical arterial cord sampling				
pH, mean (SD)	7.267 (0.08)	7.274 (0.09)	7.278 (0.08)	0.327
pCO_2, mmHg, mean (SD)	50.7 (11.4)	49.7 (11.8)	51.0 (11.3)	0.243
pO_2, mmHg, mean (SD)	15.2 (6.9)	17.6 (6.7)	14.4 (6.8)	<0.001
Bicarbonate, mmol/L, mean (SD)	22.7 (3.8)	22.3 (3.2)	22.8 (3.9)	0.227
BE(B), mmol/L, mean (SD)	−4.6 (4.1)	−4.8 (3.7)	−4.5 (4.3)	0.414
Lactate, mmol/L, mean (SD)	3.1 (1.5)	3.7 (1.6)	2.9 (1.5)	<0.001
Hemoglobin, g/dL, mean (SD)	14.96 (2.3)	15.3 (2.1)	14.4 (2.5)	<0.001
SaO_2, %, mean (SD)	33.0 (19.2)	40.4 (20.0)	30.0 (18.2)	<0.001
Veno-arterial O_2 difference, mmHg, mean (SD)	7.7 (6.4)	7.5 (7.1)	7.8 (6.2)	0.665
Fetal oxygen extraction, %, mean (SD)	35.7 (35.8)	27.3 (32.0)	38.6 (36.8)	0.007

GA = gestational age; pO_2 = partial pressure of oxygen; pCO_2 = partial pressure of carbon dioxide; BE = base excess.

To evaluate whether oxygen tension varied during pregnancy, data from venous and arterial pO_2, as well as oxygen extraction, were assessed in relation to the progression of pregnancy (Figure 2). Values from umbilical venous samples revealed a biphasic trend of oxygen level, as suggested by the significance of the quadratic regression: pO_2 levels progressively decrease until approximately 230–240 days of gestation (approximately the 33rd–34th week of gestation), when pO_2 appears to reach its lowest point; then, starting from the 34th week of pregnancy, pO_2 increases (Figure 2A), to then continue, ideally, with the positive progression observed from the 37th to the 41st weeks of gestation [13]. This tendency is better highlighted in Supplementary Figure S1, where the data were stratified according to groups of weeks of pregnancy. The arterial pO_2 level also appears to be linearly reduced with the progression of gestation (Figure 2B), with a minimal increase in the weeks near term (Supplementary Figure S1). Overall, oxygen extraction did not show significant change throughout the period considered (Figure 2C), even though a modest increase was evident comparing the groups of $27-28^{+6}$ and $29-30^{+6}$ towards the more advanced stages of pregnancy (Supplementary Figure S1).

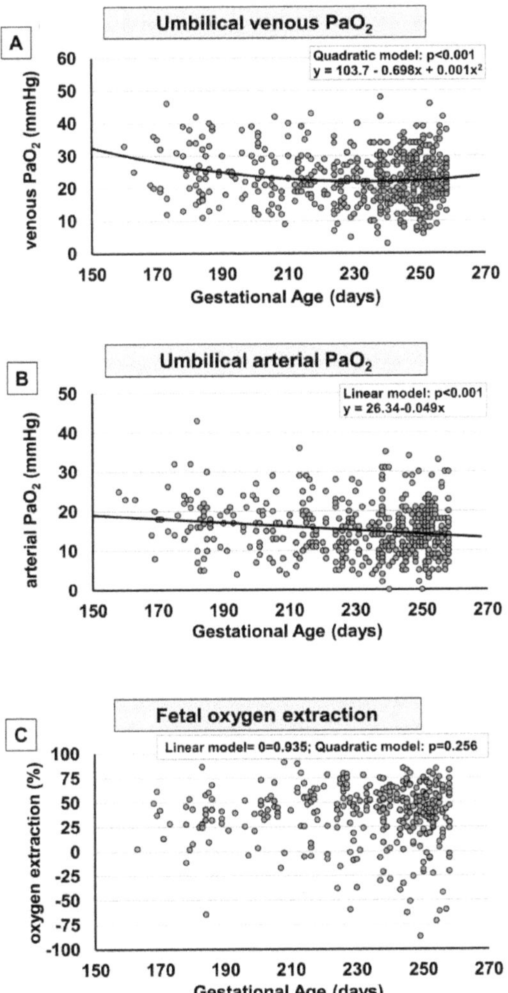

Figure 2. Scatter plots with regression lines representing umbilical cord oxygenation status [venous pO_2, panel (**A**) (n = 532); arterial pO_2, panel (**B**) (n = 500); fetal oxygen extraction, panel (**C**) (n = 399)] of the whole study population. The equation of the statistically significant model (linear or quadratic) is indicated at the top of each figure.

The levels of venous and arterial hemoglobin (Hb) progressively increased from the 23rd week onwards, reaching a plateau approximately at 230–240 days of gestation (around weeks 33–34 of gestation) (Figure 3A,B). The Hb rise appears to follow a trend opposite to that followed by pO_2, with an increase of around 1 g/dL per group of gestational age, as shown in Supplementary Figure S2. This increase explains why, despite the reduced intake of oxygen from the placenta to the fetus, the venous content of oxygen remained stable throughout the period examined (Figure 3C), excluding a minimal increase during the weeks near term (Supplementary Figure S2).

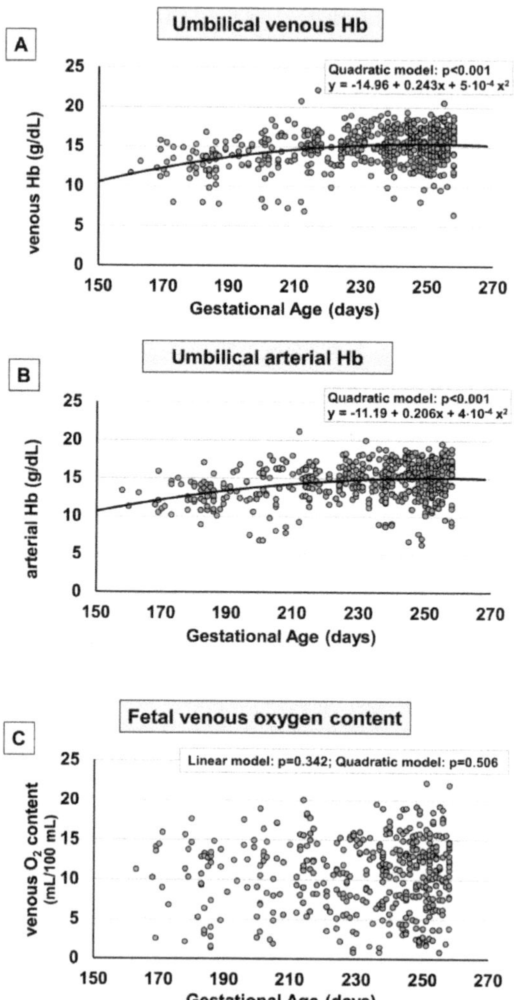

Figure 3. Scatter plots with regression lines representing umbilical cord Hb and oxygen content [venous Hb, panel (**A**) (*n* = 566); arterial Hb, panel (**B**) (*n* = 534); fetal venous oxygen content, panel (**C**) (*n* = 427)] of the whole study population. The equation of the statistically significant model (linear or quadratic) is indicated at the top of each figure.

Simultaneously with the decrease in pO_2, a linear reduction in venous and arterial pH was observed as pregnancy proceeded (Figure 4A,B), coupled with a mild reduction in venous base excess (Figure 4C), but not in arterial base excess (Figure 4D), and a mild increase in arterial bicarbonate (Figure 4F), but not in venous (Figure 4E). This tendency is emphasized in Supplementary Figure S3, where the data were stratified according to groups of weeks of pregnancy.

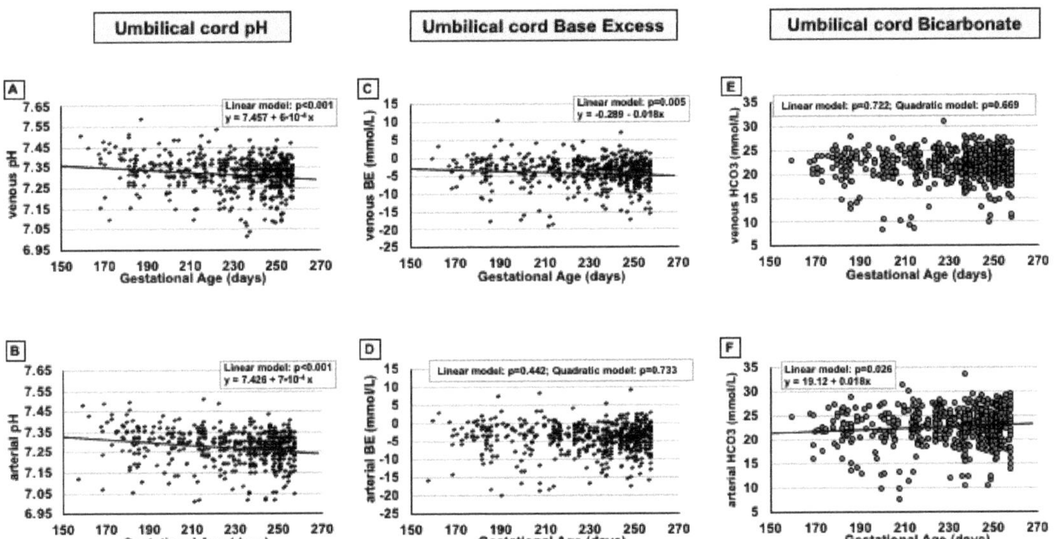

Figure 4. Scatter plots with regression lines representing umbilical cord pH, base excess, and bicarbonate [venous pH, panel (**A**) (n = 584); arterial pH, panel (**B**) (n = 552); venous base excess, panel (**C**) (n = 582); arterial base excess, panel (**D**) (n = 554); venous bicarbonate, panel (**E**) (n = 573); arterial bicarbonate, panel (**F**) (n = 541)] of the whole study population. The equation of the statistically significant model (linear or quadratic) is indicated at the top of each figure.

In analogy to what was observed in term fetuses, also in preterm fetuses, blood gas analysis displayed for the umbilical cord content of CO_2 an opposing trend if compared with the trend of oxygen, with a progressive surge both in venous (Figure 5A) and arterial samples (Figure 5B). This increase is progressive at least until the group with gestational age 31–33^{+6} weeks (Supplementary Figure S4). Starting from the 34th week of pregnancy, pCO_2 decreases and then continues, ideally, with the negative progression recently described in term fetuses [13]. Overall, CO_2 production did not seem to change significantly (Figure 5C), even though, during the more advanced weeks of gestation, we observe an apparent increase in CO_2 production (Supplementary Figure S4) that then continues, ideally, with the progression recently described from the 37th to the 41st week of gestation [13].

Between the 23rd and 36th weeks of gestation, no significant change in lactate levels was observed either in venous samples (Figure 6A) or in arterial samples (Figure 6B). A mild increase in fetal lactate production seemed evident (Figure 6C) as a consequence of a mild increase detected in newborns belonging to more advanced gestational ages (Supplementary Figure S5).

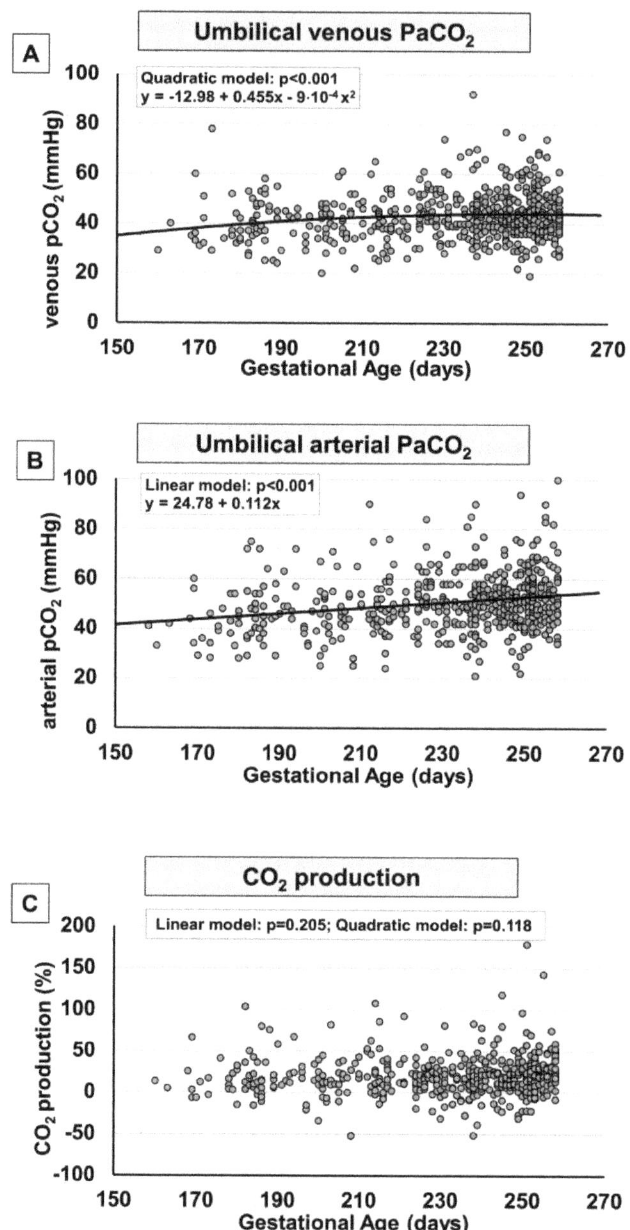

Figure 5. Scatter plots with regression lines representing umbilical cord carbon dioxide levels [venous pCO$_2$, panel (**A**) (*n* = 580); arterial pCO$_2$, panel (**B**) (*n* = 548); fetal CO$_2$ production, panel (**C**) (*n* = 527)]. The equation of the statistically significant model (linear or quadratic) is indicated at the top of each figure.

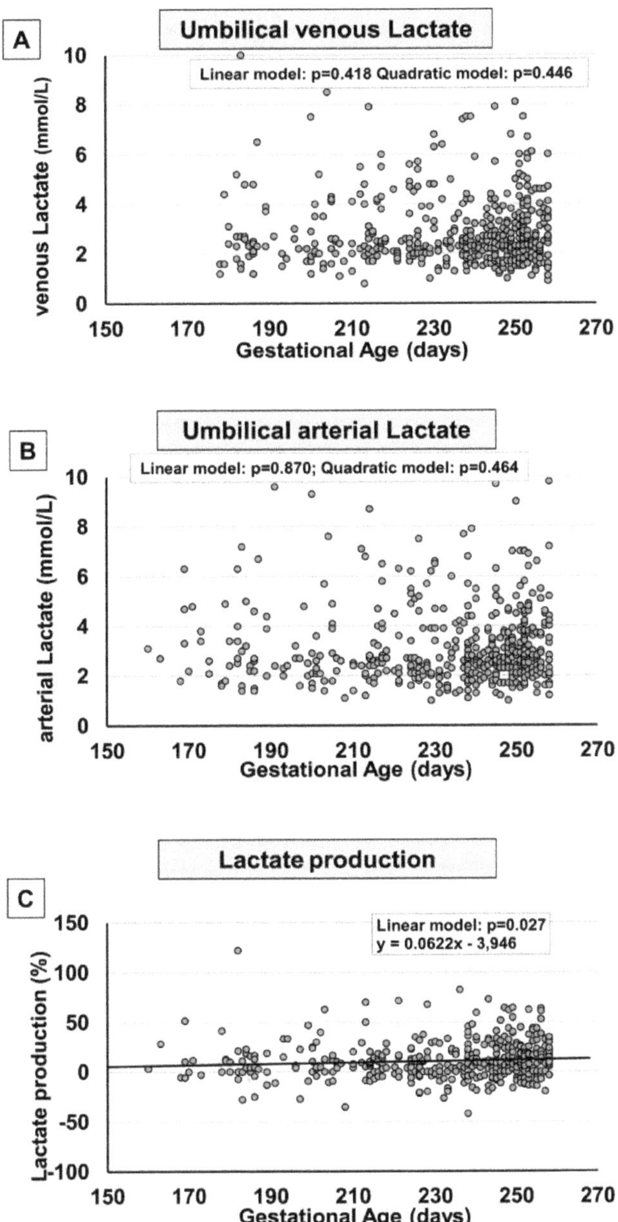

Figure 6. Scatter plots with regression lines representing umbilical cord lactate levels [venous lactate, panel (**A**) (n = 513); arterial lactate, panel (**B**) (n = 481); fetal lactate production, panel (**C**) (n = 465)]. The equation of the statistically significant model (linear or quadratic) is indicated at the top of each figure.

3. Discussion

Conception, embryogenesis, and fetal growth occur in the female reproductive tract, where the oxygen concentration is very low. A physiological hypoxia has been indicated as the key regulator of the harmonious processes of placental and embryonal development [15,16]. The importance of this peculiar environmental oxygenation is witnessed by the improved live birth rate of preimplantation embryos cultured under hypoxic conditions compared with normoxic cultures [17]. However, the embryo and the following fetoplacental unit grow and differentiate in a milieu where the levels of hypoxia undergo evident and recurring variations throughout the pregnancy.

Placental and embryonic cells react to a hypoxic environment with a series of adaptive adjustments to gene expression. The upregulation of hypoxia-inducible factors (HIFs) represents the hub through which hypoxia promotes placental development [18]. Under hypoxia, HIF-1α translocates into the nucleus and dimerizes with HIF-1β, leading to the transcription of several hundred specific target genes [19], favoring cell survival in a hypoxic environment (induction of specific enzymes involved in energy metabolism, erythropoiesis, and angiogenesis). [20]. Therefore, HIF modulation guarantees at the same time the proliferation and metabolic adaptation of embryonal cells, placental development, and trophoblast differentiation [21,22].

HIF regulation during pregnancy is complex. Although HIF is regulated also through oxygen-independent mechanisms such as hormones (angiotensin II), cytokines (interleukin-1β, tumor necrosis factor α, and NF-kβ), or growth factors (transforming growth factor-β1 and insulin-like growth factor), all of which are significantly upregulated during pregnancy [23], oxygen levels (and their fluctuations) represent the main regulator of placental HIF levels, placenta development, and normal mammalian embryo morphogenesis [24]. The pivotal role of HIF for embryonic/fetal wellness is indirectly confirmed by the high embryonic lethality, dysmorphogenesis, or severe placental defects observed in HIF-1α-/- mouse embryos [18].

Although hypoxia and HIF are indispensable for the harmonious development of the fetoplacental unit, from the first moments of pregnancy, it is evident that the levels of oxygen and HIF change following a precise dynamic.

Conception occurs when a sperm cell successfully fertilizes an egg cell in the fallopian tube, at an oxygen concentration of approximately 5–7% [25]. By this time, the morula reaches the uterine cavity, where the oxygen concentration is significantly reduced to 2% [26,27]. The transition to this more hypoxic environment is essential in the first 2–3 weeks of life to allow the proliferation, implantation, anchoring, and invasion of the blastocyst into the maternal uterus [28]. In fact, on one side, hypoxia (and the up-regulation of HIFs) is crucial to maintaining the pluripotent status of embryonic stem cells [29], as well as of cytotrophoblast cells, which are usually considered to be trophoblast stem cells [3], and to promoting their proliferation [30]. On the other side, exposure to a decreasing oxygen gradient during the travel from the oviduct to the uterus promotes a metabolic shift of cytotrophoblast cells from oxidative phosphorylation towards a glycolytic metabolism (Warburg effect) [31–33]. This metabolic adaptation confers a proliferative advantage since it promotes the production of many intermediates of the pentose phosphate pathway, such as ribose sugars necessary for nucleic acid synthesis, and induces the production of elevated lactic acid levels that are useful to promote nesting in surrounding tissues [34]. In fact, the increased production and extrusion of lactate facilitate the disaggregation of uterine tissues and promote trophoblast invasion [35].

Increasing hypoxia and high levels of HIF during the first weeks of life are also necessary to assure cytotrophoblast transmigration across the uterine epithelium and their differentiation into extravillous trophoblasts [18,36], which in turn migrate along the lumen of the spiral arterioles. These cells progressively remodel the spiral arteries, replacing the smooth muscle and the elastic lamina of the vessel. In this way, the low-flow, high-resistance original spiral arteries become high-flow, low-resistance vessels and assure, in the following weeks of pregnancy, a gradual increase in oxygen delivery [28].

In summary, the intrauterine low-oxygen tension promotes embryonic cytotrophoblast cell proliferation, their intrauterine invasion, and interconnections with the uterine spiral arterioles [37], thus laying the foundations for a subsequent increase in oxygen levels.

In the meantime, HIF upregulation promotes the concomitant activation of a series of angiogenic factors, including vascular endothelial growth factor (VEGF), basic fibroblast growth factor, platelet-derived growth factor, angiopoietin-1, angiopoietin-2, Tie2, and monocyte chemoattractant protein-1 [38]. These proangiogenic factors in turn raise the foundation for embryonal vascular and placental development and a further increase in oxygen availability [3]. In turn, such an increase in microenvironmental oxygen promotes the differentiation of cytotrophoblasts into multinucleated syncytiotrophoblasts, which are necessary to secrete several hormones for pregnancy maintenance [39].

Essentially, what happens to the embryo in the first weeks of life represents a plastic demonstration that hypoxia is indispensable; however, within this hypoxic environment, we observe a fine regulation of oxygen levels, of HIF, and consequently of the expression of its target genes aimed at modulating apparently contradictory functions, such as the expansion of stemness and differentiation.

Placentation explains why placental oxygen tension, which is less than 20 mmHg (more or less 2% O_2) during the first weeks of human gestation (approximately until the 10th week) [2,3], increases to around 60 mmHg (roughly 8% O_2) at the beginning of the second trimester of gestation [2]. The increased oxygen availability from the first weeks to the early weeks of the second trimester justifies why the fetal habitat can be considered a dynamic environment where hypoxia initially attenuates.

While recent studies performed on healthy fetal animals reported no significant variations in blood gas and acid–base status during gestation [40–47], the available data for human fetuses obtained via cordocentesis, although limited, indicate a progressive reduction in fetal oxygen levels starting from the 16–18th weeks of gestation toward its end [4–7]. The reduced oxygen supply to the fetuses is explained by the increased placental oxygen extraction with advancing gestation, as demonstrated by samples of maternal blood drawn from the subchorial lake under the chorionic plate [4]. However, these studies were performed on pregnancies complicated by fetal anomalies, a variety of prenatal pathologies, or maternal infections [4–7]; thus, such results would need to be confirmed in a healthy population. The fluctuations of oxygen levels are further confirmed by the analysis of a large amount of umbilical cord blood gas analysis, in which a clear trend reversal of oxygenation in the last weeks of gestation was demonstrated [13,14]. Although the investigation of the mechanisms underlying the increased oxygenation was beyond the scope of these studies, this phenomenon has been attributed to increased oxygen diffusion from the placenta to the fetus, probably correlated with the physiological aging of the placenta. In fact, the thinning of the placental barrier surface area due to a reduction in the distance between fetal vessels and the trophoblastic membrane over time favors an increase in placental oxygen transport [48]. In conclusion, the literature shows data apparently contradictory; while, from mid-gestation to near term, several historical studies report a progressive reduction in fetal oxygenation [4–7], other more recent studies describe increased oxygenation during the last weeks of gestation [13,14]. This contradiction can only be explained by imagining a biphasic trend of fetal oxygenation, with a first phase characterized by a progressive reduction in fetal oxygenation, followed by a phase of increased oxygenation.

The results of the present study confirm that fetuses grow in a dynamic low-oxygen environment and demonstrate for the first time that the concentration of oxygen follows a biphasic trend. In fact, at least from the 23rd week onward, we observed a progressive and linear reduction in venous pO_2 that slows and then reverses at around 33–34 weeks of gestation. A similar trend is also observed regarding the arterial samples, and therefore the oxygen extraction values do not change substantially during this time. These data suggest that the fetus is subject to a slow and progressive reduction in oxygen supply from the placenta, at least from the 23rd until the 33–34th week, which represents the inflection point between the two phases. This observation is in line with what has been observed from the second to the third trimester of pregnancy in the intervillous space of the human placentas, where mean oxygen tension measured gradually declines [3,4,49].

If we consider that the pO_2 level in mothers remains stable throughout pregnancy [50] in concomitance with the uterine blood flow [51], the gradual reduction in venous umbilical pO_2 suggests that placental growth and its increasing metabolic activity progressively make less oxygen available to the fetus. It is well known that placental oxygen consumption accounts for a large percentage of the collective fetal and placental oxygen consumption and that this percentage increases over time [15,51]. Therefore, our data suggest that the increasing consumption of oxygen by the placenta causes a progressive accentuation of fetal hypoxemia until the 33–34th week of pregnancy. In the face of a continuous reduction in the placental oxygen supply, the venous pCO_2 progressively increases until the 33–34th week of gestation, confirming the hypothesis that these changes in the supplies to the fetus are consequent to an increased metabolic activity of the placenta. The combination of the reduction in oxygenation with the increase in pCO_2 explains why the venous pH gradually decreases, at least up to the 33–34th week of pregnancy.

Increasing fetal hypoxemia is confirmed by the progressive increase in Hb concentration, the synthesis of which is known to be induced by low levels of oxygen through the activation of the HIF/erythropoietin axis [52]. The present study confirms that the Hb level gradually and progressively increases as pregnancy progresses, starting from the 23rd week of gestation, showing a trend opposite to that of pO_2, as already reported by several authors [53]. Despite the progressive decrease in pO_2, the increased Hb explains why the umbilical venous blood oxygen content remains unchanged. This strategy allows the fetus, at least from the 23rd to the 33–34th weeks of pregnancy, to simultaneously obtain two results: On the one hand, it guarantees a constant oxygen content and therefore does not deprive the peripheral tissues of the necessary oxygen, but at the same time, the concentration of oxygen dissolved in the blood is reduced. This explains why the fetus becomes progressively more hypoxemic despite the constant oxygen content that maintains the fetus's stable hypoxia status. The maintenance, from the 23rd to the 33–34th week of gestation, of this efficient balance is evidenced by the absolute stability of lactate values.

However, from the 33–34th week forward, this balance gets broken and oxygenation starts to increase, as recently described in full-term newborns [13]: This gestational age represents the watershed beyond which umbilical venous oxygenation increases, probably due to a physiologic aging of the placenta that favors an increased diffusion of oxygen [48], and consequently, the increase in fetal Hb levels stops. The oxygen availability to fetuses during the last weeks of gestation is probably even more significant because, in late gestation, the percentage of fetal Hb decreases as adult Hb (which has a lower affinity for oxygen) begins to increase [54]. A comparable reversal is observed for venous pCO_2, which increases until the 34th week; after that, pCO_2 stops increasing, ideally continuing to decrease until the end of pregnancy [13]. The arterial pCO_2 instead seems to continue to increase even after the 34th week, suggesting a possible increase in fetal pCO_2 production and therefore a possible metabolic shift toward mitochondrial activation, which will be more evident after 37 weeks [13].

Therefore, the biological role of oxygen fluctuations observed in the more advanced stages of pregnancy and also after placenta development, although difficult to interpret, may serve for tissue-specific stem cell recruitment [55] and cell differentiation [56]. In some anatomic districts, for instance, hypoxemia is essential in preserving naïve stemness potential [57,58], or modulating cell differentiation [56]. This is particularly evident in the nervous central system, where hypoxia stimulates several processes, including cell survival, proliferation, catecholaminergic differentiation of isolated neural crest stem cells [59], mesencephalic precursor cells into neurons [60], or undifferentiated astrocytes to differentiate [61,62]. In other regions, instead, it is the increased oxygenation that drives cell differentiation, as in neural retinal tissue [63], pancreatic cells [64], keratinocytes [65], hepatocyte cell lines [66], or endothelial cells [67]. Therefore, it is realistic to speculate that the intrauterine environment, which physiologically becomes more hypoxic from mid-gestation to near-term and then less hypoxic until term, guarantees tissue-specific cellular differentiation depending on oxygen tension.

As shown here, the 34th week of gestation would represent the watershed between a hypoxic first phase, characterized by increased stem cell proliferation and the differentiation of specific cell populations, and a subsequent phase in which increasing oxygenation promotes the reduction in the stem cell pool and cell differentiation. This relationship between stemness and gestational age is well evident for endothelial precursor cells (EPCs), which are significantly more expressed in the umbilical cord blood of preterm newborns if compared with term neonates [68,69]. Interestingly, the cord blood of preterm infants contains a number of EPCs that grow as gestational age progresses: very low at 24–28 weeks, and then much more expressed at 28–35 weeks [70] or until 34 weeks [71]. Considering that the differentiation of endothelial cells is favored by the increase in oxygen levels [67], we hypothesize that the expansion of EPCs is synchronous with increasing hypoxemia and stops around 33–34th when oxygen increases, promoting the completion of the vasculature.

The topic of vasculogenesis is particularly intriguing, considering that in rodents, at variance with humans, the vascularization of many organs is immature at birth. This is particularly evident in the brain, where vascularization completes after birth [8,9], and even more so in the retina, which is avascular during intrauterine life but vascularizes after birth [10,11].

Although vasculature development depends on hypoxia-driven HIF and VEGF increases [38], in the retina, vessel appearance becomes evident a few days after birth, when oxygenation increases in concomitance with lowered levels of HIF and VEGF [72]. This observation suggests that vascularization occurs in two phases. In the first intrauterine phase, hypoxia induces astrocyte maturation and the production of VEGF [73], favoring, in turn, the recruitment of EPCs. Then, over the second phase, characterized by increased oxygenation and reductions in HIF and VEGF, EPCs can differentiate into endothelial cells. A similar immaturity is present in very preterm neonates, whose retina at birth is only partially vascularized and whose brain is incompletely vascularized, especially in the germinal matrix [74]. This immaturity is believed to predispose preterm infants to an elevated risk of developing intraventricular hemorrhage. However, this risk decreases for neonates born after 33–34 weeks of pregnancy or, in highly preterm newborns, after the first days of life, suggesting that a higher oxygenated environment induces endothelial cell maturation [74].

Considering that oxygen plays an important role in the modulation of the stem population and the processes of differentiation, knowledge of its fluctuations and the consequent biological effects opens new perspectives for the studies that, in the future, will try to artificially reproduce the benefits of intrauterine life. In this regard, studies aimed at the development of artificial placenta technology [75] cannot ignore the physiological levels of oxygenation, which must be respected. But even attempts to mimic the biological effects of intrauterine hypoxemia through the stimulation of β3-adrenoceptor agonists [76] will have to take into consideration that fetal hypoxemia increases only up to the 33–34th week.

Limitations

This study has two main limitations:

1. Umbilical cord samples represent a safe, non-invasive method to obtain valuable information regarding intrauterine conditions [77]. However, the type of delivery may affect the gas analytic values: fetal oxygenation of neonates born by vaginal delivery may, in fact, be altered by the engagement of the fetus through the birth canal or by the reduction in blood flow during uterine contractions [78], as well as by cesarean section, which can affect fetal well-being due to the cardiovascular effects of anesthesia, maternal position, and maternal ventilation [79,80]. Although the impact of delivery type on the gas test result is well known [13], in this study we did not find a different incidence of vaginal or cesarean delivery at different gestational ages, and therefore the blood samples were analyzed as a whole without further stratification.
2. In this study, only umbilical cord blood samples from preterm infants were analyzed. The merging of data obtained from preterm and term newborns would have provided

a better interpretation of the fluctuations in fetal oxygenation. However, the pooling of such data would not have been methodologically suitable, as umbilical cord blood samples in preterm neonates were collected within 60 sec after birth ("early cord clamping"), while samples in term neonates were collected approximately 60 sec after birth ("delayed cord clamping") [13].

4. Materials and Methods

This retrospective study was performed at the University Hospital of Pisa, Italy, with around 1700 births per year. All the newborns born between January 1st 2016 and December 31st 2022 were eligible. Their umbilical cord blood samples were collected immediately after birth (early cord clamping). Samples collected from term newborns ($\geq$ 37 weeks) or gestations with fetal or maternal intrapartum complications (i.e., fetuses with an operative vaginal delivery involving application of forceps or a vacuum extractor, abnormal intrapartum cardiotocography requiring emergency cesarean section, meconium-stained amniotic fluid, placental abruption, cord prolapse, chorioamnionitis, maternal sepsis, convulsions, hemorrhages, uterus rupture, snapped cord) were excluded from the study. Umbilical cord blood samples for which pH, base excess (BE), or both were not available and with acidosis at birth (pH $\leq$ 7.00 and/or BE $\leq$ -12 mmol/L) [81,82] were excluded from the analysis.

Values of umbilical (venous and arterial) parameters < or >3 SD from their respective means were individually evaluated and (i) rectified if probably mid-entered, (ii) maintained unchanged if considered plausible, and (iii) excluded if considered not plausible [14]. Results of cord blood gas that did not fulfill the following criteria: (i) arterial pH < the venous pH (by at least a difference of 0.022) and (ii) arterial pCO_2 > the venous pCO_2 (by at least a difference of 5.3 mmHg) were considered unreliable and excluded from the analysis [83].

Cord blood was collected, as recently published [13]. The samples, after their labeling and identification as venous or arterial, were analyzed as soon as possible, using an automatic blood gas analyzer (GEM® Premier 4000, Instrumentation Laboratory, Lexington, MA, USA). The pH, pCO_2, pO_2, SaO_2, lactate, and Hb were measured, whereas bicarbonate and base excess were calculated, respectively, from measured pH and pCO_2 using the Henderson–Hasselbalch equation pH = 6.1 + log($[HCO_3^-]/pCO_2 \times 0.03$) [84] and the formula described by Siggaard-Anderson: $(1 - 0.014 \times Hb) \times [HCO_3^- - 24.8 + (1.43 \times Hb + 7.7) \times (pH - 7.4)]$ [85]. Oxygen content (mL/dL) was calculated using the formula $[(1.36 \times Hb (g/dL) \times SaO_2 (\%)/100) + (0.0031 \times pO_2 (mmHg)]$, where 1.36 is the volume of O_2 (mL) bound by a gram of Hb and 0.0031 is the solubility coefficient of O_2 in the blood that represents the volume of O_2 (mL) dissolved in 100 mL of plasma for each mmHg of O_2 partial pressure. Fetal oxygen extraction was determined as the difference between umbilical venous and arterial blood oxygen contents divided by umbilical venous oxygen content. Fetal CO_2 and lactate productions were calculated, respectively, as the difference between arterial blood CO_2 or lactate and venous contents, divided by the respective venous content.

Statistical Analysis

Categorical data were described with the absolute and relative (%) frequency, and continuous data were summarized with the mean and standard deviation. Gestational age was compared with different parameters of interest (umbilical venous and arterial oxygen levels, fetal oxygen extraction, Hb, oxygen content, pH, bicarbonate, base excess, CO_2, and lactate) using regression models to identify the best model between the linear and the quadratic ones. Furthermore, the curves and the equations of the best models were calculated, and the equation of the statistically significant model (linear or quadratic) was indicated at the top of each figure. When the most representative test (with a lower p) was a linear regression, the test revealed that the parameter had a linear trend (decreasing or increasing) with time. Conversely, when the most representative test (with a lower p)

was a quadratic regression, the test revealed that the parameter had a biphasic trend with time: If the quadratic model was increasing, the trend reached a maximum value and was subsequently decreasing, or if the quadratic model was decreasing, the trend reached a minimum value and was subsequently increasing. Therefore, this test suggested when, approximately, the curve reversed.

Once the data were stratified according to weeks of pregnancy (23–26^{+6}, 27–28^{+6}, 29–30^{+6}, 31–33^{+6}, 34–35^{+6}, and 36–36^{+6} weeks of gestation), a nonparametric test without Bonferroni correction was employed to evaluate the differences between the groups (Supplementary Figures). Significance was set at 0.05, and all analyses were performed by SPSS v.28 technology (IBM Corp. Released 2021. IBM SPSS Statistics for Windows, Version 28.0. IBM Corp: Armonk, NY, USA).

5. Conclusions

The present study confirms that oxygenation during fetal life undergoes important fluctuations and integrates the recently published study on oxygenation status during the last weeks of pregnancy [13]. The combination of these studies suggests that from the 23rd week onwards, the fetus becomes progressively more hypoxemic, but this condition reverses around the 33–34th week when oxygen levels progressively increase until the end of the pregnancy.

Supplementary Materials: The following supporting information can be downloaded at: https://www.mdpi.com/article/10.3390/ijms241512487/s1.

Author Contributions: Conceptualization, L.F.; methodology, L.F.; formal analysis, L.F., F.P., F.I., N.C., A.M. and R.M.; investigation, R.T.S.; data curation, L.F., F.P., F.I., N.C. and A.M.; writing—original draft preparation, L.F.; writing—review and editing, A.P., M.C. and P.B.; supervision, L.F., A.P., M.C. and P.B. All authors have read and agreed to the published version of the manuscript.

Funding: This research received no external funding.

Institutional Review Board Statement: The study was conducted in accordance with the Declaration of Helsinki and approved by the Institutional Review Board (or Ethics Committee) of the Pediatric Ethical Committee for Clinical Research of Tuscany region (protocol code 291/2022, date of approval 21 December 2021).

Informed Consent Statement: Informed consent was obtained from all subjects involved in the study.

Data Availability Statement: The datasets generated during and/or analyzed during the current study are available from the corresponding author on reasonable request.

Acknowledgments: We are most grateful to Jean-Luc Baroni and Cristina Ranzato for their support in conducting this study.

Conflicts of Interest: The authors declare no conflict of interest.

References

1. Eastman, N.J. Mount Everest in utero. *Am. J. Obstet. Gynecol.* **1954**, *67*, 701–711. [CrossRef] [PubMed]
2. Jauniaux, E.; Watson, A.; Burton, G. Evaluation of respiratory gases and acid-base gradients in human fetal fluids and uteroplacental tissue between 7 and 16 weeks' gestation. *Am. J. Obstet. Gynecol.* **2001**, *184*, 998–1003. [CrossRef] [PubMed]
3. Burton, G.J.; Cindrova-Davies, T.; Yung, H.W.; Jauniaux, E. Hypoxia and reproductive health: Oxygen and development of the human placenta. *Reproduction* **2021**, *161*, F53–F65. [CrossRef]
4. Soothill, P.W.; Nicolaides, K.H.; Rodeck, C.H.; Campbell, S. Effect of gestational age on fetal and intervillous blood gas and acid-base values in human pregnancy. *Fetal Ther.* **1986**, *1*, 168–175. [CrossRef]
5. Weiner, C.P.; Sipes, S.L.; Wenstrom, K. The effect of fetal age upon normal fetal laboratory values and venous pressure. *Obstet. Gynecol.* **1992**, *79*, 713–718.
6. Pardi, G.; Cetin, I.; Marconi, A.M.; Lanfranchi, A.; Bozzetti, P.; Ferrazzi, E.; Buscaglia, M.; Battaglia, F.C. Diagnostic value of blood sampling in fetuses with growth retardation. *N. Engl. J. Med.* **1993**, *328*, 692–696. [CrossRef]
7. Nava, S.; Bocconi, L.; Zuliani, G.; Kustermann, A.; Nicolini, U. Aspects of fetal physiology from 18 to 37 weeks' gestation as assessed by blood sampling. *Obstet. Gynecol.* **1996**, *87*, 975–980. [CrossRef]

8. Semple, B.D.; Blomgren, K.; Gimlin, K.; Ferriero, D.M.; Noble-Haeusslein, L.J. Brain development in rodents and humans: Identifying benchmarks of maturation and vulnerability to injury across species. *Prog. Neurobiol.* **2013**, *106–107*, 1–16. [CrossRef]
9. Coelho-Santos, V.; Shih, A.Y. Postnatal development of cerebrovascular structure and the neurogliovascular unit. *Wiley Interdiscip. Rev. Dev. Biol.* **2020**, *9*, e363. [CrossRef] [PubMed]
10. Stahl, A.; Connor, K.M.; Sapieha, P.; Chen, J.; Dennison, R.J.; Krah, N.M.; Seaward, M.R.; Willett, K.L.; Aderman, C.M.; Guerin, K.I.; et al. The mouse retina as an angiogenesis model. *Investig. Ophthalmol. Vis. Sci.* **2010**, *51*, 2813–2826. [CrossRef]
11. Aguilar, E.; Dorrell, M.I.; Friedlander, D.; Jacobson, R.A.; Johnson, A.; Marchetti, V.; Moreno, S.K.; Ritter, M.R.; Friedlander, M. Chapter 6. Ocular models of angiogenesis. *Methods Enzymol.* **2008**, *444*, 115–158. [CrossRef] [PubMed]
12. Wright, K.W.; Strube, Y.N. Retinopathy of Prematurity. In *Pediatric Ophthalmology and Strabismus*; Wright, K.W., Strube, Y.N., Eds.; Oxford University Press: New York, NY, USA, 2012; pp. 957–992.
13. Filippi, L.; Scaramuzzo, R.T.; Pascarella, F.; Pini, A.; Morganti, R.; Cammalleri, M.; Bagnoli, P.; Ciantelli, M. Fetal oxygenation in the last weeks of pregnancy evaluated through the umbilical cord blood gas analysis. *Front. Pediatr.* **2023**, *11*, 1140021. [CrossRef] [PubMed]
14. Richardson, B.S.; de Vrijer, B.; Brown, H.K.; Stitt, L.; Choo, S.; Regnault, T.R.H. Gestational age impacts birth to placental weight ratio and umbilical cord oxygen values with implications for the fetal oxygen margin of safety. *Early Hum. Dev.* **2022**, *164*, 105511. [CrossRef] [PubMed]
15. Carter, A.M. Placental Gas Exchange and the Oxygen Supply to the Fetus. *Compr. Physiol.* **2015**, *5*, 1381–1403. [CrossRef] [PubMed]
16. Fathollahipour, S.; Patil, P.S.; Leipzig, N.D. Oxygen Regulation in Development: Lessons from Embryogenesis towards Tissue Engineering. *Cells Tissues Organs* **2018**, *205*, 350–371. [CrossRef]
17. Houghton, F.D. Hypoxia and reproductive health: Hypoxic regulation of preimplantation embryos: Lessons from human embryonic stem cells. *Reproduction* **2021**, *161*, F41–F51. [CrossRef]
18. Cowden Dahl, K.D.; Fryer, B.H.; Mack, F.A.; Compernolle, V.; Maltepe, E.; Adelman, D.M.; Carmeliet, P.; Simon, M.C. Hypoxia-inducible factors 1alpha and 2alpha regulate trophoblast differentiation. *Mol. Cell. Biol.* **2005**, *25*, 10479–10491. [CrossRef]
19. Pringle, K.G.; Kind, K.L.; Sferruzzi-Perri, A.N.; Thompson, J.G.; Roberts, C.T. Beyond oxygen: Complex regulation and activity of hypoxia inducible factors in pregnancy. *Hum. Reprod. Update* **2010**, *16*, 415–431. [CrossRef]
20. Dengler, V.L.; Galbraith, M.; Espinosa, J.M. Transcriptional regulation by hypoxia inducible factors. *Crit. Rev. Biochem. Mol. Biol.* **2014**, *49*, 1–15. [CrossRef]
21. Genbacev, O.; Zhou, Y.; Ludlow, J.W.; Fisher, S.J. Regulation of human placental development by oxygen tension. *Science* **1997**, *277*, 1669–1672. [CrossRef]
22. Aplin, J.D. Hypoxia and human placental development. *J. Clin. Investig.* **2000**, *105*, 559–560. [CrossRef]
23. Patel, J.; Landers, K.; Mortimer, R.H.; Richard, K. Regulation of hypoxia inducible factors (HIF) in hypoxia and normoxia during placental development. *Placenta.* **2010**, *31*, 951–957. [CrossRef] [PubMed]
24. Dunwoodie, S.L. The role of hypoxia in development of the Mammalian embryo. *Dev. Cell.* **2009**, *17*, 755–773. [CrossRef] [PubMed]
25. Fischer, B.; Bavister, B.D. Oxygen tension in the oviduct and uterus of rhesus monkeys, hamsters and rabbits. *J. Reprod. Fertil.* **1993**, *99*, 673–679. [CrossRef] [PubMed]
26. Yedwab, G.A.; Paz, G.; Homonnai, T.Z.; David, M.P.; Kraicer, P.F. The temperature, pH, and partial pressure of oxygen in the cervix and uterus of woman and uterus of rats during the cycle. *Fertil. Steril.* **1976**, *27*, 304–309. [CrossRef]
27. Ottosen, L.D.; Hindkaer, J.; Husth, M.; Petersen, D.E.; Kirk, J.; Ingerslev, H.J. Observations on intrauterine oxygen tension measured by fibre-optic microsensors. *Reprod. Biomed. Online* **2006**, *13*, 380–385. [CrossRef]
28. Zhao, H.; Wong, R.J.; Stevenson, D.K. The Impact of Hypoxia in Early Pregnancy on Placental Cells. *Int. J. Mol. Sci.* **2021**, *22*, 9675. [CrossRef]
29. Forristal, C.E.; Wright, K.L.; Hanley, N.A.; Oreffo, R.O.; Houghton, F.D. Hypoxia inducible factors regulate pluripotency and proliferation in human embryonic stem cells cultured at reduced oxygen tensions. *Reproduction* **2010**, *139*, 85–97. [CrossRef]
30. Abdollahi, H.; Harris, L.J.; Zhang, P.; McIlhenny, S.; Srinivas, V.; Tulenko, T.; DiMuzio, P.J. The role of hypoxia in stem cell differentiation and therapeutics. *J. Surg. Res.* **2011**, *165*, 112–117. [CrossRef]
31. Leese, H.J. Metabolic control during preimplantation mammalian development. *Hum. Reprod. Update* **1995**, *1*, 63–72. [CrossRef]
32. Thompson, J.G.; Partridge, R.J.; Houghton, F.D.; Cox, C.I.; Leese, H.J. Oxygen uptake and carbohydrate metabolism by in vitro derived bovine embryos. *J. Reprod. Fertil.* **1996**, *106*, 299–306. [CrossRef] [PubMed]
33. Vaupel, P.; Multhoff, G. Revisiting the Warburg effect: Historical dogma versus current understanding. *J. Physiol.* **2021**, *599*, 1745–1757. [CrossRef] [PubMed]
34. Krisher, R.L.; Prather, R.S. A role for the Warburg effect in preimplantation embryo development: Metabolic modification to support rapid cell proliferation. *Mol. Reprod. Dev.* **2012**, *79*, 311–320. [CrossRef]
35. Ma, L.N.; Huang, X.B.; Muyayalo, K.P.; Mor, G.; Liao, A.H. Lactic Acid: A Novel Signaling Molecule in Early Pregnancy? *Front. Immunol.* **2020**, *11*, 279. [CrossRef]
36. Adelman, D.M.; Gertsenstein, M.; Nagy, A.; Simon, M.C.; Maltepe, E. Placental cell fates are regulated in vivo by HIF-mediated hypoxia responses. *Genes Dev.* **2000**, *14*, 3191–3203. [CrossRef]

37. Robins, J.C.; Heizer, A.; Hardiman, A.; Hubert, M.; Handwerger, S. Oxygen tension directs the differentiation pathway of human cytotrophoblast cells. *Placenta* **2007**, *28*, 1141–1146. [CrossRef] [PubMed]
38. Krock, B.L.; Skuli, N.; Simon, M.C. Hypoxia-induced angiogenesis: Good and evil. *Genes Cancer* **2011**, *2*, 1117–1133. [CrossRef]
39. Chang, C.W.; Wakeland, A.K.; Parast, M.M. Trophoblast lineage specification, differentiation and their regulation by oxygen tension. *J. Endocrinol.* **2018**, *236*, R43–R56. [CrossRef]
40. Meschia, G.; Cotter, J.R.; Breathnach, C.S.; Barron, D.H. The hemoglobin, oxygen, carbon dioxide and hydrogen ion concentrations in the umbilical bloods of sheep and goats as sampled via indwelling plastic catheters. *Q. J. Exp. Physiol. Cogn. Med. Sci.* **1965**, *50*, 185–195. [CrossRef]
41. Comline, R.S.; Silver, M. Daily changes in foetal and maternal blood of conscious pregnant ewes, with catheters in umbilical and uterine vessels. *J. Physiol.* **1970**, *209*, 567–586. [CrossRef]
42. Dawes, G.S.; Fox, H.E.; Leduc, B.M.; Liggins, G.C.; Richards, R.T. Respiratory movements and rapid eye movement sleep in the foetal lamb. *J. Physiol.* **1972**, *220*, 119–143. [CrossRef] [PubMed]
43. Clewlow, F.; Dawes, G.S.; Johnston, B.M.; Walker, D.W. Changes in breathing, electrocortical and muscle activity in unanaesthetized fetal lambs with age. *J. Physiol.* **1983**, *341*, 463–476. [CrossRef] [PubMed]
44. Fowden, A.L.; Taylor, P.M.; White, K.L.; Forhead, A.J. Ontogenic and nutritionally induced changes in fetal metabolism in the horse. *J. Physiol.* **2000**, *528*, 209–219. [CrossRef] [PubMed]
45. Giussani, D.A.; Forhead, A.J.; Fowden, A.L. Development of cardiovascular function in the horse fetus. *J. Physiol.* **2005**, *565*, 1019–1030. [CrossRef] [PubMed]
46. Yamaleyeva, L.M.; Sun, Y.; Bledsoe, T.; Hoke, A.; Gurley, S.B.; Brosnihan, K.B. Photoacoustic imaging for in vivo quantification of placental oxygenation in mice. *FASEB J.* **2017**, *31*, 5520–5529. [CrossRef]
47. Basak, K.; Luís Deán-Ben, X.; Gottschalk, S.; Reiss, M.; Razansky, D. Non-invasive determination of murine placental and foetal functional parameters with multispectral optoacoustic tomography. *Light Sci. Appl.* **2019**, *8*, 71. [CrossRef]
48. Saini, B.; Morrison, J.; Seed, M. Gas Exchange across the Placenta. In *Respiratory Disease in Pregnancy*; Lapinsky, S., Plante, L., Eds.; Cambridge University Press: Cambridge, UK, 2020; pp. 34–56.
49. Schneider, H. Oxygenation of the placental-fetal unit in humans. *Respir. Physiol. Neurobiol.* **2011**, *178*, 51–58. [CrossRef]
50. Templeton, A.; Kelman, G.R. Maternal blood-gases, (PAo2-Pao2), physiological shunt and VD/VT in normal pregnancy. *Br. J. Anaesth.* **1976**, *48*, 1001–1004. [CrossRef]
51. Assali, N.S.; Rauramo, L.; Peltonen, T. Measurement of uterine blood flow and uterine metabolism. VIII. Uterine and fetal blood flow and oxygen consumption in early human pregnancy. *Am. J. Obstet. Gynecol.* **1960**, *79*, 86–98. [CrossRef]
52. Varma, S.; Cohen, H.J. Co-transactivation of the 3′ erythropoietin hypoxia inducible enhancer by the HIF-1 protein. *Blood Cells Mol. Dis.* **1997**, *23*, 169–176. [CrossRef]
53. Jopling, J.; Henry, E.; Wiedmeier, S.E.; Christensen, R.D. Reference ranges for hematocrit and blood hemoglobin concentration during the neonatal period: Data from a multihospital health care system. *Pediatrics* **2009**, *123*, e333–e337. [CrossRef]
54. Sankaran, V.G.; Xu, J.; Orkin, S.H. Advances in the understanding of haemoglobin switching. *Br. J. Haematol.* **2010**, *149*, 181–194. [CrossRef]
55. Shahbazi, M.N.; Jedrusik, A.; Vuoristo, S.; Recher, G.; Hupalowska, A.; Bolton, V.; Fogarty, N.N.M.; Campbell, A.; Devito, L.; Ilic, D.; et al. Self-organization of the human embryo in the absence of maternal tissues. *Nat. Cell Biol.* **2016**, *18*, 700–708. [CrossRef] [PubMed]
56. Simon, M.C.; Keith, B. The role of oxygen availability in embryonic development and stem cell function. *Nat. Rev. Mol. Cell. Biol.* **2008**, *9*, 285–296. [CrossRef] [PubMed]
57. Di Mattia, M.; Mauro, A.; Citeroni, M.R.; Dufrusine, B.; Peserico, A.; Russo, V.; Berardinelli, P.; Dainese, E.; Cimini, A.; Barboni, B. Insight into Hypoxia Stemness Control. *Cells* **2021**, *10*, 2161. [CrossRef]
58. Semenza, G.L. Dynamic regulation of stem cell specification and maintenance by hypoxia-inducible factors. *Mol. Aspects Med.* **2016**, *47–48*, 15–23. [CrossRef]
59. Morrison, S.J.; Csete, M.; Groves, A.K.; Melega, W.; Wold, B.; Anderson, D.J. Culture in reduced levels of oxygen promotes clonogenic sympathoadrenal differentiation by isolated neural crest stem cells. *J. Neurosci.* **2000**, *20*, 7370–7376. [CrossRef] [PubMed]
60. Studer, L.; Csete, M.; Lee, S.H.; Kabbani, N.; Walikonis, J.; Wold, B.; McKay, R. Enhanced proliferation, survival, and dopaminergic differentiation of CNS precursors in lowered oxygen. *J. Neurosci.* **2000**, *20*, 7377–7383. [CrossRef] [PubMed]
61. Zhang, Y.; Porat, R.M.; Alon, T.; Keshet, E.; Stone, J. Tissue oxygen levels control astrocyte movement and differentiation in developing retina. *Brain Res. Dev. Brain Res.* **1999**, *118*, 135–145. [CrossRef] [PubMed]
62. Nakamura-Ishizu, A.; Kurihara, T.; Okuno, Y.; Ozawa, Y.; Kishi, K.; Goda, N.; Tsubota, K.; Okano, H.; Suda, T.; Kubota, Y. The formation of an angiogenic astrocyte template is regulated by the neuroretina in a HIF-1-dependent manner. *Dev. Biol.* **2012**, *363*, 106–114. [CrossRef]
63. Gao, L.; Chen, X.; Zeng, Y.; Li, Q.; Zou, T.; Chen, S.; Wu, Q.; Fu, C.; Xu, H.; Yin, Z.Q. Intermittent high oxygen influences the formation of neural retinal tissue from human embryonic stem cells. *Sci. Rep.* **2016**, *6*, 29944. [CrossRef]
64. Heinis, M.; Simon, M.T.; Ilc, K.; Mazure, N.M.; Pouysségur, J.; Scharfmann, R.; Duvillié, B. Oxygen tension regulates pancreatic beta-cell differentiation through hypoxia-inducible factor 1alpha. *Diabetes* **2010**, *59*, 662–669. [CrossRef] [PubMed]

65. Ngo, M.A.; Sinitsyna, N.N.; Qin, Q.; Rice, R.H. Oxygen-dependent differentiation of human keratinocytes. *J. Investig. Dermatol.* **2007**, *127*, 354–361. [CrossRef] [PubMed]

66. van Wenum, M.; Adam, A.A.A.; van der Mark, V.A.; Chang, J.C.; Wildenberg, M.E.; Hendriks, E.J.; Jongejan, A.; Moerland, P.D.; van Gulik, T.M.; Oude Elferink, R.P.; et al. Oxygen drives hepatocyte differentiation and phenotype stability in liver cell lines. *J. Cell Commun. Signal.* **2018**, *12*, 575–588. [CrossRef]

67. Berthelemy, N.; Kerdjoudj, H.; Schaaf, P.; Prin-Mathieu, C.; Lacolley, P.; Stoltz, J.F.; Voegel, J.C.; Menu, P. O_2 level controls hematopoietic circulating progenitor cells differentiation into endothelial or smooth muscle cells. *PLoS ONE* **2009**, *4*, e5514. [CrossRef] [PubMed]

68. Baker, C.D.; Ryan, S.L.; Ingram, D.A.; Seedorf, G.J.; Abman, S.H.; Balasubramaniam, V. Endothelial colony-forming cells from preterm infants are increased and more susceptible to hyperoxia. *Am. J. Respir. Crit. Care Med.* **2009**, *180*, 454–461. [CrossRef]

69. Wisgrill, L.; Schüller, S.; Bammer, M.; Berger, A.; Pollak, A.; Radke, T.F.; Kögler, G.; Spittler, A.; Helmer, H.; Husslein, P.; et al. Hematopoietic stem cells in neonates: Any differences between very preterm and term neonates? *PLoS ONE* **2014**, *9*, e106717. [CrossRef] [PubMed]

70. Javed, M.J.; Mead, L.E.; Prater, D.; Bessler, W.K.; Foster, D.; Case, J.; Goebel, W.S.; Yoder, M.C.; Haneline, L.S.; Ingram, D.A. Endothelial colony forming cells and mesenchymal stem cells are enriched at different gestational ages in human umbilical cord blood. *Pediatr. Res.* **2008**, *64*, 68–73. [CrossRef]

71. Borghesi, A.; Massa, M.; Campanelli, R.; Bollani, L.; Tzialla, C.; Figar, T.A.; Ferrari, G.; Bonetti, E.; Chiesa, G.; de Silvestri, A.; et al. Circulating endothelial progenitor cells in preterm infants with bronchopulmonary dysplasia. *Am. J. Respir. Crit. Care Med.* **2009**, *180*, 540–546. [CrossRef]

72. Cammalleri, M.; Amato, R.; Dal Monte, M.; Filippi, L.; Bagnoli, P. The β3 adrenoceptor in proliferative retinopathies: "Cinderella" steps out of its family shadow. *Pharmacol. Res.* **2023**, *190*, 106713. [CrossRef]

73. Puebla, M.; Tapia, P.J.; Espinoza, H. Key Role of Astrocytes in Postnatal Brain and Retinal Angiogenesis. *Int. J. Mol. Sci.* **2022**, *23*, 2646. [CrossRef] [PubMed]

74. Ballabh, P. Pathogenesis and prevention of intraventricular hemorrhage. *Clin. Perinatol.* **2014**, *41*, 47–67. [CrossRef] [PubMed]

75. Spencer, B.L.; Mychaliska, G.B. Milestones for clinical translation of the artificial placenta. *Semin. Fetal Neonatal Med.* **2022**, *27*, 101408. [CrossRef] [PubMed]

76. Filippi, L.; Pini, A.; Cammalleri, M.; Bagnoli, P.; Dal Monte, M. β3-Adrenoceptor, a novel player in the round-trip from neonatal diseases to cancer: Suggestive clues from embryo. *Med. Res. Rev.* **2022**, *42*, 1179–1201. [CrossRef]

77. Armstrong, L.; Stenson, B.J. Use of umbilical cord blood gas analysis in the assessment of the newborn. *Arch. Dis. Child. Fetal Neonatal Ed.* **2007**, *92*, F430–F434. [CrossRef]

78. Brotanek, V.; Hendricks, C.H.; Yoshida, T. Changes in uterine blood flow during uterine contractions. *Am. J. Obstet. Gynecol.* **1969**, *103*, 1108–1116. [CrossRef]

79. Szymanowski, P.; Szepieniec, W.K.; Zarawski, M.; Gruszecki, P.; Szweda, H.; Jóźwik, M. The impact of birth anesthesia on the parameters of oxygenation and acid-base balance in umbilical cord blood. *J. Matern. Fetal Neonatal Med.* **2020**, *33*, 3445–3452. [CrossRef]

80. Raghuraman, N.; Temming, L.A.; Doering, M.M.; Stoll, C.R.; Palanisamy, A.; Stout, M.J.; Colditz, G.A.; Cahill, A.G.; Tuuli, M.G. Maternal Oxygen Supplementation Compared with Room Air for Intrauterine Resuscitation: A Systematic Review and Meta-analysis. *JAMA Pediatr.* **2021**, *175*, 368–376. [CrossRef]

81. Ayres-de-Campos, D.; Arulkumaran, S.; FIGO Intrapartum Fetal Monitoring Expert Consensus Panel. FIGO consensus guidelines on intrapartum fetal monitoring: Physiology of fetal oxygenation and the main goals of intrapartum fetal monitoring. *Int. J. Gynaecol. Obstet.* **2015**, *131*, 5–8. [CrossRef]

82. D'Alton, M.E.; Hankins, G.D.V.; Berkowitz, R.L.; Bienstock, J.; Ghidini, A.; Goldsmith, J.; Higgins, R.; Moore, T.R.; Natale, R.; Nelson, K.B.; et al. Executive summary: Neonatal encephalopathy and neurologic outcome, second edition. Report of the American College of Obstetricians and Gynecologists' Task Force on Neonatal Encephalopathy. *Obstet. Gynecol.* **2014**, *123*, 896–901. [CrossRef]

83. Westgate, J.; Garibaldi, J.M.; Greene, K.R. Umbilical cord blood gas analysis at delivery: A time for quality data. *Br. J. Obstet. Gynaecol.* **1994**, *101*, 1054–1063. [CrossRef] [PubMed]

84. Ramsay, A.G. Clinical application of the Henderson-Hasselbalch equation. *Appl. Ther.* **1965**, *7*, 730–736. [PubMed]

85. Siggaard-Andersen, O.; Engel, K. A new acid-base nomogram. An improved method for the calculation of the relevant blood acid-base data. *Scand. J. Clin. Lab. Investig.* **1960**, *12*, 177–186. [CrossRef] [PubMed]

International Journal of
Molecular Sciences

Article

Augmented Placental Protein 13 in Placental-Associated Extracellular Vesicles in Term and Preterm Preeclampsia Is Further Elevated by Corticosteroids

Marina Marks Kazatsker [1], Adi Sharabi-Nov [2,3], Hamutal Meiri [4,5], Rami Sammour [1] and Marei Sammar [6,*]

1   Maternal and Fetal Medicine Unit, Department of Obstetrics and Gynecology, Bnai-Zion University Medical Center, Haifa 3498838, Israel; mkazatsker@gmail.com (M.M.K.); rsammour2002@gmail.com (R.S.)
2   Department of Statistics, Tel Hai Academic College, Tel Hai 122103, Israel; adi_nov@hotmail.com
3   Department of Statistics, Ziv Medical Center, Safed 131000, Israel
4   Hy-Laboratories Ltd., Rehovot 7670606, Israel; hamutal62@hotmail.com
5   TeleMarpe Ltd., 21 Beit El St., Tel Aviv 6908742, Israel
6   Prof. Ephraim Katzir Department of Biotechnology Engineering, Braude Academic College of Engineering, 51 Snunit St., Karmiel 2161002, Israel
*   Correspondence: sammar@braude.ac.il; Tel.: +972-(4)-9901769; Fax: +972-(4)-99017

Abstract: Placental protein 13 (PP13) is a regulatory protein involved in remodeling the vascular system of the pregnancy and extending the immune tolerance of the mother to the growing fetus. PP13 is localized on the surface of the syncytiotrophoblast. An ex vivo placental model shows that the PP13 is released via placental-associated extracellular vesicles (PEVs) to the maternal uterine vein. This exploratory study aimed to determine PEV-associated PP13 in the maternal circulation as compared to the known soluble fraction since each has a specific communication pathway. Patients admitted to Bnai Zion Medical Center for delivery were recruited, and included 19 preeclampsia (PE) patients (7 preterm PE gestational age < 37 weeks' gestation), 16 preterm delivery (PTD, delivery at GA < 37 weeks' gestation), and 15 matched term delivery controls. Treatment by corticosteroids (Celestone), which is often given to patients with suspected preterm PE and PTD, was recorded. The PEV proteome was purified from the patients' plasma by size exclusion chromatography (SEC) to separate the soluble and PEV-associated PP13. The total level of PP13 (soluble and PEV-associated) was determined using mild detergent that depleted the PEV proteome. PP13 fractions were determined by ELISA with PP13 specific antibodies. ELISA with alkaline phosphatase (PLAP)- and cluster differentiation 63 (CD63)-specific antibodies served to verify the placental origin of the PEVs. SPSS was used for statistical analysis. The patients' medical, pregnancy, and delivery records in all groups were similar except, as expected, that a larger number of PE and PTD patients had smaller babies who were delivered earlier, and the PE patients had hypertension and proteinuria. The SEC analysis detected the presence of PP13 in the cargo of the PEVs and on their surface, in addition to the known soluble fraction. The median soluble PP13 was not significantly different across the PE, PTD, and term delivery control groups. However, after depleting the PEV of their proteome, the total PP13 (soluble and PEV-associated) was augmented in the cases of preterm PE, reaching 2153 pg/mL [IQR 1866–2838] but not in cases of PTD reaching 1576 pg/mL [1011–2014] or term delivery groups reaching 964 pg/mL [875–1636]), $p < 0.01$. On the surface of the circulating PEV from PTD patients, there was a decrease in PP13. Corticosteroid treatment was accompanied by a massive depletion of PP13 from the PEV, especially in preterm PE patients. This exploratory study is, thus, the first to determine PEV-associated PP13 in maternal circulation, providing a quantitative determination of the soluble and the PEV-associated fractions, and it shows that the latter is the larger. We found an increase in the amount of PP13 carried via the PEV-associated pathway in PE and PTD patients compared to term delivery cases, which was further augmented when the patients were treated with corticosteroids, especially in preterm PE. The signal conveyed by this novel communication pathway warrants further research to investigate these two differential pathways for the liberation of PP13.

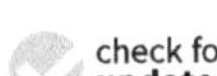
check for
updates

Citation: Kazatsker, M.M.; Sharabi-Nov, A.; Meiri, H.; Sammour, R.; Sammar, M. Augmented Placental Protein 13 in Placental-Associated Extracellular Vesicles in Term and Preterm Preeclampsia Is Further Elevated by Corticosteroids. *Int. J. Mol. Sci.* **2023**, *24*, 12051. https://doi.org/10.3390/ijms241512051

Academic Editor: Udo Jeschke

Received: 29 May 2023
Revised: 24 July 2023
Accepted: 25 July 2023
Published: 27 July 2023

Keywords: extracellular vesicles; placental protein 13 (PP13); galectin 13; preeclampsia (PE); preterm delivery (PTD); size exclusion chromatography (SEC); corticosteroids

1. Introduction

The interphase between the placenta and the maternal circulation has been extensively investigated in normal pregnancy and in pregnancy complications to obtain greater insights into the processes of placentation, fetal growth, maternal body adaptation to the pregnancy, and signal transmission between the pregnancy and the maternal organs [1,2]. Placental-specific galectins [1,3] and galectin 13, or placental protein 13 (PP13) in particular, are major regulatory proteins released by the placenta that communicate signals to the maternal organs about the well-being of the pregnancy. PP13 is specifically expressed by the syncytiotrophoblast and is first detected at GA 5–6 in the soluble maternal circulation. It is then gradually increased during gestation along with the increase in placental size (3–5). Studies have suggested that in preeclampsia (PE), and mainly in preterm PE requiring delivery before gestational week 37 (GA < 37 weeks' gestation), there is a lower level of maternal blood PP13. Accordingly, the determination of PP13 in the first trimester was used as a marker for predicting the subsequent development of PE later in pregnancy. Near delivery with PE, there is a shedding of PP13 from the syncytiotrophoblast and the maternal blood circulation shows an increase in the level of PP13, at least in some studies [3–5].

Animal studies have found that PP13 expands the uterine and vascular arteries and veins, thus, priming the pregnancy for the required increase in the supply of oxygen and nutrients to the pregnancy [4,5]. PP13 has also been shown to be responsible for the mother's immune system tolerance to the growing fetus [3,4]. Hence, PP13 is likely to play an important role in the fetal–maternal interactions during the course of normal and complicated pregnancies [3,5–7].

In the last ten years, studies have shown that extracellular vesicles (EVs) constitute the main communication vehicles between activated cell types, hormonal glands, tumors, fetal and placental cells, and between the immune system and remote organs [8,9]. These nanometer-sized membrane vesicles (exosomes and microvesicles) are released by the surface of certain cells and organs to convey signals (including possible complications at their site of origin) to remote organs. Their small size enables them to pass through small capillaries to enter the circulation, thus, transporting their cargo (RNA, proteins) and surface markers from their place of origin to remote organs. The study of EVs has, thus, contributed to major advances in the clinical management of diseases such as cancer, Alzheimer's, COVID-19 vaccines, and central nervous system diseases. Advanced purification and subsequent size and surface characterization of EVs have led to the development of the current nomenclature of their size and surface shape [7–9].

This exploratory study focused on the evaluation of the communication pathway of the release of placental EVs (PEVs) that carry regulatory proteins, such as PP13, into the maternal circulation. Unlike our previous study, where we determined PP13 levels on the surface and in the internal cargo of PEVs collected after the ex vivo perfusion of the isolated placenta [6], here we focused on PEVs in the maternal circulation [7–9]. We compared the PP13 of the PEV in PE and preterm delivery (PTD) to term delivery.

PE is a major hypertensive disorder of pregnancy (affecting 2–8% of all pregnancies) [10,11]. Advances in the past decade have enabled better prediction of PE, especially preterm PE (PE requiring delivery before 37 week gestation) by first-trimester biomarkers and its prevention by low-dose aspirin [12,13]. Recent studies described the use of pro- and anti-angiogenic factors in identifying women at risk for developing PE in the third trimester [14]. Currently, most biomarkers of these complications are mainly determined using serum, whereas a plasma evaluation is less often conducted.

The current study focused on exploring the PEV-associated PP13, and determining changes in its maternal circulation level in PE, thus, providing an innovative diagnostic approach to the prediction of PE. In addition, we also evaluated the PEV-associated PP13 in cases of preterm delivery (PTD) [15,16], as a control for the preterm PE cases.

When we collected the data, we entered information on treatment with corticosteroids routinely used in the clinical management of PTD and preterm PE. The impact of corticosteroids was previously shown to affect the level of various regulatory proteins. Corticosteroids are mainly used to facilitate the maturation of the fetal respiratory system when preterm birth is suspected. There are reports that indicate various side effects following the use of corticosteroids [17–22]. We have previously reported that following corticosteroids' treatment, there is a transient surge of maternal blood PP13 at different GAs, but the topic was not systematically followed [23,24].

Overall, this study had three goals: (1) identify the relative fractions of soluble and PEV-associated PP13 in the maternal circulation; (2) analyze the differential changes of the above in PE and PTD compared to term delivery controls; and (3) estimate the impact of corticosteroids on the levels of PP13 in each fraction.

2. Results

2.1. Cohort Characterization

We enrolled pregnant women attending the delivery clinic at Bnai Zion Medical Center (BZMC) with suspected PE and PTD compared to term delivery controls. The cohort included 19 PE cases (7 cases of preterm PE (delivery at GA < 37 weeks' gestation), of whom 3 were delivered at GA 34 and below), and 16 cases of preterm delivery (all delivered at GA < 37 weeks' gestation) of whom 3 were delivered before 34 weeks' gestation compared to 15 cases of term delivery controls (Figure 1).

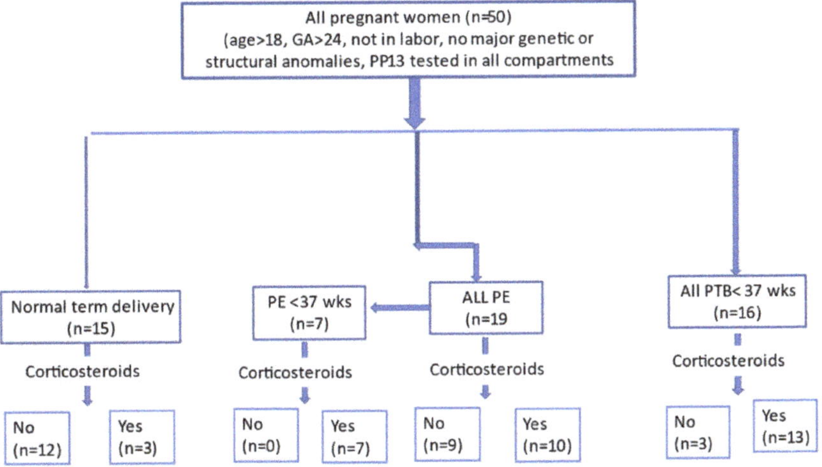

Figure 1. Study flow chart. We enrolled 50 patients—19 cases of PE, of whom 7 delivered at GA < 37 weeks' gestation due to PE severity (PE < 37 wks), 16 PTD delivered at GA < 37 weeks' gestation (PTD < 37 wks) unrelated to PE or other iatrogenic causes, and 15 term delivery controls. The number of women treated and untreated with corticosteroids (+/−) is shown.

As shown in Table 1, the patients in the PE and PTD groups have a higher frequency of conception by in vitro fertilization (IVF), they deliver earlier, their babies are smaller, and a larger proportion of these newborns remain in the newborn intensive care unit (NICU) for at least one week. As expected, patients in the PE group have hypertension and proteinuria, but there are no other significant differences between the groups.

Table 1. Characterization of the study population—pregnancy and delivery features.

Parameter	Term Control Delivery at GA> 37 wks (n = 15)	All PE (n = 19)	Preterm PE Delivery < 37 wks (n = 7)	PTD Delivery < 37 wks (n = 16)	*p*
Enrolment					
Maternal age (years)	30.0 [27.0–33.0]	32.0 [30.0–36.0]	32.0 [30.0–36.0]	30.5 [28.0–35.0]	0.244
BMI (kg/h^2)	25.0 [22.2–32.3]	32.0 [26.0–36.0]	27.0 [22.0–34.0]	27.0 [26.0–29.0]	0.449
BMI > 30 n (%)	4 (33.3)	11 (57.9)	2 (28.6)	2 (13.3)	0.189
Smoker n (%)	4 (26.7)	10 (52.6)	3 (42.9)	9 (56.3)	0.592
Previous PE n (%)	0 (0)	5 (26.3)	4 (57.1)	5 (31.3)	0.151
Null parity n (%)	2 (13.3)	0 (0)	0 (0)	0 (0)	0.247
Conception, IVF n (%)	0 (0)	1 (5.3)	0 (0)	0 (0)	0.798
Previous GDM n (%)	2 (13.3)	2 (10.5)	0 (0)	2 (12.5)	0.765
Chronic hypertension 0, (%)	0 (0)	3 (15.8)	1 (14.3)	0 (0)	0.301
Diabetes mellitus 0 (%)	2 (13.3)	1 (5.3)	0 (0)	0 (0)	0.503
Systolic BP (mm Hg)	109 [b] [105–119]	162 [a] [150–180]	180 [a] [133–195]	123 [b] [115–128]	**<0.001**
Diastolic BP (mm Hg)	71 [b] [69–75]	100 [a] [90–107]	107 [a] [82–110]	77 [b] [73–80]	<0.001
MAP (mm Hg)	84 [b] [81–90]	118 [a] [108–133]	133 [a] [99–136]	91 [b] [88–96]	<0.001
Creatinine (mg/dL)	0.50 [b] [0.4–0.6]	0.70 [a] [0.6–0.8]	0.8 [a] [0.6–0.8]	0.5 [b] [0.5–0.6]	0.004
Aspirin n (%)	0 (0)	3 (17.6)	2 (33.3)	1 (6.3)	0.020
Progesterone [#] n (%)	3 (20.0)	0 (0)	0 (0)	1 (6.3)	0.115
Delivery					
GA at delivery (wks)	39.1 [a] [38.7–39.9]	37.1 [b] [35.0–37.4]	34.3 [c] [32.9–35.6]	36.2 [b] [34.7–36.6]	<0.001
Infant's birthweight (gr)	3290 [a] [2840–3620]	2730 [b] [2180–3125]	2100 [c] [1830–2240]	2498 [b] [2235–3068]	<0.001
Infant gender (male) n (%)	10 (66.7)	10 (55.6)	5 (71.4)	16 (100)	0.036
NICU days	39.1 [a] [38.7–39.9]	37.1 [b] [35.0–37.4]	34.3 [c] [32.9–35.6]	36.2 [b] [34.7–36.6]	<0.001

Continuous variables are shown as medians and the interquartile range [IQR], and categorical variables are shown as frequencies—n, and percentages (%). [#] Progesterone was given for a short cervix. The letter letters "a" to "c" represent significant differences between the groups' medians using the Kruskal–Wallis non-parametric test. The letter "a" is significantly higher, "b" is significantly lower than "a", and "c" is lower than all. PE—preeclampsia, PTD—preterm delivery. BMI—body mass index, MAP—mean arterial blood pressure. IVF—in vitro fertilization, NICU—newborn intensive care unit, GA—gestational week, BP—blood pressure, GDM- gestational diabetes mellitous, NICU—newborn intensive care unit.

2.2. Quantitative ELISA Analysis of PP13 in Different Maternal Blood Fractions

In this exploratory study, we used size exclusion chromatography (SEC) to separate the soluble and PEV-associated PP13. The placental origin of the PEV was verified using the specific placental marker placental-associated alkaline phosphatase (PLAP). The identity of the EVs was further verified by cluster differentiation 63 (CD63). Mild detergent was used to deplete the PEVs from their PP13 cargo that was released to the plasma, thus, creating the total PP13 (PEVs associated and soluble combined). The PP13 on the surface of the PEVs was determined using small columns and SEC without detergent, a procedure that kept the PEVs intact to determine the surface PP13. We used ELISA to quantify the level of PP13 in each fraction (Supplementary Figure S1).

As shown in Figure 2 and Table 2, the main finding is the significant increase in the total PP13 level compared to the soluble level. In the term delivery controls, the level of total PP13 is 964 pg/mL [IQR 875–1636], which is 2.5 times higher than the level of soluble PP13 fraction. In the group of All PE, the level of total PP13 is 1598 pg/mL [1070–1981], and it further augments to 2153 pg/mL [1866–2938, *p* < 0.01] in the preterm PE cases. The latter is higher than the total PP13 level in the PTD group, which is 1576 pg/mL [1011–2014].

Since the soluble levels of preterm PE cases compared to the PTD and term deliver control cases are not significantly different, the increase in the level of total PP13 in the preterm PE group is attributed to the increase in the level of PP13 in the PEV-associated PP13 fraction, which is what the PP13-associated fraction level shows (Table 3).

In the three cases of early PE (cases requiring delivery at GA < 34 weeks' gestation, not shown), the level increases from 455 pg/mL [421–861] in the soluble fraction to 1981 pg/mL [1866–2838] in the total PP13 fraction. The increase in the total PP13 in the early PTD (cases delivered at GA < 34 weeks' gestation) is only twice the soluble level. However, neither is significant (Table 2).

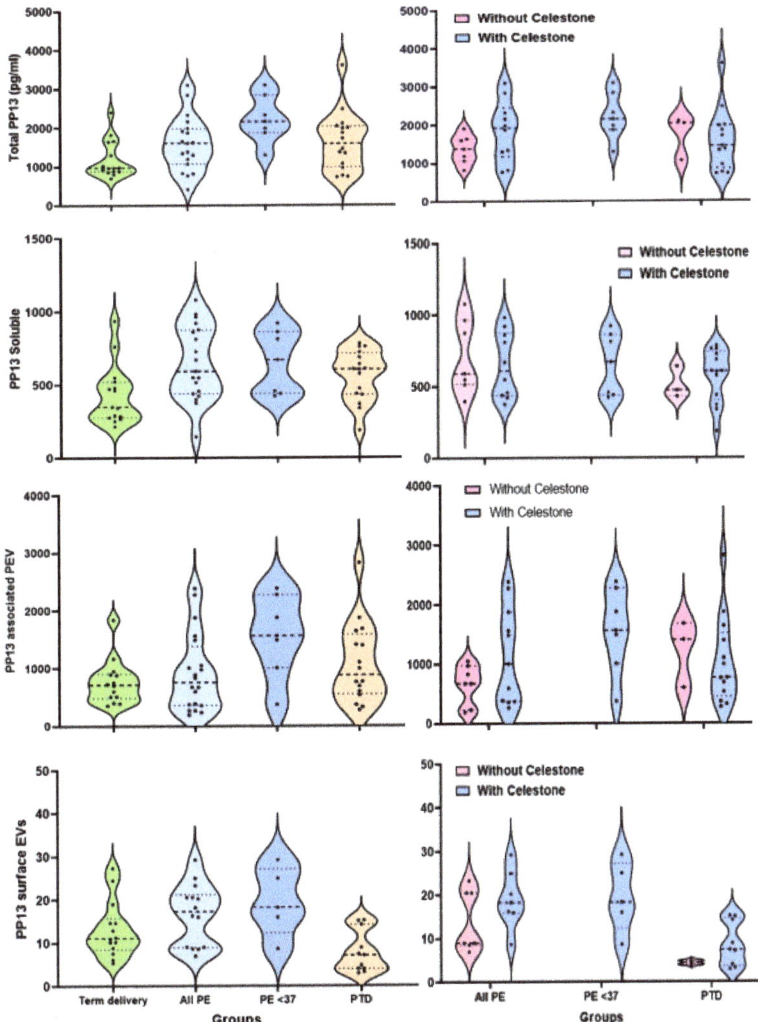

Figure 2. Violin plot of PP13 levels as a function of the different fractions. The four panels on the right side are the violin plots of the level of PP13 in the total fraction (top), the solubilized fraction (below), the PEV-associated fraction (below), while the very bottom plot is the PP13 of the surface of PEV. The four panels on the left side are the violin plots of the same PP13 fractions for the groups of patients treated with (pink) and without (blue) corticosteroids. In each violin plot, the horizontal dashed line represents the lower quartile, the median, and the third quartile. Dots along the violin show different patients' values within the violin. The longitudinal lines on both violin apexes represent the minimum and the maximum values. PE—preeclampsia, PTD—preterm delivery, PE < 37—preterm PE, PEV—placental-associated extracellular vesicles.

The results show that the cargo of PP13 in the PEV-associated PP13 is 699 pg/mL [IQR 511–891] in the term delivery controls compared to 830 pg/mL [355–1485] in all cases of PE, and 877 pg/mL [564–1519] in the PTD group (Table 2). In preterm PE, the PRV-associated cargo is the highest compared to the term delivery and the PTD groups. Accordingly, the PEV-associated pathway appears to be a major route for PP13 liberation to the maternal circulation in PE (Table 2). The PP13 level on the PEV surface is higher in PE, especially in preterm PE, whereas in the PTD group, the level decreases (Table 2, Figure 2).

Table 2. Placenta protein 13 (PP13) levels according to groups and fractions.

Fractions	Groups	Term Control (Delivery > 37wks)	All Preeclampsia	Preterm PE (Delivery < 37 wks)	PTD (Delivery < 37 wks)
		(n = 15)	(n = 19)	(n = 7)	(n = 16)
PP13 soluble		349 [276–522]	591 [437–875]	668 [437–861]	602 [431–701]
Total PP13		964 [875–1636]	1598 [1070–1981]	2153 * [1886–2838]	1576 [1011–2014]
PP13-associated PEV		699 [511–891]	830 [355–1485]	1560 * [1004–2277]	877 [564–1519]
Surface PEVs PP13		11.1 [8.8–14.6]	17.2 [9.0–20.5]	18.2 * [15.9–25.0]	7.0 * [3.9–14.1]

Results are shown as medians and the interquartile range [IQR]. * Significantly different from the control group using a Mann–Whitney non-parametric test ($p < 0.01$). PE: preeclampsia, PTD: preterm delivery, PEV: placental-associated extracellular vesicles. PP13-associated PEV was obtained by subtracting the soluble PP13 from the total PP13. Surface PEV PP13 was determined on aliquots of mini-SEC columns.

Table 3. Placenta protein 13 (PP13) levels as a function of treatment with corticosteroids.

Group / Treatment / Fraction	Term Delivery (n = 15) Celestone +	Celestone −	PE (n = 19) Celestone +	Celestone −	Preterm PE (n = 7) Celestone +	Celestone −	PTD (n = 16) Celestone +	Celestone −	*p* Celestone +	*p* Celestone −
	(n = 3)	(n = 12)	(n = 10)	(n = 9)	(n = 7)	(n = 0)	(n = 13)	(n = 3)		
Total PP13	911 [875–1636]	967 [866–1475]	1923 [1293–2316]	1372 [1070–1640]	2153 [1866–2838]		1442 [655–1973]	2037 [1068–2102]	0.129	0.150
PP13 soluble	471 [276–522]	345 [b] [274–513]	608 [437–861]	591 [a] [514–966]	668 [437–861]		604 [435–720]	471 [ab] [427–636]	0.425	0.041
PP13-associated PEV	599 [390–1164]	711 [511–884]	1245 [372–1879]	676 [233–975]	1560 [1004–2277]		763 [531–1387]	1401 [597–1676]	0.365	0.318
Surface PEVs PP13	10.1 [ab] [5.3–27.3]	11.2 [a] [8.8–14.6]	18.2 [a] [15.9–25.0]	14.7 [a] [8.5–20.5]	18.2 [a] [15.9–25.0]		7.3 [b] [3.9–14.1]	4.3 [b] [3.9–4.7]	0.017	0.075

Results are shown as medians and the interquartile range [IQR] using the Kruskal–Wallis non-parametric test. The letter "a" stands for significantly higher, "ab" is also significantly higher but lower than the values marked "a", "b" is significantly lower than any. PE: preeclampsia, PTD: preterm delivery. Enzyme-linked immunosorbent assay (ELISA) was used for the quantification of PP13 in each fraction. Total PP13 was obtained after treatment with mild detergent. Soluble PP13 was determined directly. The PEV-associated PP13 is the total PP13 minus the soluble PP13. The surface PEVs PP13 was determined on aliquots of mini-SEC columns.

2.3. The Impact of Corticosteroids

A decade ago, we found a temporary increase in the level of PP13 in patients treated with various corticosteroids [23,24]. Since cases of preterm PE and PTD were often treated with corticosteroids, we compared the relative fraction of PP13 levels between patients treated or untreated with corticosteroids (Table 3).

The results show a larger PP13 in PE patients treated with corticosteroids as compared to non-treated patients, which is true for all fractions (Table 3 and the right side of Figure 2). No comparison was available for preterm PE since they were all treated. In PTD, the surface of the PEV has higher PP13 and the soluble fraction appears higher but there is no significant difference in the total PP13 or the PEV-associated PP13 (Table 3, Figure 2, right side for each compartment).

3. Discussion

This exploratory study is the first to quantify the levels of PEV-associated PP13 in the maternal circulation in term delivery controls, in PE, and in PTD. This pathway is generated by the syncytiotrophoblast that sheds PEVs into the maternal circulation in addition to the previously reported pathway of soluble PP13 [3,25]. Previous studies have focused on the soluble PP13, and although the majority of these studies reported consistent results, the use of different analyzers and antibodies, and different methods of blood processing resulted in inconsistencies as to the usefulness of PP13 as a PE marker [3,25]. Here, it is found that the fraction of PEV-associated PP13 is the major pathway that carries the largest amount of PP13 from the placental origin into the maternal circulation. The amount carried by this pathway is higher in PE, especially in preterm PE cases, compared to the term or the preterm delivery cases.

While SEC is an analytical methodology that is not widely used in clinical labs, the same results were obtained by blood treatment with a mild detergent to generate the total PP13. The latter can easily be used in clinical labs. Hence, this study brings the PP13 marker back into the arena of predicting PE, with a verified overview of its actual presence in the maternal circulation, and a justified way to analyze it properly. This is important since PP13 is shown to be important for rendering the mother immune-suppressive to the growing fetus [25,26]. It is also important given there are studies that show how PP13 primes the uterine arteries and veins to increase the delivery of oxygen and nutrients to the placenta and CO_2 and metabolite removal [27,28].

In a previous study, we used isolated placentae tested ex vivo and determined the level of PP13 in exosomes and microvesicles that were purified from the uterine vein after perfusing [6]. In that model, the levels of PP13 were normalized to the protein level and the data indicated that for a given amount of protein, the level of PP13 was lower on the surface and in the cargo of both the exosomes and microvesicles. Here, we show that near delivery, there is an increase in PEV-associated PP13 in PE cases, and primarily in preterm PE, compared to term delivery controls. This discrepancy can be resolved by the previously reported higher number of PEVs conveyed to the maternal circulation in PE versus term delivery controls [29,30]. Thus, it appears that while each individual PEV may carry less PP13, there is an overall increase due to the larger number of PEVs that are delivered into the maternal circulation.

At the research level, we are now developing a method for mounting the PEVs from the maternal circulation onto a glass surface of 96 well microplates and developing multiplex PEV arrays to fluorescently determine their amount and distribution by visualizing immune-labeled complexes and optical nanoscopy. The aim is to generate a novel analyzer of the risk of developing PE, which will be based on the PEV pathway between the placenta and the maternal circulation.

In the last ten years, EVs were shown to communicate signals carried internally or on their surface between remote organs in many differentiation and pathological conditions [29–33]. Here, we provide additional evidence that a specific type of EV, the PEVs, which are delivered from the placenta into the maternal circulation, should be further analyzed not only for PP13 but also for their RNA content [29] and other proteins such as placental growth factors (PlGF) and soluble FMS-like protein kinase-1 (sFlt-1), which are widely used in the prediction of PE near delivery according to their negative predictive values [14,34].

Corticosteroids are given to women attending a delivery admission clinic with suspected PTD or preterm PE. They are used to facilitate fetal lung maturation [17–23]. Here, we find that in PE, they augment the maternal blood levels of PP13. This may be linked to the cytokine storm or to the changes in the levels of TNF-alpha in PE [34–36], which are linked to the loss of immune tolerance in PE.

Limitations—This is an exploratory study, and we had no prior estimates of the anticipated level of PEV-associated PP13. We aimed for approximately the same number of patients in each group, which resulted in the fact that the study was underpowered. Although significant statistical differences were found, the small size of the groups may reflect an over-impact of certain individual patients. Thus, larger cohorts are needed to validate our findings in prospective studies.

Another limitation stems from the use of the mini-column for PEV purification, which is considered to be more analytical and not fully quantitative, since only a small amount of plasma could be loaded onto the mini-SEC columns. Subsequent studies are, thus, required to fully explore this more hidden reservoir of the PP13 on the surface of PEVs.

4. Materials and Methods

4.1. Sample and Patients

This study is exploratory. We had no previous information on the amount of PP13 in the PEV fraction, and, thus, aimed to have approximately the same number of patients in each group. Pregnant women attending the delivery clinic of Bnai Zion Medical Center (BZMC) in Haifa were invited to take part between August 2020 and May 2022. The term delivery controls were enrolled on the same or the next day as the study groups (to avoid bias). The enrollment criteria were GA of 24 weeks and above, and not being in labor when enrolled in the study. Other inclusion criteria were a maternal age of 18 years and above, viable singleton pregnancies without major fetal structural and genetic abnormalities, and the patients' agreement to undergo all test procedures and deliver at the medical center. The exclusion criteria were multiple pregnancies, fetal abnormalities, preexisting renal, hematological, autoimmune, or severe cardiovascular conditions, or the inability to sign the informed consent due to mental disabilities. GA was determined from the last menstrual period and was verified by evaluating the records of the routine first-trimester ultrasound of the fetal crown–rump length [37].

The study was approved by the Ethics Committee of BZMC (approval # BZMC-0107-19), and informed written consent was obtained from all participants. Demographics, medical, and pregnancy history, and delivery records were extracted from the hospital's electronic medical records. These included the drugs taken during pregnancy such as low-dose aspirin, vaginal progesterone, tocolysis, and corticosteroids.

Blood pressure was measured at the time of enrollment with a Welch Allyn Nonin SPo2 device. These devices are calibrated regularly as per protocol. Measurements were made according to the guidelines of the Fetal Medicine Foundation (FMF), which advises measuring the diastolic and systolic blood pressure twice, 20 min apart, and calculating the diastolic, systolic, and mean arterial blood pressure [38].

The pregnancy complications are outlined below.

Preeclampsia (PE)—data on pregnancy outcomes were obtained from hospital records. PE was diagnosed according to the guidelines of the International Society for the Study of Hypertension in Pregnancy (ISSHP). According to this definition, the diagnosis of PE requires the presence of new-onset hypertension (systolic blood pressure $\geq$ 140 mmHg or diastolic blood pressure $\geq$ 90 mmHg) at GA $\geq$ 20 weeks' gestation, or chronic hypertension and either proteinuria ($\geq$300 mg/24 h or protein-to-creatinine ratio $\geq$ 30 mg/mmol or $\geq$2+ on dipstick testing), or evidence of renal dysfunction (serum creatinine > 97 µmol/L), hepatic dysfunction (transaminases $\geq$ 65 IU/L), or hematological dysfunction (platelet count < 100,000/µL). Outcome measures were all PE, preterm, and early PE, with delivery at any gestation, or at <37, and <34 weeks' gestation [11–13].

Preterm delivery (PTD) was defined as delivery before 37 weeks' gestation [14,16] that was not related to fetal growth restriction or PE, chorioamnionitis, placenta abruption, placenta previa, or placenta accreta.

4.2. Blood Drawing and Processing

At the time of enrollment, 10 mL of whole blood was drawn into K_2EDTA tubes (BD, Heidelberg, Germany), turned upside down several times to assure a good mixture of the blood with the solution, then centrifuged at $1500\times g$ for 10 min at RT. The clear plasma was aspired and stored in 0.5 mL cryovials at $-80\ °C$ until use, with labels listing the patient's code and the date of sample collection.

4.3. The Different Fractions of PP13

We developed a step-wise method to determine PP13 in the different fractions of maternal blood (Figure 3). Soluble PP13 was measured directly. PEV-associated PP13 was obtained by SEC. Treatment with mild detergent yielded a total PP13. PP13 of the surface PEVs was determined on mini-SEC columns (Figure 3).

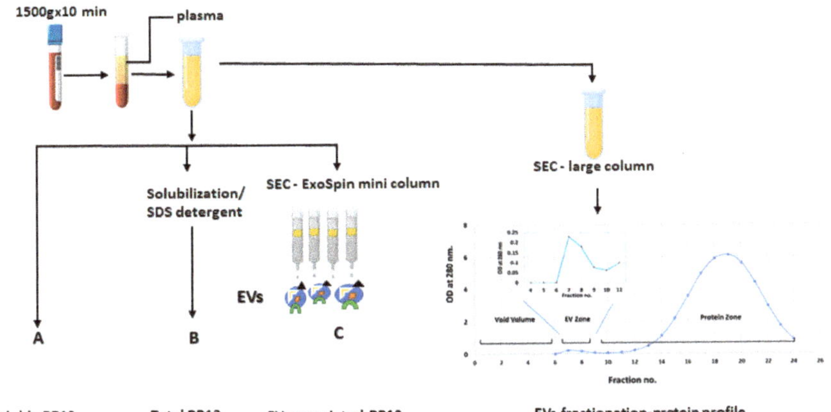

Figure 3. Sub-fractionation of maternal plasma. Starting from whole blood, the plasma was collected by EDTA-containing tubes. A–C: PP13 level was determined by ELISA in the soluble compartment (A). The total PP13 was determined by ELISA after treating the plasma with mild detergent to deplete PP13 from the PEV (B). Exo Spin mini-size exclusion chromatography (SEC) columns were used to exclusively isolate the PEV and determine PP13 on their surface (C). Note that in C, the level was limited by column capacity. To the right side, we describe the entire process of SEC while the actual SEC profile is shown in the Supplementary Figure S1. The schematic profile at the lower right side below the SEC large column tube indicates the zone of fractions 6–8 of the SEC includes the PEVs while the zone of fractions 12–24 include the soluble proteins (further details are demonstrated in Supplementary Figure S1).

4.4. Size Exclusion Chromatography

The extracellular vesicles (EVs) from maternal plasma were isolated by size exclusion chromatography (SEC) according to the manufacturer's instructions (Izon Science, Lyon, France). Plasma samples were thawed on ice and centrifuged at $2000 \times g$ for 10 min to remove aggregates generated during the freeze/thaw cycle. The supernatant was filtered using a 0.22 μm constant well filtration system after centrifugation. Then, 0.5 mL of plasma was loaded on the top of the pre-equilibrated EV column (Izon Science). The column was washed with 3 mL PBS to collect 0.5 mL fractions of the void volume. Filtered PBS was added into the column and 24 fractions of 0.5 mL each were collected as eluents. The protein content was monitored in each fraction by measuring the optical density OD at 280 nm. Protein concentration was determined by BCA assay (Thermo-Fisher, Waltham, MA, USA). All fractions were stored at −80 °C for downstream analyses.

4.5. Isolation and Characterization of EVs

The isolation of EVs from maternal blood was performed by Exo-spin™ 96 according to the manufacturer's instructions (Cell Guidance Systems, Cambridge, UK). Briefly, the plasma was centrifuged at $16,000 \times g$ for 30 min to remove any remaining cell debris and large aggregates. Aliquots of 100 μL of the centrifuged plasma were loaded on each pre-equilibrated column and allowed to enter the column. The flow-through was then collected into the waste plate and discarded. PBS (180 μL) was applied to each column and the exosomes were eluted into the collection plate. The protein content in the eluate was monitored by measurement of the OD at 280 nm. Samples were stored at −70 °C until use.

4.6. Solubilization of PEVs

To quantify the total PP13 in the plasma (soluble and PEV-associated PP13), the plasma was treated with 0.1% sodium dodecyl sulphate-SDS/PBS as a detergent for 30 min on ice to solubilize the EVs and release the PP13 content in the EV compartments (inside and outside) [39]. Total PP13 was determined in the solubilized proteins by ELISA as described in Section 4.6 below. The profile of the SEC before and after detergent treatment is shown in Supplementary Figure S1.

4.7. Quantification of PP13 by ELISA

Determination of the total PP13, soluble PP13, and EV-associated PP13 concentrations was performed by a competitive enzyme-associated immunosorbent assay (ELISA) according to the manufacturer's instructions (Cusabio, Wuhan, China; cat. #: CSB-E12733h). All the samples were analyzed in duplicate. Briefly, the wells of the ELISA plates precoated by the manufacturer with anti-PP13 antibodies were incubated for 2 h, with the diluted samples together with PP13 conjugated to HRP to allow for the competition between PP13 in the plasma samples labeled PP13 to reach equilibrium. Access reagents were washed and 3,3′,5,5′ tetramethylbenzidine (TMB) chromogenic substrate was added. The reaction was stopped by 2 N HCL and the optical density of the colored developer was determined at 450 nm using an ELISA plate reader. PP13 concentration was calculated based on the standard curve generated in the same experiment.

Indirect ELISA was used to determine the PP13, PLAP, and CD63 of the SEC-eluted fractions [1–24]. For this assay, the SEC fractions were diluted 1:50 in a 50 mM carbonate buffer at pH 9.6 and were coated in duplicate on the wells of flat-bottom ELISA plates (Falcon, BD Heidelberg, Germany). The SEC diluted fractions were left for overnight incubation at 4 °C followed by blocking of the nonspecific binding sites with 2% bovine serum albumin (BSA) in phosphate-buffered saline (PBS). The three pairs of SEC fractions were incubated with either anti-PP13 antibodies at 0.1 microgram /mL of monoclonal anti-PP13 (clones 534 and 215-28-3) [26,29–38,40–46], or anti-PLAP antibody (NDOGE2, a generous gift from Dr. Manu Vatish, Oxford University [6,9]) or the commercially available anti-CD63 (Thermos Fischer Scientific, Waltham, MA, USA) for 2 h at room temperature (RT). After washing off excess reagents, the plates were incubated for 1 h with goat anti-mouse IgG conjugated with horseradish peroxidase, HRP (Dianova, Königswinter, Germany) for 1 h at RT. Extensive washing with PBS containing 0.05% Tween was performed between steps. The reaction product was developed with TMB substrate (Thermo Fisher Scientific, Waltham, MA, USA), stopped with 2 N HCl, and the optical density was measured using a microplate spectrophotometer reader (BioTeck instruments Inc., Santa Clara, CA, USA) at 450 nm.

4.8. Statistics

The data were analyzed using the SPSS version 28 (IBM, Chicago, IL, USA). For descriptive statistics, the categorical variables are presented as frequencies (n) and percentages, whereas the continuous variables are presented as the median or medians and interquartile range [IQR]. For the inferential statistics, differences between groups for the continuous variables were examined using a Kruskal–Wallis or a Mann–Whitney non-parametric test. Relationships between groups and the categorical variables, and were calculated using chi-square tests or the Fisher exact test, depending on sample size.

A violin plot was used to describe the medians, quartiles, and the upper and lower range. The advantage of the violin plot over the box plot is that it not only shows the medians, quartiles, and ranges but also depicts the cases included in each quartile, which is more informative than box plots where the box size is fixed.

5. Conclusions

This exploratory study was designed to quantify the pathway of PEVs carrying PP13 on their surface and in their inside cargo into the maternal circulation (Figure 4). PP13 carried via this pathway in PE and PTD patients is higher than in the term delivery controls, especially among preterm PE cases. When the patients were treated with corticosteroids, a further increase in PP13 liberation was found. These results illustrate that the PEVs create an important communication pathway between the placenta and maternal circulation, and this finding warrants further research.

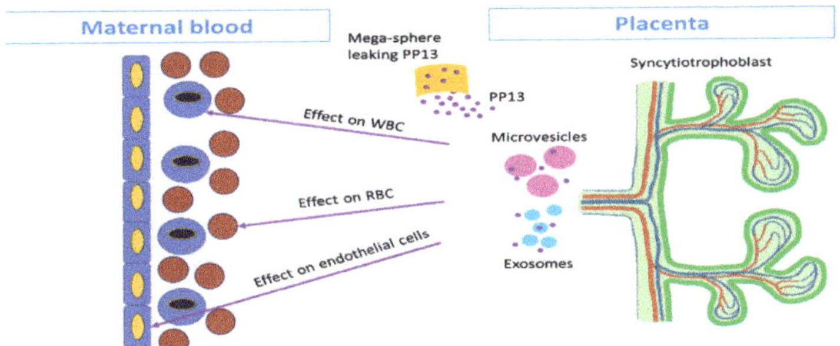

Figure 4. PP13 pathways to the maternal circulation. The placenta syncytiotrophoblast (green to the right) releases PP13 through the uterine vein into the maternal circulation (the blue blood vessel to the left). The PEVs serve as the communication pathway. The microvesicles (pink) and the exosomes (turquoise) carry PP13 (purple) on their surface and inside. There is also a leakage of soluble PP13 (purple) either into the intracellular space or through necrotic vesicles (orange) that are too big to pass through the capillaries, thus, leaking their content into the soluble fraction of the maternal circulation. PP13 is, thus, liberated either as a soluble protein or in association with the PEVs. Clinical pathologies affect PP13 mainly via the PEV pathway. When reaching the maternal circulation, PP13 binds to the ABO antigen on the red blood cells (RBC, [47]), the white blood cells (WBC [3,26]), or the endothelial layer [27,28,48], Sammar et al. Elements of this figure were modified from our former review Sammar et al. [7] *Int. J. Mol. Sci.* **2019**, *20*, 3192. https://doi.org/10.3390/ijms20133192.

Supplementary Materials: The following supporting information can be downloaded at: https://www.mdpi.com/article/10.3390/ijms241512051/s1.

Author Contributions: This study was initiated by M.S. who conceptualized the study on all fronts and designed its flow. He conducted all the immune-biochemical experiments and marker determinations. R.S. and M.M.K. enrolled the patients, obtained their informed consent, collected all the patients' medical, demographic, and delivery records, drew blood, and processed it into plasma. H.M. prepared the clinical study protocol, the informed consent, and the clinical study report forms (CRF) into which the hospital medical records were entered by M.M.K. H.M. also verified the study database and worked with A.S.-N., who conducted the statistical analyses. All authors participated in the manuscript writing and editing, and also read and approved the final version of the manuscript. All authors have read and agreed to the published version of the manuscript.

Funding: This project was supported by the internal fund of Braude Engineering College.

Data Availability Statement: Not applicable.

Acknowledgments: The authors thanks Roni Cohen Zisman for producing Figure 4.

Conflicts of Interest: The authors declare no conflict of interest. No significant financial support for this work was received. There was no any source, financial or any other, that could have influenced its outcome.

Abbreviations

ACOG	American College of Obstetricians and Gynecologists
BMI	Body mass index
BP	Blood pressure
BZMC	Bnai Zion Medical Center
CD63	Cluster Differentiation 63
EVs	Extracellular Vesicles
ELISA	Enzyme-linked Immuno Sorbent Assay
FGR	Fetal Growth Restriction
GA	Gestational Age
GDM	Gestational Diabetes Mellitous
ISSHP	International Society for the Study of Hypertension Disorders in Pregnancy
ISUOG	International Society for Ultrasound in Obstetrics and Gynecology
IVF	In Vitro Fertilization
MAP	Mean arterial blood pressure
NICU	Newborn Intensive Care Unit
NPV	Negative Predictive value
PE	Preeclampsia
PEVs	Placental-associated Extracellular Vesicles
PlGF	Placental Growth Factor
PLAP	Placental-derived Alkaline Phosphatase.
PP13	Placental Protein 13
PTD	Preterm Delivery
sFlt-1	Soluble FMS (oncogene for Feline McDonough Sarcoma)-like tyrosine kinase 1
SEC	Size Exclusion Chromatography

References

1. Burton, G.J.; Jauniaux, E. What is the placenta? *Am. J. Obstet. Gynecol.* **2015**, *213*, S6.e1–S6.e4. [CrossRef]
2. Goldstein, J.A.; Gallagher, K.; Beck, C.; Kumar, R.; Gernand, A.D. Maternal-Fetal Inflammation in the Placenta and the Developmental Origins of Health and Disease. *Front. Immunol.* **2020**, *11*, 531543. [CrossRef]
3. Than, N.G.; Balogh, A.; Romero, R.; Kárpáti, É.; Erez, O.; Szilágyi, A.; Kovalszky, I.; Sammar, M.; Gizurarson, S.; Matkó, J.; et al. Placental Protein 13 (PP13)—A Placental Immunoregulatory Galectin Protecting Pregnancy. *Front. Immunol.* **2014**, *5*, 348. [CrossRef] [PubMed]
4. Than, N.G.; Rahman, O.A.; Magenheim, R.; Nagy, B.; Fule, T.; Hargitai, B.; Sammar, M.; Hupuczi, P.; Tarca, A.L.; Szabo, G.; et al. Placental protein 13 (galectin-13) has decreased placental expression but increased shedding and maternal serum concentrations in patients presenting with preterm pre-eclampsia and HELLP syndrome. *Virchows Arch.* **2008**, *453*, 387–400. [CrossRef] [PubMed]
5. Sammar, M.; Nisemblat, S.; Fleischfarb, Z.; Golan, A.; Sadan, O.; Meiri, H.; Huppertz, B.; Gonen, R. Placenta-bound and Body Fluid PP13 and its mRNA in Normal Pregnancy Compared to Preeclampsia, HELLP and Preterm Delivery. *Placenta* **2011**, *32*, S30–S36. [CrossRef] [PubMed]
6. Sammar, M.; Dragovic, R.; Meiri, H.; Vatish, M.; Sharabi-Nov, A.; Sargent, I.; Redman, C.; Tannetta, D. Reduced placental protein 13 (PP13) in placental derived syncytiotrophoblast extracellular vesicles in preeclampsia—A novel tool to study the impaired cargo transmission of the placenta to the maternal organs. *Placenta* **2018**, *66*, 17–25. [CrossRef]
7. Sammar, M.; Drobnjak, T.; Mandala, M.; Gizurarson, S.; Huppertz, B.; Meiri, H. Galectin 13 (PP13) Facilitates Remodeling and Structural Stabilization of Maternal Vessels during Pregnancy. *Int. J. Mol. Sci.* **2019**, *20*, 3192. [CrossRef]
8. Ortega, M.A.; Fraile-Martínez, O.; García-Montero, C.; Paradela, A.; Sánchez-Gil, M.A.; Rodriguez-Martin, S.; De León-Luis, J.A.; Pereda-Cerquella, C.; Bujan, J.; Guijarro, L.G.; et al. Unfolding the role of placental-derived Extracellular Vesicles in Pregnancy: From homeostasis to pathophysiology. *Front. Cell Dev. Biol.* **2022**, *10*, 1060850. [CrossRef]
9. Awoyemi, T.; Cerdeira, A.S.; Zhang, W.; Jiang, S.; Rahbar, M.; Logenthiran, P.; Redman, C.; Vatish, M. Preeclampsia and syncytiotrophoblast membrane extracellular vesicles (STB-EVs). *Clin. Sci.* **2022**, *136*, 1793–1807. [CrossRef]
10. American College of Obstetricians and Gynecologists. Gestational Hypertension and Preeclampsia: ACOG Practice Bulletin Summary, Number 222. *Obstet. Gynecol.* **2020**, *135*, 1492–1495. [CrossRef]
11. Magee, L.A.; Brown, M.A.; Hall, D.R.; Gupte, S.; Hennessy, A.; Karumanchi, S.A.; Kenny, L.C.; McCarthy, F.; Myers, J.; Poon, L.C.; et al. The 2021 International Society for the Study of Hypertension in Pregnancy classification, diagnosis & management recommendations for international practice. *Pregnancy Hypertens.* **2021**, *27*, 148–169. [CrossRef] [PubMed]
12. *Managing Complications in Pregnancy and Childbirth: A Guide for Midwives and Doctors*, 2nd ed.; World Health Organization: Geneva, Switzerland, 2017. Available online: http://apps.who.int/iris/betstream/handle/10665/255760/9789241565493-eng.pdf;sequence=1 (accessed on 28 May 2023).

13. Rolnik, D.L.; Wright, D.; Poon, L.C.; O'Gorman, N.; Syngelaki, A.; de Paco Matallana, C.; Akolekar, R.; Cicero, S.; Janga, D.; Singh, M.; et al. Aspirin versus Placebo in Pregnancies at High Risk for Preterm Preeclampsia. *N. Engl. J. Med.* **2017**, *377*, 613–622. [CrossRef]

14. Stepan, H.; Galindo, A.; Hund, M.; Schlembach, D.; Sillman, J.; Surbek, D.; Vatish, M. Clinical utility of sFlt-1 and PlGF in screening, prediction, diagnosis and monitoring of pre-eclampsia and fetal growth restriction. *Ultrasound Obstet. Gynecol.* **2023**, *61*, 168–180. [CrossRef] [PubMed]

15. American College of Obstetricians and Gynecologists' Committee on Practice Bulletins—Obstetrics Practice Bulletin No. 171: Management of Preterm Labor. *Obstet. Gynecol.* **2016**, *128*, e155–e164. [CrossRef] [PubMed]

16. Goldenberg, R.L.; Culhane, J.F.; Iams, J.D.; Romero, R. Epidemiology and causes of preterm birth. *Lancet* **2008**, *371*, 75–84. [CrossRef] [PubMed]

17. Gyamfi-Bannerman, C.; Thom, E.A.; Blackwell, S.C.; Tita, A.T.; Reddy, U.M.; Saade, G.R.; Rouse, D.J.; McKenna, D.S.; Clark, E.A.; Thorp, J.M.; et al. Antenatal Betamethasone for Women at Risk for Late Preterm Delivery. *N. Engl. J. Med.* **2016**, *374*, 1311–1320. [CrossRef]

18. Bauer, M.E.; Price, L.K.; MacEachern, M.P.; Housey, M.; Langen, E.S.; Bauer, S.T. Maternal leukocytosis after antenatal corticosteroid administration: A systematic review and meta-analysis. *J. Obstet. Gynaecol.* **2017**, *38*, 210–216. [CrossRef]

19. Williams, M.J.; Ramson, J.A.; Brownfoot, F.C. Different corticosteroids and regimens for accelerating fetal lung maturation for babies at risk of preterm birth. *Cochrane Database Syst. Rev.* **2022**, *8*, CD006764. [CrossRef]

20. Shanks, A.L.; Grasch, J.L.; Quinney, S.K.; Haas, D.M. Controversies in antenatal corticosteroids. *Semin. Fetal Neonatal Med.* **2019**, *24*, 182–188. [CrossRef]

21. Chollat, C.; Marret, S. Magnesium sulfate and fetal neuroprotection: Overview of clinical evidence. *Neural Regen. Res.* **2018**, *13*, 2044–2049. [CrossRef]

22. Shennan, A.; Suff, N.; Jacobsson, B.; the FIGO Working Group for Preterm Birth; Simpson, J.L.; Norman, J.; Grobman, W.A.; Bianchi, A.; Mujanja, S.; Valencia, C.M.; et al. Abstracts of the XXIII FIGO World Congress of Gynecology & Obstetrics. *Int. J. Gynecol. Obstet.* **2021**, *155*, 31–74. [CrossRef]

23. Huppertz, B.; Meiri, H.; Neumaier-Wagner, P.; Hebisch, G.; Sammar, M. Tocolysis increases the release of placental protein 13 (PP13) from the placenta into the maternal blood. *Placenta* **2005**, *26*, A40.

24. Bartz, C.; Meiri, H.; Hebisch, G.; Sammar, M.; Werner, R.; Huppertz, B. Tocolytic substances damage syncytiotrophoblast demonstrated by the release of placental protein 13 into the maternal circulation. *Hypertens. Pregnancy* **2006**, *28*, 25–26.

25. Than, N.G.; Romero, R.; Goodman, M.; Weckle, A.; Xing, J.; Dong, Z.; Xu, Y.; Tarquini, F.; Szilagyi, A.; Gal, P.; et al. A primate subfamily of galectins expressed at the maternal–fetal interface that promote immune cell death. *Proc. Natl. Acad. Sci. USA* **2009**, *106*, 9731–9736. [CrossRef] [PubMed]

26. Kliman, H.J.; Sammar, M.; Grimpel, Y.I.; Lynch, S.K.; Milano, K.M.; Pick, E.; Bejar, J.; Arad, A.; Lee, J.J.; Meiri, H.; et al. Placental Protein 13 and Decidual Zones of Necrosis: An Immunologic Diversion That May be Linked to Preeclampsia. *Reprod. Sci.* **2012**, *19*, 16–30. [CrossRef] [PubMed]

27. Drobnjak, T.; Gizurarson, S.; Gokina, N.I.; Meiri, H.; Mandalá, M.; Huppertz, B.; Osol, G. Placental protein 13 (PP13)-induced vasodilation of resistance arteries from pregnant and nonpregnant rats occurs via endothelial-signaling pathways. *Hypertens. Pregnancy* **2017**, *36*, 186–195. [CrossRef]

28. Drobnjak, T.; Jónsdóttir, A.M.; Helgadóttir, H.; Runólfsdóttir, M.S.; Meiri, H.; Sammar, M.; Osol, G.; Mandalà, M.; Huppertz, B.; Gizurarson, S. Placental protein 13 (PP13) stimulates rat uterine vessels after slow subcutaneous administration. *Int. J. Women's Health* **2019**, *11*, 213–222. [CrossRef]

29. Miranda, J.; Paules, C.; Nair, S.; Lai, A.; Palma, C.; Scholz-Romero, K.; Rice, G.E.; Gratacos, E.; Crispi, F.; Salomon, C. Placental exosomes profile in maternal and fetal circulation in intrauterine growth restriction—Liquid biopsies to monitoring fetal growth. *Placenta* **2018**, *64*, 34–43. [CrossRef]

30. Knight, M.; Redman, C.W.G.; Linton, E.A.; Sargent, I.L. Shedding of syncytiotrophoblast microvilli into the maternal circulation in pre-eclamptic pregnancies. *BJOG: Int. J. Obstet. Gynaecol.* **1998**, *105*, 632–640. [CrossRef]

31. Dragovic, R.; Collett, G.; Hole, P.; Ferguson, D.; Redman, C.; Sargent, I.; Tannetta, D. Isolation of syncytiotrophoblast microvesicles and exosomes and their characterisation by multicolour flow cytometry and fluorescence Nanoparticle Tracking Analysis. *Methods* **2015**, *87*, 64–74. [CrossRef]

32. Rädler, J.; Gupta, D.; Zickler, A.; EL Andaloussi, S. Exploiting the biogenesis of extracellular vesicles for bioengineering and therapeutic cargo loading. *Mol. Ther.* **2023**, *31*, 1231–1250. [CrossRef]

33. Ghafourian, M.; Mahdavi, R.; Jonoush, Z.A.; Sadeghi, M.; Ghadiri, N.; Farzaneh, M.; Salehi, A.M. The implications of exosomes in pregnancy: Emerging as new diagnostic markers and therapeutics targets. *Cell Commun. Signal.* **2022**, *20*, 51. [CrossRef]

34. Sharabi-Nov, A.; Kumar, K.; Vodušek, V.F.; Sršen, T.P.; Tul, N.; Fabjan, T.; Meiri, H.; Nicolaides, K.H.; Osredkar, J. Establishing a Differential Marker Profile for Pregnancy Complications Near Delivery. *Fetal Diagn. Ther.* **2019**, *47*, 471–484. [CrossRef] [PubMed]

35. Danielli, M.; Thomas, R.C.; Gillies, C.L.; Hu, J.; Khunti, K.; Tan, B.K. Blood biomarkers to predict the onset of pre-eclampsia: A systematic review and meta-analysis. *Heliyon* **2022**, *8*, e11226. [CrossRef] [PubMed]

36. Hornaday, K.K.; Wood, E.M.; Slater, D.M. Is there a maternal blood biomarker that can predict spontaneous preterm birth prior to labour onset? A systematic review. *PLoS ONE* **2022**, *17*, e0265853. [CrossRef]

37. Hadlock, F.P.; Shah, Y.P.; Kanon, D.J.; Lindsey, J.V. Fetal crown-rump length: Reevaluation of relation to menstrual age (5–18 weeks) with high-resolution real-time US. *Radiology* **1992**, *182*, 501–505. [CrossRef]

38. Poon, L.C.Y.; Kametas, N.A.; Chelemen, T.; Leal, A.; Nicolaides, K.H. Maternal risk factors for hypertensive disorders in pregnancy: A multivariate approach. *J. Hum. Hypertens.* **2009**, *24*, 104–110. [CrossRef] [PubMed]

39. Burger, O.; Pick, E.; Zwickel, J.; Klayman, M.; Meiri, H.; Slotky, R.; Mandel, S.; Rabinovitch, L.; Paltieli, Y.; Admon, A.; et al. Placental protein 13 (PP-13): Effects on cultured trophoblasts, and its detection in human body fluids in normal and pathological pregnancies. *Placenta* **2004**, *25*, 608–622. [CrossRef]

40. Than, N.G.; Pick, E.; Bellyei, S.; Szigeti, A.; Burger, O.; Berente, Z.; Janaky, T.; Boronkai, A.; Kliman, H.; Meiri, H.; et al. Functional analyses of placental protein 13/galectin. *JBIC J. Biol. Inorg. Chem.* **2004**, *271*, 1065–1078. [CrossRef]

41. Nicolaides, K.H.; Bindra, R.; Turan, O.M.; Chefetz, I.; Sammar, M.; Meiri, H.; Tal, J.; Cuckle, H.S. A novel approach to first-trimester screening for early pre-eclampsia combining serum PP-13 and Doppler ultrasound. *Ultrasound Obstet. Gynecol.* **2005**, *27*, 13–17. [CrossRef]

42. Chafetz, I.; Kuhnreich, I.; Sammar, M.; Tal, Y.; Gibor, Y.; Meiri, H.; Cuckle, H.; Wolf, M. First-trimester placental protein 13 screening for preeclampsia and intrauterine growth restriction. *Am. J. Obstet. Gynecol.* **2007**, *197*, 35.e1–35.e7. [CrossRef] [PubMed]

43. Romero, R.; Kusanovic, J.P.; Than, N.G.; Erez, O.; Gotsch, F.; Espinoza, J.; Edwin, S.; Chefetz, I.; Gomez, R.; Nien, J.K.; et al. First-trimester maternal serum PP13 in the risk assessment for preeclampsia. *Am. J. Obstet. Gynecol.* **2008**, *199*, 122.e1–122.e11. [CrossRef] [PubMed]

44. Gonen, R.; Shahar, R.; Grimpel, Y.; Chefetz, I.; Sammar, M.; Meiri, H.; Gibor, Y. Placental protein 13 as an early marker for pre-eclampsia: A prospective longitudinal study. *BJOG Int. J. Obstet. Gynaecol.* **2008**, *115*, 1465–1472. [CrossRef]

45. Huppertz, B.; Sammar, M.; Chefetz, I.; Neumaier-Wagner, P.; Bartz, C.; Meiri, H. Longitudinal Determination of Serum Placental Protein 13 during Development of Preeclampsia. *Fetal Diagn. Ther.* **2008**, *24*, 230–236. [CrossRef]

46. Osteikoetxea, X.; Sódar, B.; Németh, A.; Szabó-Taylor, K.; Pálóczi, K.; Vukman, K.V.; Tamási, V.; Balogh, A.; Kittel, Á.; Pállinger, É.E.; et al. Differential detergent sensitivity of extracellular vesicle subpopulations. *Org. Biomol. Chem.* **2015**, *13*, 9775–9782. [CrossRef] [PubMed]

47. Than, N.G.; Romero, R.; Meiri, H.; Erez, O.; Xu, Y.; Tarquini, F.; Barna, L.; Szilagyi, A.; Ackerman, R.; Sammar, M.; et al. PP13, Maternal ABO Blood Groups and the Risk Assessment of Pregnancy Complications. *PLoS ONE* **2011**, *6*, e21564. [CrossRef]

48. Drobnjak, T.; Meiri, H.; Mandalá, M.; Huppertz, B.; Gizurarson, S. Pharmacokinetics of placental protein 13 after intravenous and subcutaneous administration in rabbits. *Drug Des. Dev. Ther.* **2018**, *12*, 1977–1983. [CrossRef]

International Journal of
Molecular Sciences

Article

The Role of Annexin A1 in DNA Damage Response in Placental Cells: Impact on Gestational Diabetes Mellitus

Jusciele Brogin Moreli [1,2], Mayk Ricardo dos Santos [3], Iracema de Mattos Paranhos Calderon [4], Cristina Bichels Hebeda [5], Sandra Helena Poliselli Farsky [5], Estela Bevilacqua [6] and Sonia Maria Oliani [1,3,7,*]

1 Post-Graduation in Structural and Functional Biology, Federal University of São Paulo (UNIFESP), São Paulo 04023-062, Brazil; juscielemoreli@gmail.com
2 Faceres School of Medicine (FACERES), São José do Rio Preto 15090-305, Brazil
3 Department of Biology, School of Biosciences, Humanities and Exact Sciences, São Paulo State University (UNESP), São José do Rio Preto 15054-000, Brazil; mayk.kd@gmail.com
4 Graduate Program in Gynecology, Obstetrics and Mastology, Botucatu Medical School, São Paulo State University (UNESP), Botucatu 18618-687, Brazil; iracema.calderon@gmail.com
5 Department of Clinical and Toxicological Analyses, Faculty of Pharmaceutical Sciences, University of Sao Paulo (USP), São Paulo 05508-000, Brazil; crisbh@gmail.com (C.B.H.); sfarsky@usp.br (S.H.P.F.)
6 Department of Cell and Developmental Biology, Institute of Biomedical Sciences, University of São Paulo (USP), São Paulo 05508-000, Brazil; bevilacq@usp.br
7 Advanced Research Center in Medicine (CEPAM), União das Faculdades dos Grandes Lagos (Unilago), São José do Rio Preto 15030-070, Brazil
* Correspondence: sonia.oliani@unesp.br

Abstract: The functions of annexin A1 (ANXA1), which is expressed on membranes and in cytoplasmic granules, have been fully described. Nonetheless, the role of this protein in protecting against DNA damage in the nucleus is still emerging and requires further investigation. Here, we investigated the involvement of ANXA1 in the DNA damage response in placental cells. Placenta was collected from ANXA1 knockout mice (AnxA1$^{-/-}$) and pregnant women with gestational diabetes mellitus (GDM). The placental morphology and ANXA1 expression, which are related to the modulation of cellular response markers in the presence of DNA damage, were analyzed. The total area of AnxA1$^{-/-}$ placenta was smaller due to a reduced labyrinth zone, enhanced DNA damage, and impaired base excision repair (BER) enzymes, which resulted in the induction of apoptosis in the labyrinthine and junctional layers. The placentas of pregnant women with GDM showed reduced expression of AnxA1 in the villous compartment, increased DNA damage, apoptosis, and a reduction of enzymes involved in the BER pathway. Our translational data provide valuable insights into the possible involvement of ANXA1 in the response of placental cells to oxidative DNA damage and represent an advancement in investigations into the mechanisms involved in placental biology.

Keywords: apoptosis; oxidative stress; base excision repair; annexin A1 knockout mice; high-risk pregnancy

Citation: Moreli, J.B.; Santos, M.R.d.; Calderon, I.d.M.P.; Hebeda, C.B.; Farsky, S.H.P.; Bevilacqua, E.; Oliani, S.M. The Role of Annexin A1 in DNA Damage Response in Placental Cells: Impact on Gestational Diabetes Mellitus. *Int. J. Mol. Sci.* **2023**, 24, 10155. https://doi.org/10.3390/ijms241210155

Academic Editor: Giovanni Tossetta

Received: 10 May 2023
Revised: 5 June 2023
Accepted: 6 June 2023
Published: 15 June 2023

1. Introduction

Annexin A1 (ANXA1) is a member of the annexin superfamily of calcium- and phospholipid-binding proteins [1]. The protein core comprises 346 amino acids (37 kDa) with C- and N-terminal domains, the latter of which confers specificity and physiological activity to each annexin [2]. Besides being characterized as a glucocorticoid-regulated anti-inflammatory protein [3], ANXA1 has also been reported to be involved in critical pathophysiological processes, including cell proliferation, differentiation, and apoptosis of epithelial and cancer cells, which implicates this protein in tissue repair and cancer metastasis [4–6].

ANXA1 is found in three distinct subcellular locations: the cytoplasm, nucleus, and plasma membrane. Although none of the annexins contain nuclear-targeting sequences, nuclear localization of ANXA1 has been reported under certain conditions [7].

The translocation of ANXA1 from the cytoplasm to the nucleus is believed to start with mitogenic/proliferative signals or after a DNA-damaging stimulus, such as oxidative stress [7]. In breast cancer cells, nuclear localization of ANXA1 was associated with protection against heat-induced DNA damage [8].

ANXA1 contains a DNA- and/or RNA-binding sequence [9] and has been proposed to perform helicase activity in the nucleus [10,11]. Because helicase activity is required in DNA replication and repair, nuclear ANXA1 may participate in tumorigenesis [6]. DNA lesions, such as double-strand breaks and oxidation of DNA bases, especially guanine, are also induced by hyperglycemia-mediated oxidative stress [12–15]. We previously demonstrated that maternal hyperglycemia levels are directly proportional to DNA damage and inversely proportional to the expression of base excision repair (BER) enzymes in peripheral blood cells [15]. However, this relationship was not observed in the same cell types in newborns because of the regulation of expression of BER enzymes [15]. BER is the most efficient mechanism for repairing endogenous DNA damage, in which DNA glycosylase (8-oxo guanine DNA glycosylase [OGG1]) removes the damaged base, resulting in an apurinic/apyrimidinic (AP) site. The AP endonuclease (AP endonuclease 1 [APE1]) cleaves the AP site, allowing DNA polymerase to synthesize the repair patch. The latter is re-ligated using the DNA ligase III activity [16].

As a determinant condition of hyperglycemia, GDM, defined as any degree of glucose intolerance with onset or first recognition during pregnancy, is among the most common complications associated with pregnancy [17]. Maternal metabolic changes can, directly and indirectly, be reflected in the placenta, interposed between the maternal and fetal circulation [18], and can lead to impaired embryonic and fetal development [19–21]. Cell death [22], inflammation [23,24], and oxidative stress [25] are important hallmarks of placental cell damage in mothers with GDM.

Because nuclear ANXA1 is apparently involved in DNA replication and repair, we tested the hypothesis that nuclear ANXA1 is associated with the placental cellular response induced by oxidative DNA damage. To this end, we used placental samples from AnxA1 knockout animals (AnxA1$^{-/-}$) and from mothers with GDM, a classical condition of oxidative stress and inflammation during pregnancy.

2. Results

2.1. Deficiency of AnxA1 Gene Modifies the Placental Morphology

The labyrinthine and junctional zones of mouse placenta were identified and measured using histological sections (Figure 1). The mice placentas comprised the following distinct layers: the labyrinth, junctional zone, and maternal decidua (Figure 1A,B). The junctional zone contained spongiotrophoblasts, glycogen, and giant trophoblasts (Figure 1C,D). The labyrinth (Figure 1E,F)—the site of gas and nutrient exchange—was formed by a layer of giant trophoblasts in contact with the maternal blood and syncytiotrophoblast layers facing the fetal capillaries.

The placentas of AnxA1$^{-/-}$ mice showed a smaller total area (Figure 1G), no difference in junctional zone (Figure 1H) and reduction of labyrinthine zone (Figure 1I) when compared with WT placentas.

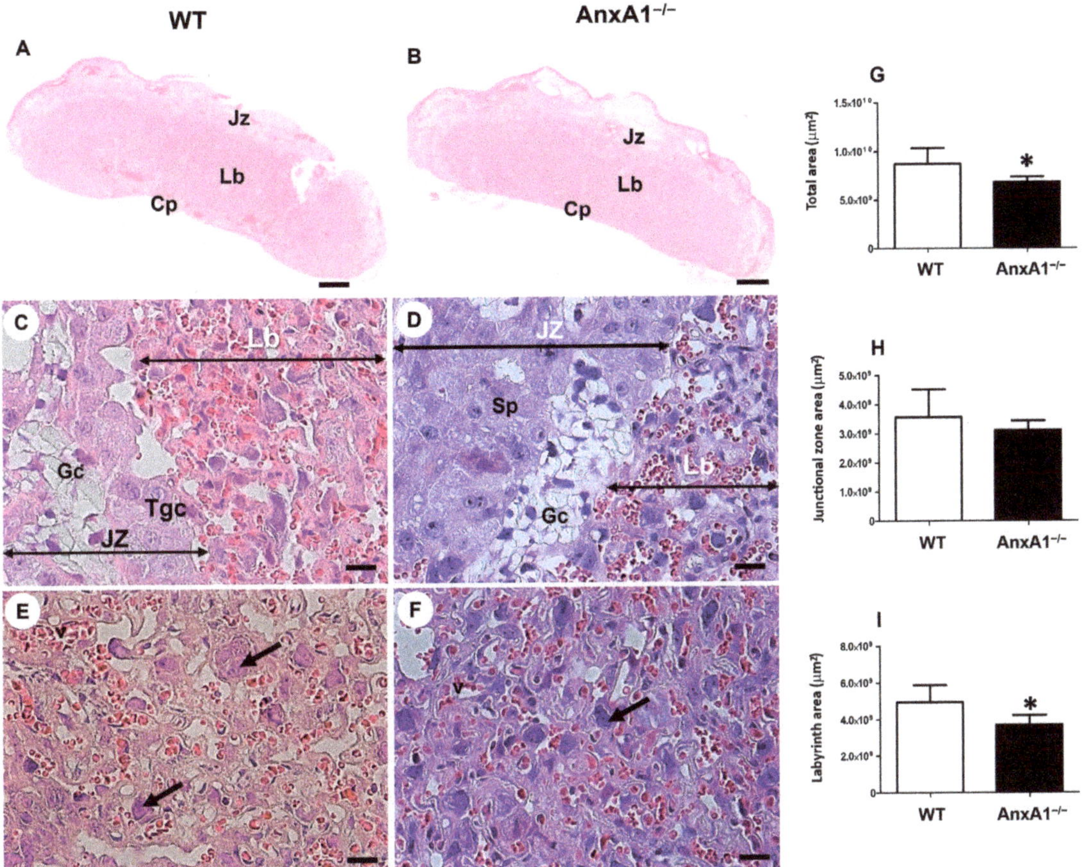

Figure 1. Representative images of placental sections from wild-type (WT) (**A**) and AnxA1 knockout animals (AnxA1$^{-/-}$, (**B**)). The histologic sections show the junctional (Jz) (**C,D**) and labyrinth (Lb) (**E,F**) placental zones. Scale bars = 20 μm. Jz: junctional zone; Lb: labyrinth; Cp: chorionic plate; Gc: glycogen cells; Sp: spongiotrophoblast; Tgc: junctional zone giant trophoblast cells; arrows: labyrinthine giant trophoblast cells. The sections were stained with hematoxylin and eosin Morphometric analyses of total area (**G**), junctional zone (**H**) and labyrinth (**I**) of placentas from WT and AnxA1$^{-/-}$ animals. Values are shown as mean ± SD. * $p < 0.05$, n = 6/group.

2.2. Oxidative DNA Damage Is Augmented, and the Expression of Repair Enzymes Is Impaired in Placenta from AnxA1$^{-/-}$ Mice

In placenta from AnxA1$^{-/-}$ mice, the oxidative DNA damage, evaluated based on positive nuclear staining for 8-oxoguanine, was increased in the junctional zone (Figure 2A,B,F) and in the labyrinth (Figure 2C,D,G). The expression of the OGG1 DNA repair enzyme was less in AnxA1$^{-/-}$ in the placental nuclei of both junctional (Figure 2H,I,M) and labyrinthine (Figure 2J,K,N) zones than in the placenta from WT mice. Similarly, the expression of the other component in the DNA repair pathway, namely APE-1, was also reduced in the placenta of AnxA1$^{-/-}$ mice (Figure 2O–U).

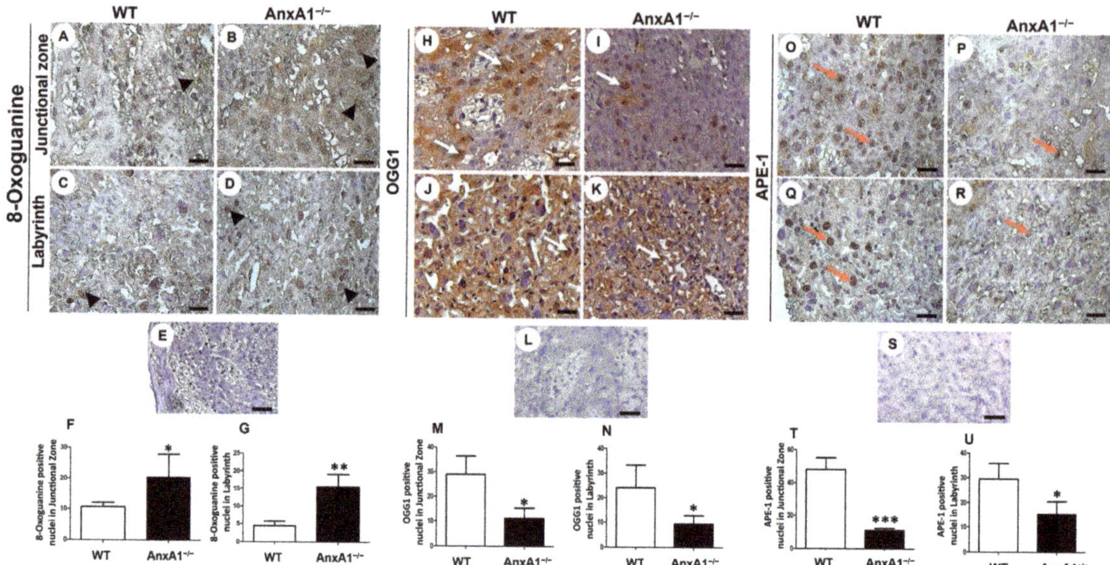

Figure 2. Immunolocalization of 8-Oxoguanine (oxidative DNA damage, (**A–G**)) and OGG1 (**H–N**) and APE-1 (DNA repair enzymes, (**O–U**)) in placenta sections from WT and AnxA1$^{-/-}$ animals. Placental cells reactive to 8-Oxoguanine (arrowheads; (**A–D**)), OGG-1 (arrows; (**H–K**)), and APE-1 (red arrows; (**O–R**)) are found in both junctional and labyrinthine placental zones with a similar pattern of immunoreactivity. (**E,L,S**) Buffer was used instead of the primary antibody as a negative control, in panels. Bars = 50 μm. (**A–E,H–L,O–S**) Immunoperoxidase and hematoxylin counterstaining. (**F,G,M,N,T,U**) Quantification of 8 hydroxyguanosine- (**F,G**), OGG-1- (**M,N**), and APE-1- (**T,U**) positive nuclei per 10,000 μm^2 of placental area. Values are shown as mean $\pm$ SD; * $p < 0.05$; ** $p < 0.01$, *** $p < 0.001$, n = 6/group.

2.3. DNA Double-Strand Breaks and Apoptosis Are Enhanced in Placenta from AnxA1$^{-/-}$ Mice

The placental cells in the AnxA1$^{-/-}$ group showed increased expression of gamma H2AX in the junctional zone (Figure 3A,B,F) and labyrinth (Figure 3C,D,G) layers. The number of cells showing the expression of cleaved caspase 3 was increased only in the labyrinth (Figure 3J,K,N).

2.4. GDM Alters Maternal Clinical Parameters

Pregnant women with GDM presented with higher levels of HbA1c and their newborns were heavier than those of ND women. Other clinical parameters, such as age, BMI, pregnancy weight gain, and placental weight, showed no differences between the GDM and ND pregnant women (Table 1).

Table 1. Clinical data.

	ND (n = 10)	**GDM (n = 10)**
Maternal age (years)	30.85 $\pm$ 6.11	29.43 $\pm$ 5.36
BMI (Kg/m^2)	34.88 $\pm$ 8.44	38.05 $\pm$ 4.37
Pregnancy weight gain (Kg)	12.83 $\pm$ 7.50	12.34 $\pm$ 6.94
HbA1c (%)	5.36 $\pm$ 0.63	6.32 $\pm$ 0.82 *
Placental weight (g)	623.10 $\pm$ 78.38	632.50 $\pm$ 179.80
Newborn weight (g)	3062.00 $\pm$ 276.60	3540.00 $\pm$ 574.50 *

Data presented as means $\pm$ standard deviation. BMI: body mass index evaluated in the third trimester of pregnancy; HbA1c: Glycated hemoglobin evaluated in the third trimester of pregnancy; * ($p < 0.05$).

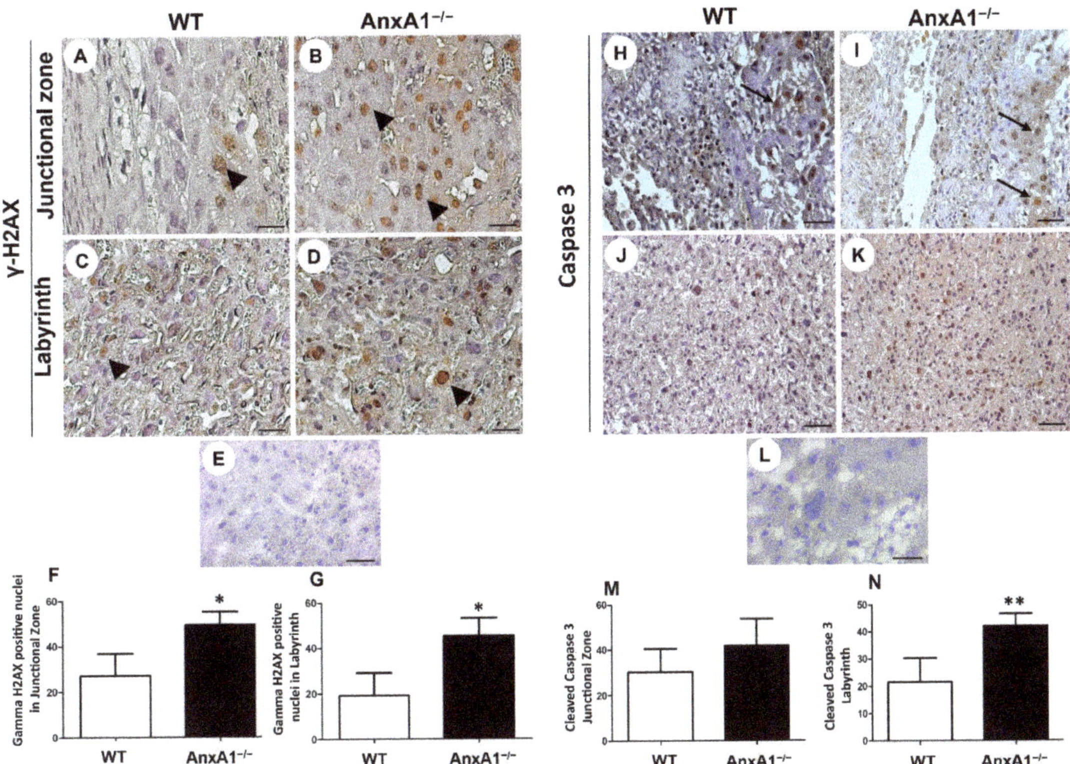

Figure 3. Immunolocalization of γ-H2AX (DNA double-strand breaks, **(A–G)**) and cleaved caspase-3 (apoptosis, **(H–N)**) in placenta sections from WT and AnxA1$^{-/-}$ animals. **(A–E,H–L)** Placental cells reactive to γ-H2AX (arrowheads; **(A–E)**) and caspase-3 (arrows; **(H–L)**) are found in both junctional and labyrinthine placental zones. **(E,L)** Negative control for immunohistochemical analysis. **(A–E,H–L)** Immunoperoxidase and hematoxylin counterstaining. Bars = 50 μm. **(F,G,M,N)** Quantification of γ-H2AX-stained nuclei **(F,G)** and caspase-3-positive cells **(M,N)** per 10,000 μm^2 of placental area. Values are shown as mean ± SD; * $p < 0.05$; ** $p < 0.01$, n = 6/group.

2.5. GDM Changes the Villous Morphology, Leads to Oxidative DNA Damage, and Impairs the Expression of Nuclear DNA Repair Enzymes

As observed in the placentas from ND women (Figure 4A,C,E), the intermediate villus components were preserved in the placentas from women with GDM (Figure 4B,D,F). Syncytiotrophoblast lined the villi, in contact with the intervillous space (filled with maternal blood), which contained vessels and mesenchyme cells. Nuclear aggregates in syncytiotrophoblast (syncytial knots) and fibrin deposits were observed, especially in the GDM group (Figure 4B; Table S1).

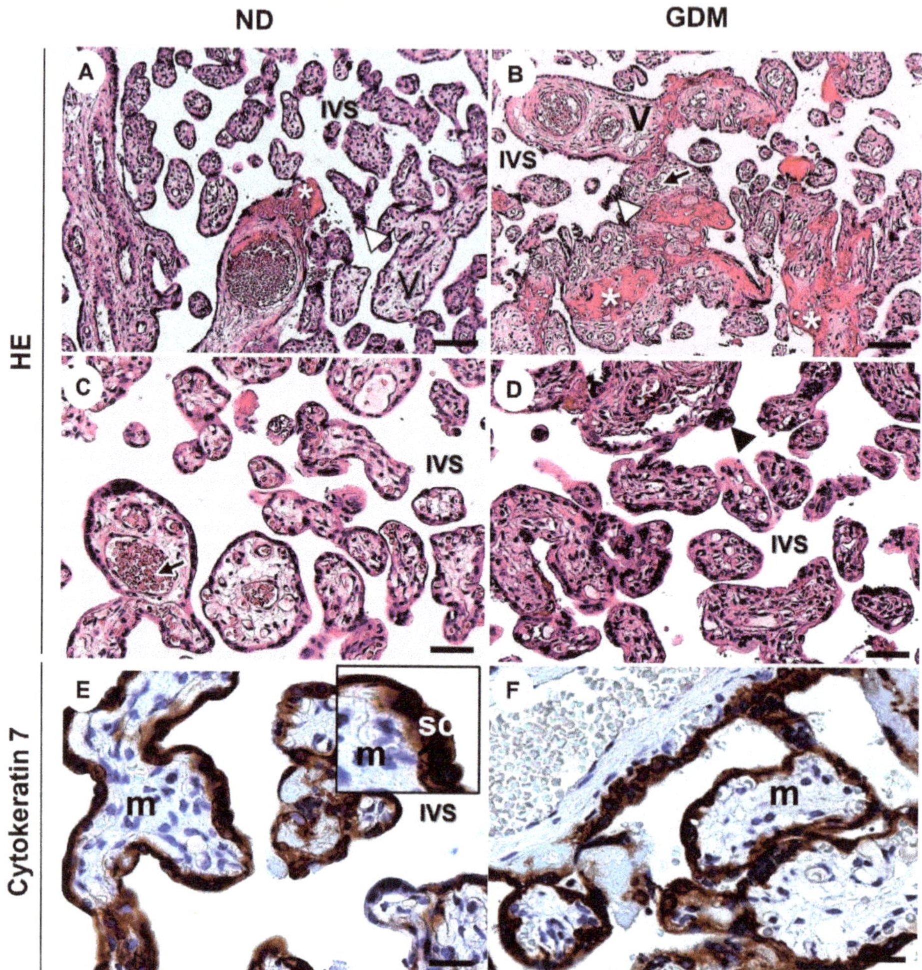

Figure 4. Representative images of villous compartment of term placentas stained with hematoxylin and eosin (**A–D**) and immunolabeling of cytokeratin 7 for identification of the syncytiotrophoblast layer (SC) around the mesenchyme (m) from pregnant women with no diabetes (ND; (**A,C,E**)) or gestational diabetes mellitus (GDM; (**B,D,F**)). (**A–C**) Note the aggregations of syncytiotrophoblast nuclei (arrowhead) and increased intramural fibrinoid (asterisk) mainly in GDM villi (**B**). V: chorionic villous; black arrows: fetal vessels; IVS: intervillous space. Bars = 50 μm.

Oxidative DNA damage was evaluated using nuclear 8-Oxoguanine detection. Cells positive for nuclear staining of 8-Oxoguanine were observed in the syncytiotrophoblast layer, and mesenchymal and endothelial cells in both groups (Figure 5A,B). Quantitative analysis revealed increased DNA damage in placenta samples, particularly in the syncytiotrophoblast from the GDM group (Figure 5B,C).

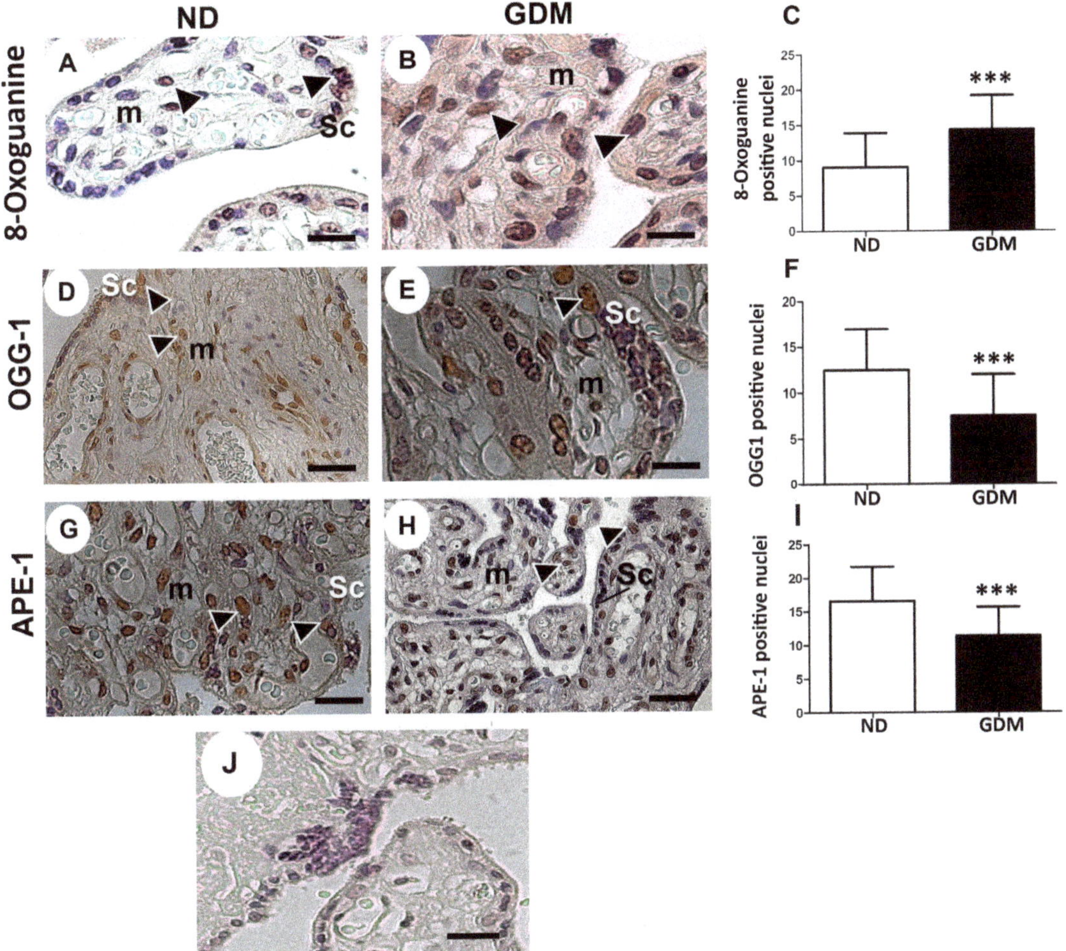

Figure 5. Immunolocalization of oxidative DNA damage (**A–C**) and DNA repair enzymes (**D–I**) in placentas from pregnant women with no diabetes (ND; (**A,D,G**)) or gestational diabetes mellitus (GDM; (**B,E,H**)). Nuclei of the syncytiotrophoblast (arrowheads) and mesenchyme cells (m) were reactive for 8-Oxoguanine (**A,B**), OGG-1 (**D,E**) and APE-1 (**G,H**) in ND and GDM placentas. (**J**) Buffer was used instead of the primary antibody as a negative control. Immunoperoxidase and hematoxylin counterstaining. Bars = 50 μm. (**C,F,I**) Quantification of 8-Oxoguanine (**C**), OGG-1 (**F**), and APE-1 (**I**) reactive nuclei per 10,000 μm^2 of placental area. Values are shown as mean ± SD; *** $p < 0.001$, n = 10/group.

OGG1 (Figure 5D,E) and APE-1 (Figure 5G,H), the enzymes in the BER pathway, were expressed in syncytiotrophoblast, mesenchymal cells, and endothelial cells in the ND group. Compared with the ND group, lower expression of nuclear BER enzymes (Figure 5F,I) was found in placentas from the GDM group, mainly because of reduced expression in the syncytiotrophoblast (Figure 5E,H).

2.6. Placenta from Women with GDM Presents Augmented DNA Double-Strand Breaks and Apoptosis

Double-stranded DNA breaks were evaluated by detecting H2AX phosphorylation at serine 139. Placenta cells exposed to hyperglycemia showed increased DNA double-strand breaks (Figure 6A–C) and apoptotic cells (Figure 6D–F) compared with those in the ND group, as evaluated using activated caspase 3.

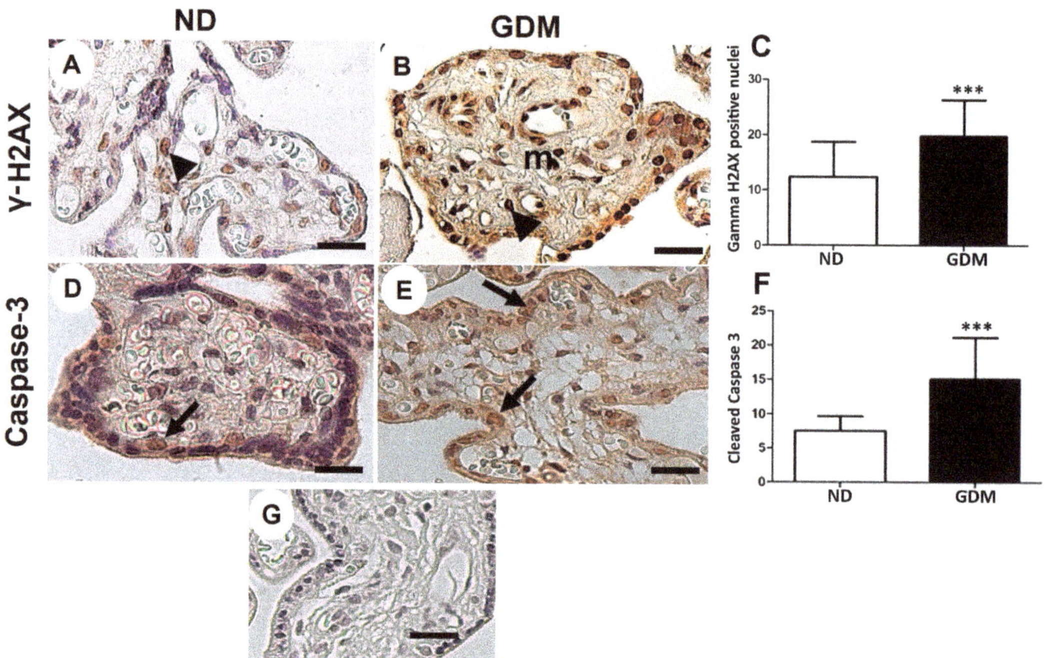

Figure 6. Immunolocalization of γ-H2AX (DNA double-strand breaks, (**A–C**)) and cleaved caspase-3 (apoptosis, (**D–F**)) in placenta from pregnant women with no diabetes (ND; (**A,D**)) or gestational diabetes mellitus (GDM; (**B,E**)). (**A,B,D,E**) Nuclei of the syncytiotrophoblast (arrows) and mesenchymal cells (m) are reactive to γ-H2AX (arrowheads). Caspase-3 reactive cells follow a similar pattern of γ-H2AX immunoreactivity (arrows). (**G**) Negative control used in immunohistochemical analysis. Immunoperoxidase and hematoxylin counterstaining. Bars = 50 μm. (**C,F**) Quantification of γ-H2AX-positive nuclei (**C**) and caspase-3-reactive cells (**F**) per 10,000 μm^2 of placental area. Values are shown as mean ± SD; *** $p < 0.001$, n = 10/group.

2.7. ANXA1 Expression Is Impaired in the Placenta from Women with GDM

In both the groups, ANXA1-positive cells were observed to be present in different cellular components of the placental sections (Figure 7). In chorionic villi, ANXA1 was strongly immunoreactive in the cytoplasm and nuclei of syncytiotrophoblast, and weakly stained in mesenchymal cells (Figure 7A–D). In both syncytiotrophoblast and mesenchymal cells, quantitative analysis of the staining in cytoplasm and nuclei indicated reduced ANXA1 expression in the placentas of GDM women (Figure 7E,F).

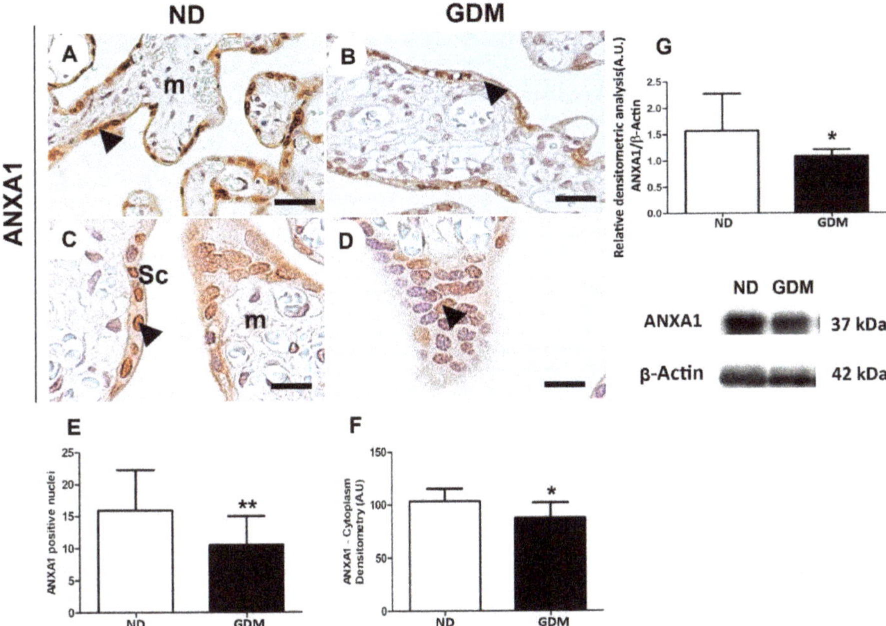

Figure 7. ANXA1 expression in placenta from pregnant women with no diabetes (ND; (**A,C,G**)) or gestational diabetes mellitus (GDM; (**B,D,G**)) evaluated by immunohistochemistry (**A–F**) and western blot (**G**). Placental reactive components (arrowheads) are mainly nuclei and cytoplasm of syncytiotrophoblast (Sc) and some cells found in mesenchyme (m). The sample were counter-stained with hematoxylin. Bars = 50 μm. 40× (**A,B**); 100× (**C,D**) Quantification of positive nuclei in 10,000 μ^2 (**E**) and cytoplasm densitometry (**F**). The relative band intensities from western blot experiments were normalized to β-actin and analyzed with ImageJ software (**G**). Values as mean ± SD; * $p < 0.05$; ** $p < 0.01$, n = 10/group.

Immunoreactive bands (~37 kDa) were detected using the ANXA1 antibody in the extracts of villous placental tissue of pregnant women in both the ND and GDM groups. Densitometric analyses of these bands confirmed a reduction in the expression of ANXA1 in placentas from the GDM group (Figure 7G).

3. Discussion

ANXA1 is synthesized by immune, epithelial, and cancer cells via the action of different chemical mediators, such as glucocorticoids and cytokines [26,27]. ANXA1, stored in cytoplasmic granules and released into the extracellular compartment, binds to membrane formyl peptides for downstream intracellular signaling of anti-inflammatory, proliferative, apoptosis, and migration processes [26,28,29]. Moreover, ANXA1 is also found in the nucleus, where a protective action against DNA damage has recently been proposed [30]. This study also highlights the nuclear localization of ANXA1. Here, the translational data showed that the reduced expression of ANXA1 in the placenta was associated with increased apoptosis, which may reflect a failure to protect DNA from oxidative damage. In this context, modulation of BER enzymes is a possible underlying mechanism.

Placentas deficient in AnxA1 showed increased oxidative DNA damage, reduced expression of nuclear BER enzymes with DNA double-strand breaks, and apoptosis induction. This study found gH2Ax was detected both in the labyrinth and junctional zone. However, increased apoptosis and structural reduction were only detected in the labyrinth. Although we have not studied the possible bases of this difference, a study using giant trophoblastic cells obtained from trophoblastic stem cells showed that these cells can resist apoptosis induced by DNA damage [31]. This resistance is associated mainly with the selective upregulation of p21. The presence of p21 in the cytoplasm of giant cells prevented apoptosis induced by DNA damage, similar to that found in cancer cells able to escape apoptosis. Giant cells constitute a significant component of the junctional zone, which may be an explanation for why lesser apoptosis and maintenance of the junctional layer were found in our KO animals.

Analysis of placentas exposed to an adverse environment, such as hyperglycemia, showed similar outcomes associated with ANXA1 deficiency in the nucleus and cytoplasm of GDM samples.

It is important to emphasize that pregnant women in the GDM group, although treated with a combination of insulin and diet, showed increased HbA1c levels and weight of newborn compared with those in the ND group. The exaggerated supply of glucose from the mother to the fetus explains the weight of the newborns observed in this study [32]. The degree of maternal glucose tolerance is not reflected in clinical aspects. Changes in placental morphology and physiology have been observed in different degrees of hyperglycemic disorders, including GDM [33]. Previous studies have demonstrated that GDM placentas show a higher expression of inflammasome pathway components [24], with an exacerbation in the production of inflammatory cytokines, oxidative stress, DNA damage response, and apoptotic cells [15,22,24,25]. As these effects were detected in the placenta samples from $AnxA1^{-/-}$ mice, our data strongly suggest a protective role for ANXA1 in placental physiology, which can be mitigated in certain adverse conditions, such as hyperglycemia.

An imbalance in the expression of ANXA1 in the placenta of high-risk pregnancies has been previously described. ANXA1 showed protective activity against *Toxoplasma gondii* infection, based on the observation that the third-trimester placentas expressing lesser ANXA1 were more permissive to this infection than the first-trimester placentas expressing more ANXA1. In the third-trimester placental villous explant culture, parasite reduction was observed after treatment with an ANXA1 peptide mimetic (Ac2-26) [34]. Reduced expression of placental AnxA1 was observed in mothers post ZIKA infection. The placentas of these mothers showed an increased inflammatory response and impaired tissue repair [35]. These data may be related to a deficiency in ANXA1, which consolidates the importance of ANXA1 in the placental response to aggression.

As discussed in a review by Boudhraa [6], ANXA1 can translocate to the nucleus in response to mitogenic/proliferative and DNA-damaging stimuli, such as hydrogen peroxide. In the nuclear compartment, ANXA1 binds to DNA, where helicase activity has been proposed, allowing it to include ANXA1 in DNA synthesis and repair mechanisms [10]. In MCF-7 breast cancer cells, the loss of ANXA1 upon stress led to an increased susceptibility to DNA damage and mutation [8]. However, experiments performed in nasopharyngeal carcinoma cells, suggest that knockdown of ANXA1 inhibits DNA damage by decreasing the generation of intracellular reactive oxygen species and the formation of γ-H2AX and promotes DNA repair by increasing DNA-dependent protein kinase activity [36]. Moreover, prevention of nuclear translocation of ANXA1 using the small peptide Tat-NTS inhibited cellular proliferation (G2/M phase arrest), migration, and invasion of glioblastoma cells [37]. Thus, the DNA damage response related to ANXA1 is tissue-dependent and the helicase activity proposed to ANXA1 may explain the involvement of this protein with DNA repair mechanisms.

Our data indicate crucial functions of ANXA1 in placental cells, especially in the nuclear compartment. Under hyperglycemic conditions (a classical oxidative stress condition), we observed a reduction in nuclear expression of ANXA1 in GDM placentas. As observed in MCF7 breast adenocarcinoma cells, our data indicate that the absence or reduction of ANXA1 in placental cells might underlie the cell death response and impair placental function.

Overall, our translational data indicate that differential expression of ANXA1 in the placenta alters the cellular response to oxidative DNA damage, leading to apoptosis. We focused on the reduction in BER enzymes in the nuclear compartment to explain apoptotic signaling. These findings demonstrate the relevance of ANXA1 in placental cell responses and shed light on the mechanisms that regulate the functions of this critical protein in placental biology.

4. Materials and Methods

4.1. Ethical Statement

This cross-sectional study included placental samples from AnxA1 knockout (AnxA1$^{-/-}$) mice and mothers with GDM. All animal procedures were performed according to the Brazilian Society for the Science of Laboratory Animals (SBCAL) and were approved by the Institutional Animal Care and Use Committee of the Faculty of Pharmaceutical Sciences at the University of Sao Paulo (Protocol 521). Placental samples from pregnant women were obtained from the Diabetes and Pregnancy Service of Botucatu Medical School/UNESP, Brazil, with the approval of the Research Ethics Committee (protocol # 48609715.0.0000.5505). Written informed consent was obtained from all participants according to the principles of the Declaration of Helsinki.

4.2. AnxA1 Knockout (AnxA1$^{-/-}$) Placental Samples

Male and female wild type (WT) and AnxA1$^{-/-}$ BALB/c mice, aged 5–6 weeks, were maintained and reproduced at the animal house of the Faculty of Pharmaceutical Sciences, University of Sao Paulo (Brazil). The deficiency of AnxA1 knockout in placentas was demonstrated by immunohistochemical analysis and showed in Supplemental Figure S1. Animals were provided chow (Nuvilab) and water ad libitum. All animals were housed in a temperature-controlled room (22–25 °C and 70% relative humidity) with a 12 h light-dark cycle. Female mice were caged overnight with males (3:1) and successful mating was verified the following morning by the presence of a vaginal plug (day 0.5 of pregnancy). On day 18.5 of pregnancy, mice were euthanized by cervical dislocation or were anesthetized with xylazine and ketamine (i.p., 7 and 77 mg/kg, respectively; Vetbrands, Jacarei, SP, Brazil) [38] for collection of placenta samples (six from each group: WT and AnxA1$^{-/-}$). The placenta samples (n = 6/group) were processed for morphological (hematoxylin-eosin staining) and immunohistochemical analyses.

4.3. Population Characterization and Collection of Human Placenta Samples

GDM was diagnosed using a 75 g glucose tolerance test (75 g-GTT), as recommended by the American Diabetes Association [17], between the 24th and 28th gestational weeks. Twenty placenta samples were used in this study, ten each from the nondiabetic (ND; normal 75 g-GTT) and GDM (abnormal 75 g-GTT) groups.

Population characteristics included age, body mass index (BMI) in the third trimester of pregnancy, weight gain during pregnancy, gestational age at delivery, glycemic mean (GM), and glycated hemoglobin (HbA1c) levels. GM was calculated from the arithmetic mean of plasma glucose levels measured in all glucose profiles obtained during treatment (diet or diet + insulin). Plasma glucose levels were measured using the oxidase method (Glucose Analyzer II Beckman, Fullerton, CA, USA) and HbA1c levels were measured using high-performance liquid chromatography (D10™ Hemoglobin Testing System, Bio-Rad Laboratories, Hercules, CA, USA). Placental and fetal weights were also included.

The inclusion criteria were as follows: (i) hyperglycemia defined at a minimum gestational age of 28–30 weeks; (ii) prenatal and delivery care at the service; (iii) absence of clinically diagnosed infections and negative serology for HIV and syphilis, absence of multiple pregnancies, overt diabetes, fetal malformations, fetal death, alcohol consumption, or illicit drug habits; and (iv) deliveries before the 36th week of gestation.

Placentas were collected immediately after delivery (cesarean section) and cut into smaller fragments from different cotyledons. Decidual and villous areas were dissected and rinsed in phosphate-buffered saline. A part of the placental villous area was subjected to routine morphological and immunohistochemical procedures, and the remaining part was frozen and stored at $-80\ °C$ for western blotting.

4.4. Preparation of Placenta Samples for Histological Assessment

Placenta samples from pregnant women with GDM and from $AnxA1^{-/-}$ mice were fixed in 4% buffered paraformaldehyde for 24 h, dehydrated in a graded ethanol series, and embedded in paraffin (Merck, Darmstadt, Germany). Three-micrometer thick sections were obtained for hematoxylin-eosin staining and immunohistochemistry.

4.5. Morphological and Morphometric Analyses of Placenta Samples from $AnxA1^{-/-}$ Mice

Morphometric analysis was performed on formalin-fixed placental samples stained with hematoxylin and eosin and scanned using the Image-Pro Plus version 4.5 for Windows software—Zeiss-Jenaval (Zeiss-Jenaval, Jena, Germany). Placental slides from six dams in each group were analyzed using the ImageJ software (National Institutes of Health, Bethesda, MD, USA). The thickest point of the labyrinth or junctional zone was first identified to measure the thickness of the placental layer. The thickness of each layer was measured and the ratio of the thickness of each layer to the total thickness of the placenta was calculated.

4.6. Immunohistochemical Detection of Oxidative DNA Damage, DNA-Double Strand Breaks, DNA Repair Enzymes, and Apoptosis in GDM and $AnxA1^{-/-}$ Placenta Samples

The immunohistochemical detection of 8-Oxoguanine (oxidative DNA damage marker), gamma H2AX (DNA-double strand breaks), APE-1 and OGG1 (DNA repair enzymes), and cleaved caspase-3 (apoptosis) in placental sections from GDM and $AnxA1^{-/-}$ mice was performed. Endogenous peroxide activity was blocked using 3% hydrogen peroxide for 30 min. The tissue sections were then incubated with the following primary antibodies overnight at $4\ °C$: 8-Oxoguanine (Abcam, Cambridge, UK), gamma H2AX (Novus Biologicals, Littleton, CO, USA), OGG1 (Novus Biologicals), APE-1 (Novus Biologicals), polyclonal rabbit anti-caspase-3 (Abcam). For negative control, sections were incubated with 10% TBS-BSA (Sigma–Aldrich, St. Louis, MO, USA) instead of the primary antibody. After washing with TBS, the sections were incubated with horseradish peroxidase-conjugated secondary antibodies (Abcam). Staining was visualized using a 3,3$'$-diaminobenzidine substrate (Invitrogen, Waltham, MA, USA). Finally, the sections were counterstained with hematoxylin (Inlab Confiança, São Paulo, Brazil) and mounted using Entellan (Merck, Germany).

4.7. Nuclear OGG1, APE-1, and Caspase-3 Analysis in Placenta from $AnxA1^{-/-}$ Mice

Cells with positive nuclear staining for OGG1 and APE-1 in the immunohistochemical analysis were counted in ten fields ($\times 40$) in each placental zone (labyrinthine or junctional) for all placenta samples. The ImageJ software (NIH) was used to quantify the area and number of positive nuclei to calculate the ratio of positive nuclei per placental zone.

Data are presented as the number of positive cells per $10^4\ \mu m^2$ area. Cleaved caspase-3 analysis was performed similarly, considering the number of positive cells. Images were obtained using an Axioskop2-Mot Plus Microscope (Carl Zeiss, Jena, Germany) with the AxioVision software (SE64 Rel. 4.9.1).

4.8. Morphological Analysis of Placenta from Pregnant Women with GDM

Morphological analysis was performed on formalin-fixed placenta samples stained with hematoxylin-eosin and cytokeratin 7. The expression of cytokeratin 7 (a specific marker for trophoblast cells) was detected immunohistochemically using polyclonal rabbit IgG anti-cytokeratin 7 (Abcam; 1:50). Images were obtained using an Axioskop 2-Mot Plus Microscope (Carl Zeiss) with the AxioVision software.

4.9. Analysis of Nuclear OGG1, APE-1, and Caspase-3 in Placenta from Pregnant Women with GDM

Cells with positive nuclear staining for OGG1 and APE-1 in the villous area were counted in ten fields ($\times 40$) of each placenta. As described for mouse placentas, the ImageJ software (NIH) was used to quantify the villous area, and the ratio between the number of positive nuclei and villous area was calculated. Data are presented as the number of positive cells per 10^4 μm^2 area. Cleaved caspase-3 analysis was performed considering positive cells.

4.10. Detection and Analysis of Cytoplasmic and Nuclear ANXA1 in GDM Placentas

The expression of ANXA1 was determined by immunohistochemical staining as described above, using the following primary antibody: polyclonal rabbit anti-AnxA1 (Zymed Laboratories, Cambridge, UK) at 1:5000 for 1 h.

The intensity of cytoplasmic ANXA1 staining in villous cells was measured densitometrically using 10^3 random points from ten fields ($\times 40$) from each placenta on an arbitrary scale from 0 to 255.

Cells with positive nuclear staining for ANXA1 in the villous area were counted in ten fields ($\times 40$) of each placenta. The ImageJ software (NIH) was used to quantify the villous area and the ratio of positive nuclei to the villous area was calculated. Data are presented as the number of positive cells per 10^4 μm^2 area. Images were obtained using an Axioskop 2-Mot Plus Microscope (Carl Zeiss, Jena, Germany) with the AxioVision software (SE64 Rel. 4.9.1).

4.11. Western Blot Analysis of ANXA1 Using Placental Extracts from Patients with GDM

Villous fragments of frozen human placental tissues were transferred to propylene tubes containing the lysis buffer (Merck, Darmstadt, Germany) and a cocktail of protease inhibitors (Complete Mini, EDTA-free protease inhibitor cocktail tablets, Roche, Switzerland). Samples were homogenized on ice using an electric homogenizer. The homogenates were centrifuged at 12,000 rpm for 15 min at 4 °C. The protein concentration in the supernatant was measured using a BCA protein assay (Pierce^{TM} BCA Protein Assay Kit, Thermo Scientific, Waltham, MA, USA) and then stored at -80 °C for western blot analysis.

Equal amounts of total protein from each group were separated electrophoretically on a 15% SDS-polyacrylamide gel and then transferred to a 0.45 μm nitrocellulose membrane (Millipore, Burlington, MA, USA). The transfer of proteins was confirmed by staining the membranes with 10% Ponceau S solution (Sigma Aldrich, St. Louis, MO, USA). The blotted membranes were blocked with 3% TBS-T-milk for 1 h, washed three times with TBS buffer, and then incubated overnight at 4 °C with anti-β-actin (1:2000; Novus Biologicals, Centennial, ON, USA) and anti-ANXA1 (1:5000; Invitrogen, Carlsbad, CA, USA) antibodies in 3% TBS-T-milk and washed three times with TBS buffer. The membranes were exposed to a horseradish peroxidase-conjugated secondary antibody (1:1000; Abcam) in 3% TBS-T-milk for 1 h and washed three times with TBS buffer. Immunoreactive bands (indicative of peroxidase activity) were detected using the enhanced chemiluminescence method. Quantitative analysis of ANXA1 was performed by densitometry using the ImageJ software (NIH, Bethesda, MD, USA). β-Actin was used as a loading control.

4.12. Statistical Analysis

Data are presented as mean $\pm$ standard deviation (SD). Statistical analyses were performed using the GraphPad software version 6.00. First, we performed the Kolmogorov–Smirnov normality test to determine whether the data distribution was parametric or nonparametric. Student's *t*-test or the Mann–Whitney *U* test was used for group comparisons. Statistical significance was set at $p < 0.05$.

Supplementary Materials: The supporting information can be downloaded at: https://www.mdpi.com/article/10.3390/ijms241210155/s1. Reference [39] is cited in the supplementary materials Table S1.

Author Contributions: Study design: J.B.M. and S.M.O.; Study conduct: J.B.M. and M.R.d.S.; Data collection: J.B.M., I.d.M.P.C., C.B.H. and S.H.P.F.; Data analysis: J.B.M., M.R.d.S., E.B. and S.M.O.; Data interpretation: J.B.M., S.H.P.F., E.B. and S.M.O.; Drafting manuscript: J.B.M., M.R.d.S., E.B., S.H.P.F. and S.M.O.; Revising manuscript content: All authors. Approving the final version of the manuscript: all authors take responsibility for the integrity of the data analysis. All authors have read and agreed to the published version of the manuscript.

Funding: Fundação de Amparo à Pesquisa do Estado de São Paulo—FAPESP (SMO, grant number 2019/19949-7 and SRPF, grant number 2014/07328-4) for financial support and Coordenação de Aperfeiçoamento de Pessoal de Nível Superior (CAPES) for M.R.d.S. and J.B.M. fellowship.

Institutional Review Board Statement: This cross-sectional study included placental samples from AnxA1 knockout (AnxA1$^{-/-}$) mice and mothers with GDM. All animal procedures were performed according to the Brazilian Society for the Science of Laboratory Animals (SBCAL) and were approved by the Institutional Animal Care and Use Committee of the Faculty of Pharmaceutical Sciences at the University of Sao Paulo (Protocol 521). Placental samples from pregnant women were obtained from the Diabetes and Pregnancy Service of Botucatu Medical School/UNESP, Brazil, with the approval of the Research Ethics Committee (protocol # 48609715.0.0000.5505). Written informed consent was obtained from all participants according to the principles of the Declaration of Helsinki.

Informed Consent Statement: Informed consent was obtained from all subjects involved in the study.

Data Availability Statement: The data presented in this study are available on request from the corresponding author.

Conflicts of Interest: The authors declare no conflict of interest.

References

1. Flower, R.J. Lipocortin and the mechanism of action of the glucocorticoids. *Br. J. Pharmacol.* **1988**, *94*, 987–1015. [CrossRef] [PubMed]
2. Crumpton, M.J.; Dedman, J.R. Protein terminology tangle. *Nature* **1990**, *345*, 212. [CrossRef] [PubMed]
3. Perretti, M.; D'Acquisto, F. Annexin A1 and glucocorticoids as effectors of the resolution of inflammation. *Nat. Rev. Immunol.* **2009**, *9*, 62–70. [CrossRef] [PubMed]
4. Lim, L.H.; Pervaiz, S. Annexin 1: The new face of an old molecule. *FASEB J.* **2007**, *21*, 968–975. [CrossRef]
5. Chittajallu, R.; Vignes, M.; Dev, K.K.; Barnes, J.M.; Collingridge, G.L.; Henley, J.M. Regulation of glutamate release by presynaptic kainate receptors in the hippocampus. *Nature* **1996**, *379*, 78–81. [CrossRef]
6. Boudhraa, Z.; Bouchon, B.; Viallard, C.; D'Incan, M.; Degoul, F. Annexin A1 localization and its relevance to cancer. *Clin. Sci.* **2016**, *130*, 205–220. [CrossRef]
7. Rhee, H.J.; Kim, G.Y.; Huh, J.W.; Kim, S.W.; Na, D.S. Annexin I is a stress protein induced by heat, oxidative stress and a sulfhydryl-reactive agent. *Eur. J. Biochem.* **2000**, *267*, 3220–3225. [CrossRef]
8. Nair, S.; Hande, M.P.; Lim, L.H. Annexin-1 protects MCF7 breast cancer cells against heat-induced growth arrest and DNA damage. *Cancer Lett.* **2010**, *294*, 111–117. [CrossRef]
9. Hirata, A.; Hirata, F. Lipocortin (Annexin) I heterotetramer binds to purine RNA and pyrimidine DNA. *Biochem. Biophys. Res. Commun.* **1999**, *265*, 200–204. [CrossRef]
10. Hirata, A.; Hirata, F. DNA chain unwinding and annealing reactions of lipocortin (annexin) I heterotetramer: Regulation by Ca^{2+} and Mg^{2+}. *Biochem. Biophys. Res. Commun.* **2002**, *291*, 205–209. [CrossRef] [PubMed]
11. Lin, S.; Hickey, R.; Malkas, L. The biochemical status of the DNA synthesome can distinguish between permanent and temporary cell growth arrest. *Cell Growth Differ.* **1997**, *8*, 1359–1369. [PubMed]
12. Bokhari, B.; Sharma, S. Stress Marks on the Genome: Use or Lose? *Int. J. Mol. Sci.* **2019**, *20*, 364. [CrossRef]
13. Jackson, S.P.; Bartek, J. The DNA-damage response in human biology and disease. *Nature* **2009**, *461*, 1071–1078. [CrossRef] [PubMed]
14. Moreli, J.B.; Santos, J.H.; Rocha, C.R.; Damasceno, D.C.; Morceli, G.; Rudge, M.V.; Bevilacqua, E.; Calderon, I.M. DNA damage and its cellular response in mother and fetus exposed to hyperglycemic environment. *BioMed Res. Int.* **2014**, *2014*, 676758. [CrossRef]
15. Moreli, J.B.; Santos, J.H.; Lorenzon-Ojea, A.R.; Corrêa-Silva, S.; Fortunato, R.S.; Rocha, C.R.; Rudge, M.V.; Damasceno, D.C.; Bevilacqua, E.; Calderon, I.M. Hyperglycemia Differentially Affects Maternal and Fetal DNA Integrity and DNA Damage Response. *Int. J. Biol. Sci.* **2016**, *12*, 466–477. [CrossRef]
16. Whitaker, A.M.; Schaich, M.A.; Smith, M.R.; Flynn, T.S.; Freudenthal, B.D. Base excision repair of oxidative DNA damage: From mechanism to disease. *Front. Biosci.* **2017**, *22*, 1493–1522.
17. American Diabetes Association Professional Practice Committee. 2. Classification and Diagnosis of Diabetes: Standards of Medical Care in Diabetes-2022. *Diabetes Care* **2022**, *45* (Suppl. 1), S17–S38. [CrossRef]

18. Desoye, G.; Wells, J.C.K. Pregnancies in Diabetes and Obesity: The Capacity-Load Model of Placental Adaptation. *Diabetes* **2021**, *70*, 823–830. [CrossRef]
19. Damasceno, D.C.; Netto, A.O.; Iessi, I.L.; Gallego, F.Q.; Corvino, S.B.; Dallaqua, B.; Sinzato, Y.K.; Bueno, A.; Calderon, I.M.; Rudge, M.V. Streptozotocin-induced diabetes models: Pathophysiological mechanisms and fetal outcomes. *BioMed Res. Int.* **2014**, *2014*, 819065. [CrossRef]
20. Hoch, D.; Gauster, M.; Hauguel-de Mouzon, S.; Desoye, G. Diabesity-associated oxidative and inflammatory stress signalling in the early human placenta. *Mol. Asp. Med.* **2019**, *66*, 21–30. [CrossRef]
21. Bedell, S.; Hutson, J.; de Vrijer, B.; Eastabrook, G. Effects of Maternal Obesity and Gestational Diabetes Mellitus on the Placenta: Current Knowledge and Targets for Therapeutic Interventions. *Curr. Vasc. Pharmacol.* **2021**, *19*, 176–192. [CrossRef]
22. Sgarbosa, F.; Barbisan, L.F.; Brasil, M.A.; Costa, E.; Calderon, I.M.; Gonçalves, C.R.; Bevilacqua, E.; Rudge, M.V. Changes in apoptosis and Bcl-2 expression in human hyperglycemic, term placental trophoblast. *Diabetes Res. Clin. Pract.* **2006**, *73*, 143–149. [CrossRef]
23. Moreli, J.B.; Corrêa-Silva, S.; Damasceno, D.C.; Sinzato, Y.K.; Lorenzon-Ojea, A.R.; Borbely, A.U.; Rudge, M.V.; Bevilacqua, E.; Calderon, I.M. Changes in the TNF-alpha/IL-10 ratio in hyperglycemia-associated pregnancies. *Diabetes Res. Clin. Pract.* **2015**, *107*, 362–369. [CrossRef] [PubMed]
24. Corrêa-Silva, S.; Alencar, A.P.; Moreli, J.B.; Borbely, A.U.; de SLima, L.; Scavone, C.; Damasceno, D.C.; Rudge, M.V.C.; Bevilacqua, E.; Calderon, I.M.P. Hyperglycemia induces inflammatory mediators in the human chorionic villous. *Cytokine* **2018**, *111*, 41–48. [CrossRef] [PubMed]
25. Myatt, L.; Cui, X. Oxidative stress in the placenta. *Histochem. Cell Biol.* **2004**, *122*, 369–382. [CrossRef]
26. Prieto-Fernández, L.; Menéndez, S.T.; Otero-Rosales, M.; Montoro-Jiménez, I.; Hermida-Prado, F.; García-Pedrero, J.M.; Álvarez-Teijeiro, S. Pathobiological functions and clinical implications of annexin dysregulation in human cancers. *Front. Cell Dev. Biol.* **2022**, *10*, 1009908. [CrossRef]
27. Ganesan, T.; Sinniah, A.; Ibrahim, Z.A.; Chik, Z.; Alshawsh, M.A. Annexin A1: A Bane or a Boon in Cancer? A Systematic Review. *Molecules* **2020**, *25*, 3700. [CrossRef] [PubMed]
28. Petrella, A.; Festa, M.; Ercolino, S.F.; Zerilli, M.; Stassi, G.; Solito, E.; Parente, L. Induction of annexin-1 during TRAIL-induced apoptosis in thyroid carcinoma cells. *Cell Death Differ.* **2005**, *12*, 1358–1360. [CrossRef] [PubMed]
29. Alldridge, L.C.; Bryant, C.E. Annexin 1 regulates cell proliferation by disruption of cell morphology and inhibition of cyclin D1 expression through sustained activation of the ERK1/2 MAPK signal. *Exp. Cell Res.* **2003**, *290*, 93–107. [CrossRef] [PubMed]
30. Oliani, S.M.; Perretti, M. Cell localization of the anti-inflammatory protein annexin 1 during experimental inflammatory response. *Ital. J. Anat. Embryol.* **2001**, *106*, 69–77.
31. de Renty, C.; DePamphilis, M.; Ullah, Z. Cytoplasmic localization of p21 protects trophoblast giant cells from DNA damage induced apoptosis. *PLoS ONE* **2014**, *9*, e97434. [CrossRef] [PubMed]
32. Hay, W.W., Jr. Placental-fetal glucose exchange and fetal glucose metabolism. *Trans. Am. Clin. Climatol. Assoc.* **2006**, *117*, 321–340. [PubMed]
33. Desoye, G.; Cervar-Zivkovic, M. Diabetes Mellitus, Obesity, and the Placenta. *Obstet. Gynecol. Clin. N. Am.* **2020**, *47*, 65–79.
34. de Oliveira Cardoso, M.F.; Moreli, J.B.; Gomes, A.O.; de Freitas Zanon, C.; Silva, A.E.; Paulesu, L.R.; Ietta, F.; Mineo, J.R.; Ferro, E.A.; Oliani, S.M. Annexin A1 peptide is able to induce an anti-parasitic effect in human placental explants infected by Toxoplasma gondii. *Microb. Pathog.* **2018**, *123*, 153–161. [CrossRef] [PubMed]
35. Molás, R.B.; Ribeiro, M.R.; Ramalho Dos Santos, M.J.C.; Borbely, A.U.; Oliani, D.V.; Oliani, A.H.; Nadkarni, S.; Nogueira, M.L.; Moreli, J.B.; Oliani, S.M. The involvement of annexin A1 in human placental response to maternal Zika virus infection. *Antiviral. Res.* **2020**, *179*, 104809. [CrossRef]
36. Liao, L.; Yan, W.J.; Tian, C.M.; Li, M.Y.; Tian, Y.Q.; Zeng, G.Q. Knockdown of Annexin A1 Enhances Radioresistance and Inhibits Apoptosis in Nasopharyngeal Carcinoma. *Technol. Cancer Res. Treat.* **2018**, *17*, 1533034617750309. [CrossRef]
37. Luo, Z.; Liu, L.; Li, X.; Chen, W.; Lu, Z. Correction to: Tat-NTS Suppresses the Proliferation, Migration and Invasion of Glioblastoma Cells by Inhibiting Annexin-A1 Nuclear Translocation. *Cell. Mol. Neurobiol.* **2022**, *42*, 2727–2732. [CrossRef]
38. Hebeda, C.B.; Machado, I.D.; Reif-Silva, I.; Moreli, J.B.; Oliani, S.M.; Nadkarni, S.; Perretti, M.; Bevilacqua, E.; Farsky, S.H.P. Endogenous annexin A1 (AnxA1) modulates early-phase gestation and offspring sex-ratio skewing. *J. Cell. Physiol.* **2018**, *233*, 6591–6603. [CrossRef]
39. Meyerholz, D.K.; Beck, A.P. Fundamental Concepts for Semiquantitative Tissue Scoring in Translational Research. *ILAR J.* **1998**, *59*, 13–17. [CrossRef]

International Journal of
Molecular Sciences

Editorial

Special Issue "Physiology and Pathophysiology of the Placenta"

Giovanni Tossetta

Department of Experimental and Clinical Medicine, Università Politecnica delle Marche, 60126 Ancona, Italy;
g.tossetta@univpm.it

Citation: Tossetta, G. Special Issue
"Physiology and Pathophysiology of
the Placenta". *Int. J. Mol. Sci.* **2024**, *25*,
3594. https://doi.org/10.3390/
ijms25073594

Received: 14 March 2024
Accepted: 20 March 2024
Published: 22 March 2024

The placenta is a transient but essential organ for normal in utero development, playing several essential functions in normal pregnancy [1–5]. The important role of the placenta during pregnancy is highlighted when placental development is impaired, leading to the development of pregnancy complications such as gestational trophoblastic diseases (GTD), gestational diabetes mellitus (GDM), preeclampsia (PE), preterm birth (PTB) and intrauterine growth restriction (IUGR) [6–9].

In this Special Issue, articles (6) and reviews (5) addressing the major problems in the physiology and pathophysiology of the placenta have been selected for publication.

The study by Peñailillo and colleagues evaluated the role of forkhead box M1 (FOXM1), a transcription factor involved in angiogenesis and cell migration [10–12] in trophoblast invasion, identifying that FOXM1 expression is significantly higher in trophoblast cells exposed to hypoxia (3% O_2), while its expression significantly decreases under more tight hypoxic conditions (1% O_2). FOXM1 overexpression in HTR-8/SVneo cells increases their migration and tubule formation ability. Moreover, trophospheres obtained from a 3D culture show higher FOXM1 expression compared to pre-invasion trophospheres. Furthermore, FOXM1-depletion in HTR-8/SVneo cells leads to the downregulation of Polo-Like Kinase 4 (PLK4), Vascular Endothelial Growth Factor (VEGF), and Matrix Metallopeptidase 2 (MMP-2) mRNA expression demonstrating how FOXM1 participates in embryo implantation by favoring early trophoblast invasion.

Placenta accreta spectrum (PAS) is an important pregnancy complication caused by an excessive invasion of cytotrophoblast cells in the endometrial–myometrial interface [13,14]. Timofeeva and colleagues reported how the evaluation of the serum levels of miR-26a-5p, miR-17-5p, and miR-101-3p in the first trimester demonstrated 100% sensitivity in detecting PAS. Thus, the detection of these miRNAs can serve as an auxiliary method for the first-trimester screening of pregnant women. The importance of evaluating miRNAs during pregnancy has also been highlighted in the review article published by Giannubilo and colleagues.

PE occurs in 5–7% of pregnancies and is generally diagnosed in the second half of pregnancy when its clinical manifestations (proteinuria and hypertension) occur [15–17]. Shallow trophoblast invasion in the endometrium/myometrium characterizing PE is also the cause of the occurrence of a hypoxic environment, which leads to increased oxidative stress [18,19]. Oxidative stress is involved in several complications and diseases, including cancer, neurodegenerative diseases and endothelial dysfunction [20–26].

Antihypertensive therapy is essential for the management of patients with PE, and methyldopa (Dopegyt®) and nifedipine (Cordaflex®) are the common drugs used to stabilize blood pressure [27,28]. The study by Ziganshina and colleagues analyzed the effect of antihypertensive therapy on the expression of fucosylated glycans in the fetal capillaries of placental terminal villi in patients with early-onset PE (EOPE; onset < 34 weeks of gestation) and late-onset PE (LOPE; onset ≥ 34 weeks of gestation). The authors found that the expression patterns of fucosylated glycans in endothelial glycocalyx (eGC) in the terminal villi of EOPE and LOPE are characterized by the predominant expression of structures with a type-2 core due to one or both being antihypertensive drugs. These changes in eGC fucoglycans were in accordance with maternal hemodynamics, fetoplacental hemodynamics, and humoral factors associated with eGC damage, demonstrating the effects of antihypertensive therapy on placental eGC in women with PE.

The study by Filippi and colleagues evaluated umbilical venous and arterial oxygen levels, fetal oxygen extraction, oxygen content, CO_2, and lactate in healthy newborns with gestational age < 37 weeks and found a progressive decrease in oxygen levels associated with a concomitant increase in CO_2 levels and reduction in pH starting from the 23rd week to the 33rd–34th week of gestation. From the 33rd–34th week onwards, fetal oxygenation increased, demonstrating that oxygenation during intrauterine life continues to vary even after placenta development.

Placental protein 13 (PP13) is a key protein involved in vascular remodeling and immune tolerance [29–31]. Kazatsker and colleagues evaluated soluble and placental-associated extracellular vesicles (PEV-associated PP13) in the maternal uterine vein in normal, PE and preterm conditions. The authors found that soluble PP13 was not significantly altered across PE, preterm and term delivery, but after depleting the PEV of their proteome, the total PP13 (soluble and PEV-associated PP13) was increased in preterm PE (but not in cases of preterm or term delivery). Corticosteroid treatment caused a depletion of PP13 from the PEV, especially in preterm PE patients.

The functions of annexin A1 (ANXA1) have a key role in protecting cells against DNA damage [32,33]. The study by Moreli and colleagues found that, in ANXA1 knockout mice (AnxA1−/−), the labyrinth zone of the placenta was reduced, and there was increased DNA damage, causing apoptosis in the labyrinthine and junctional layers. In pregnant women with gestational diabetes mellitus (GDM), placenta AnxA1 expression was reduced, and there was increased DNA damage and apoptosis, suggesting the possible involvement of ANXA1 in oxidative DNA damage response under hyperglycemia.

In addition to these insightful research articles, two reviews in this Special Issue highlight the role of uterine receptivity and point mutations in recurrent miscarriage (reviewed by Günther and colleagues) and pregnancy loss (reviewed by Maksiutenko and colleagues), respectively. Furthermore, Berna-Erro and colleagues review the main platelet modifications during hypoxia in order to highlight new platelet markers of hypoxic conditions during labor.

Finally, the review by Benagiano and colleagues highlights the nature of maternal–embryonic communication and the major mechanisms active during the pre-implantation period in order to better understand how the placenta forms during pregnancy.

The eleven articles published in this Special Issue prove the growing interest in finding new molecular targets and mechanisms involved in the regulation of placental development and new markers to predict pregnancy complications. We hope to provide our readers with a new representative and useful snapshots of the current problems in placental development in order to inspire new studies in this field. I personally acknowledge all the contributors of this Special Issue, the Editorial Board, and the assistant editors of the *International Journal of Molecular Sciences* for their support.

Conflicts of Interest: The author declares no conflict of interest.

References

1. Cardaropoli, S.; Paulesu, L.; Romagnoli, R.; Ietta, F.; Marzioni, D.; Castellucci, M.; Rolfo, A.; Vasario, E.; Piccoli, E.; Todros, T. Macrophage migration inhibitory factor in fetoplacental tissues from preeclamptic pregnancies with or without fetal growth restriction. *Clin. Dev. Immunol.* **2012**, *2012*, 639342. [CrossRef] [PubMed]
2. Marzioni, D.; Crescimanno, C.; Zaccheo, D.; Coppari, R.; Underhill, C.B.; Castellucci, M. Hyaluronate and CD44 expression patterns in the human placenta throughout pregnancy. *Eur. J. Histochem.* **2001**, *45*, 131–140. [CrossRef] [PubMed]
3. Marzioni, D.; Fiore, G.; Giordano, A.; Nabissi, M.; Florio, P.; Verdenelli, F.; Petraglia, F.; Castellucci, M. Placental expression of substance P and vasoactive intestinal peptide: Evidence for a local effect on hormone release. *J. Clin. Endocrinol. Metab.* **2005**, *90*, 2378–2383. [CrossRef] [PubMed]
4. Tossetta, G.; Avellini, C.; Licini, C.; Giannubilo, S.R.; Castellucci, M.; Marzioni, D. High temperature requirement A1 and fibronectin: Two possible players in placental tissue remodelling. *Eur. J. Histochem.* **2016**, *60*, 2724. [CrossRef]

5. Muhlhauser, J.; Marzioni, D.; Morroni, M.; Vuckovic, M.; Crescimanno, C.; Castellucci, M. Codistribution of basic fibroblast growth factor and heparan sulfate proteoglycan in the growth zones of the human placenta. *Cell Tissue Res.* **1996**, *285*, 101–107. [CrossRef] [PubMed]

6. Fantone, S.; Giannubilo, S.R.; Marzioni, D.; Tossetta, G. HTRA family proteins in pregnancy outcome. *Tissue Cell* **2021**, *72*, 101549. [CrossRef] [PubMed]

7. Marzioni, D.; Quaranta, A.; Lorenzi, T.; Morroni, M.; Crescimanno, C.; De Nictolis, M.; Toti, P.; Muzzonigro, G.; Baldi, A.; De Luca, A.; et al. Expression pattern alterations of the serine protease HtrA1 in normal human placental tissues and in gestational trophoblastic diseases. *Histol. Histopathol.* **2009**, *24*, 1213–1222. [CrossRef] [PubMed]

8. Cecati, M.; Sartini, D.; Campagna, R.; Biagini, A.; Ciavattini, A.; Emanuelli, M.; Giannubilo, S.R. Molecular analysis of endometrial inflammation in preterm birth. *Cell. Mol. Biol.* **2017**, *63*, 51–57. [CrossRef] [PubMed]

9. Marzioni, D.; Todros, T.; Cardaropoli, S.; Rolfo, A.; Lorenzi, T.; Ciarmela, P.; Romagnoli, R.; Paulesu, L.; Castellucci, M. Activating protein-1 family of transcription factors in the human placenta complicated by preeclampsia with and without fetal growth restriction. *Placenta* **2010**, *31*, 919–927. [CrossRef]

10. Zhang, Z.; Li, M.; Sun, T.; Zhang, Z.; Liu, C. FOXM1: Functional Roles of FOXM1 in Non-Malignant Diseases. *Biomolecules* **2023**, *13*, 857. [CrossRef]

11. Yi, D.; Liu, B.; Wang, T.; Liao, Q.; Zhu, M.M.; Zhao, Y.Y.; Dai, Z. Endothelial Autocrine Signaling through CXCL12/CXCR4/FoxM1 Axis Contributes to Severe Pulmonary Arterial Hypertension. *Int. J. Mol. Sci.* **2021**, *22*, 3182. [CrossRef] [PubMed]

12. Wang, R.T.; Miao, R.C.; Zhang, X.; Yang, G.H.; Mu, Y.P.; Zhang, Z.Y.; Qu, K.; Liu, C. Fork head box M1 regulates vascular endothelial growth factor-A expression to promote the angiogenesis and tumor cell growth of gallbladder cancer. *World J. Gastroenterol.* **2021**, *27*, 692–707. [CrossRef] [PubMed]

13. Ghosh, A.; Lee, S.; Lim, C.; Vogelzang, R.L.; Chrisman, H.B. Placenta Accreta Spectrum: An Overview. *Semin. Interv. Radiol.* **2023**, *40*, 467–471. [CrossRef] [PubMed]

14. Dar, P.; Doulaveris, G. First-trimester screening for placenta accreta spectrum. *Am. J. Obstet. Gynecol. MFM* **2024**, 101329. [CrossRef] [PubMed]

15. Marin, R.; Chiarello, D.I.; Abad, C.; Rojas, D.; Toledo, F.; Sobrevia, L. Oxidative stress and mitochondrial dysfunction in early-onset and late-onset preeclampsia. *Biochim. Biophys. Acta Mol. Basis Dis.* **2020**, *1866*, 165961. [CrossRef] [PubMed]

16. Hurrell, A.; Duhig, K.; Vandermolen, B.; Shennan, A.H. Recent advances in the diagnosis and management of pre-eclampsia. *Fac. Rev.* **2020**, *9*, 10. [CrossRef] [PubMed]

17. Aneman, I.; Pienaar, D.; Suvakov, S.; Simic, T.P.; Garovic, V.D.; McClements, L. Mechanisms of Key Innate Immune Cells in Early- and Late-Onset Preeclampsia. *Front. Immunol.* **2020**, *11*, 1864. [CrossRef] [PubMed]

18. Yang, M.; Wang, M.; Li, N. Advances in pathogenesis of preeclampsia. *Arch. Gynecol. Obstet.* **2024**, 1–9. [CrossRef] [PubMed]

19. Grzeszczak, K.; Lanocha-Arendarczyk, N.; Malinowski, W.; Zietek, P.; Kosik-Bogacka, D. Oxidative Stress in Pregnancy. *Biomolecules* **2023**, *13*, 1768. [CrossRef]

20. Campagna, R.; Mateuszuk, L.; Wojnar-Lason, K.; Kaczara, P.; Tworzydlo, A.; Kij, A.; Bujok, R.; Mlynarski, J.; Wang, Y.; Sartini, D.; et al. Nicotinamide N-methyltransferase in endothelium protects against oxidant stress-induced endothelial injury. *Biochim. Biophys. Acta Mol. Cell Res.* **2021**, *1868*, 119082. [CrossRef]

21. Emanuelli, M.; Sartini, D.; Molinelli, E.; Campagna, R.; Pozzi, V.; Salvolini, E.; Simonetti, O.; Campanati, A.; Offidani, A. The Double-Edged Sword of Oxidative Stress in Skin Damage and Melanoma: From Physiopathology to Therapeutic Approaches. *Antioxidants* **2022**, *11*, 612. [CrossRef] [PubMed]

22. Ekundayo, B.E.; Obafemi, T.O.; Adewale, O.B.; Obafemi, B.A.; Oyinloye, B.E.; Ekundayo, S.K. Oxidative Stress, Endoplasmic Reticulum Stress and Apoptosis in the Pathology of Alzheimer's Disease. *Cell Biochem. Biophys.* **2024**, 1–21. [CrossRef]

23. Bacchetti, T.; Campagna, R.; Sartini, D.; Cecati, M.; Morresi, C.; Bellachioma, L.; Martinelli, E.; Rocchetti, G.; Lucini, L.; Ferretti, G.; et al. *C. spinosa* L. subsp. rupestris Phytochemical Profile and Effect on Oxidative Stress in Normal and Cancer Cells. *Molecules* **2022**, *27*, 6488. [CrossRef] [PubMed]

24. Sartini, D.; Campagna, R.; Lucarini, G.; Pompei, V.; Salvolini, E.; Mattioli-Belmonte, M.; Molinelli, E.; Brisigotti, V.; Campanati, A.; Bacchetti, T.; et al. Differential immunohistochemical expression of paraoxonase-2 in actinic keratosis and squamous cell carcinoma. *Hum. Cell* **2021**, *34*, 1929–1931. [CrossRef] [PubMed]

25. Campagna, R.; Pozzi, V.; Giorgini, S.; Morichetti, D.; Goteri, G.; Sartini, D.; Serritelli, E.N.; Emanuelli, M. Paraoxonase-2 is upregulated in triple negative breast cancer and contributes to tumor progression and chemoresistance. *Hum. Cell* **2023**, *36*, 1108–1119. [CrossRef] [PubMed]

26. Campagna, R.; Belloni, A.; Pozzi, V.; Salvucci, A.; Notarstefano, V.; Togni, L.; Mascitti, M.; Sartini, D.; Giorgini, E.; Salvolini, E.; et al. Role Played by Paraoxonase-2 Enzyme in Cell Viability, Proliferation and Sensitivity to Chemotherapy of Oral Squamous Cell Carcinoma Cell Lines. *Int. J. Mol. Sci.* **2022**, *24*, 338. [CrossRef] [PubMed]

27. Easterling, T.; Mundle, S.; Bracken, H.; Parvekar, S.; Mool, S.; Magee, L.A.; von Dadelszen, P.; Shochet, T.; Winikoff, B. Oral antihypertensive regimens (nifedipine retard, labetalol, and methyldopa) for management of severe hypertension in pregnancy: An open-label, randomised controlled trial. *Lancet* **2019**, *394*, 1011–1021. [CrossRef] [PubMed]

28. Odigboegwu, O.; Pan, L.J.; Chatterjee, P. Use of Antihypertensive Drugs During Preeclampsia. *Front. Cardiovasc. Med.* **2018**, *5*, 50. [CrossRef]

29. Sammar, M.; Drobnjak, T.; Mandala, M.; Gizurarson, S.; Huppertz, B.; Meiri, H. Galectin 13 (PP13) Facilitates Remodeling and Structural Stabilization of Maternal Vessels during Pregnancy. *Int. J. Mol. Sci.* **2019**, *20*, 3192. [CrossRef]
30. Meiri, H.; Osol, G.; Cetin, I.; Gizurarson, S.; Huppertz, B. Personalized Therapy against Preeclampsia by Replenishing Placental Protein 13 (PP13) Targeted to Patients with Impaired PP13 Molecule or Function. *Comput. Struct. Biotechnol. J.* **2017**, *15*, 433–446. [CrossRef]
31. Huppertz, B. Maternal-fetal interactions, predictive markers for preeclampsia, and programming. *J. Reprod. Immunol.* **2015**, *108*, 26–32. [CrossRef] [PubMed]
32. Sousa, S.O.; Santos, M.R.D.; Teixeira, S.C.; Ferro, E.A.V.; Oliani, S.M. ANNEXIN A1: Roles in Placenta, Cell Survival, and Nucleus. *Cells* **2022**, *11*, 2057. [CrossRef] [PubMed]
33. Araujo, T.G.; Mota, S.T.S.; Ferreira, H.S.V.; Ribeiro, M.A.; Goulart, L.R.; Vecchi, L. Annexin A1 as a Regulator of Immune Response in Cancer. *Cells* **2021**, *10*, 2245. [CrossRef] [PubMed]

MDPI

St. Alban-Anlage 66

4052 Basel

Switzerland

www.mdpi.com

International Journal of Molecular Sciences Editorial Office

E-mail: ijms@mdpi.com

www.mdpi.com/journal/ijms